4-24-78

INTRODUCTION to COMBINATORICS

INTRODUCTION
TO
COMBINATORICS

by

IOAN TOMESCU

Translated from the Romanian

by

I. TOMESCU and S. RUDEANU

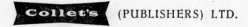 (PUBLISHERS) LTD.

London and Wellingborough

1975

Collet's (Publishers) Ltd.
Denington Estate, Wellingborough,
England

This book is a translation of
INTRODUCERE ÎN COMBINATORICĂ
by Ioan Tomescu

First English edition 1975

© First published by Editura tehnică, Bucharest, 1972

© *Collet's (Publishers) Ltd.*
Printed in Romania

Contents

Editor's Foreword

During the last two decades the subject of combinatorics has grown from being a mathematical backwater to an important and rapidly developing field. A number of books on the subject have been published during that period but many of them are proceedings of conferences or treat a small and specialised area of the subject. Textbooks covering a broad range of combinatorial mathematics are still comparatively few in number and so the present book, dealing with a variety of topics, should prove to be of value.

The book was published first in a Romanian edition and in his preface to it Acad. Gr. C. Moisil refers to the lack of axioms and deductive rules for combinatorics. As the years go by this becomes less and less the case; axioms for certain areas of the subject are being developed, as are techniques for solving certain classes of problems. Nevertheless, the variety of means to which Moisil refers remains a feature of the subject, particularly in enumerative combinatorics where a little thought may well produce a solution to a particular problem more easily than a general technique designed to handle any of a whole family of problems.

The English edition of this book was written by the author himself, and to it he has added some problems at the end of each chapter. These vary and are not ordered in their degree of difficulty. Some of them are exercises on the material in the chapter; others give additional results with references to where they have been solved in the literature. This, of course, increases considerably the information in the book. A number of the references are to Romanian papers, the results of which deserve to be more widely known amongst English speaking mathematicians and the publication of this book should help to make them so.

Topics covered in the fourteen chapters of the book include certain properties of numbers occuring in combinatorial problems, enumeration (Pólya-de Bruijn methods, inclusion-exclusion and Möbius functions), existence of various configurations (distinct representatives and Ramsey's theorem) and optimization problems (shortest paths and distances in graphs and chromatic numbers). Each chapter is supplied with a bibliography and where appropriate I have added details of English editions of books. I have also brought the abbreviations for modern periodical titles into line with those used in Mathematical Reviews.

Combinatorics, as Moisil tells us, is a subject used in various fields and so this book is of interest not only to mathematicians specialising in combinatorics but also to chemists, economists, operational researchers, engineers, etc. It is also suitable for students at undergraduate level.

The graph theory terminology used in the book generally follows that of the French School and the reader is referred to Berge (*Graphes et Hypergraphes*, Dunod, Paris, 1970; English edition: North-Holland, Amsterdam, 1973) for any definitions not explicitly given here. A knowledge of Boolean algebra is assumed in some chapters.

E. KEITH LLOYD

Preface

I am very pleased that a young and gifted Romanian mathematician has written a book on combinatorics. The readers of this book will thus have the opportunity to find out what this so much disputed combinatorics is. They will also understand, especially when reading chapters 12 and 13, the novelty which the existence of computers has brought to various branches of mathematics, possibly to all of them.

Not long ago, combinatorics, as well as graph theory, was considered as a strange (but not unworthy) concern of a few mathematicians; though not of unimportant ones. It should be emphasized that some of them bore famous names, such as Cayley, Euler, Hamilton, Sylvester.

Today graph theory has become a major discipline, even if it is not included in a dogmatic classification of the branches of mathematics.

The theory of graphs is an aspect of the logic of relations which comprises both a logistic and a purely algebraical one: the study of the algebra of matrices with elements in a Boolean algebra or, more generally, in a distributive lattice.

The use of graph theory in various fields, from chemistry to economics, from the study of electrical networks to the analysis of texts and to politics, endow it with a prestige that must be taken into account by those who classify sciences.

This is not the case with combinatorics. A branch of mathematics has its axioms, but its deductive rules are, of course, those common to all the deductive sciences. The few scientists who enumerate them are said to deal with mathematical logic rather than with mathematics itself. Combinatorics has no characteristic axioms of its own. Dealing with natural numbers, its axioms seem to be those of number theory.

But neither the theory of numbers nor combinatorics deals with the axiomatics of natural numbers, Moreover, a branch of mathematics has its own concepts and structure, an evolution of its own and of its concepts and structure, and its revolutions. We never find such things in combinatorics. Does this mean that combinatorics is not a part of pure mathematics?

Or is combinatorics a concern only of applied mathematicians?

Mechanics, both celestial and terrestrial, show us how to understand a certain kind of phenomena; electromagnetic theory, quantum mechanics, operational research, and mathematical linguistics, help us to understand other kinds of phenomena. They have no systems of axioms like that of Zermelo for set theory or that of Peano for arithmetic; nor like that of group theory, of topological spaces or of categories, for which an axiomatics of set theory is implicitly presupposed.

In all the above-mentioned fields the main concern centres round the phenomena that are to be studied: the movement of the Earth round the Sun or the grammatical category of the case.

Obviously, it is interesting to be able to count some sets and such counting problems have been raised in various fields. But objections could arise as to whether the following problem is one of applied mathematics: in how many ways can we arrange n husbands and their n wives round a circular table so that each man be seated between two women neither of them being his wife? Can that be considered a serious problem?

Curiously enough, this actually is a serious mathematical problem (see page 56 and the following).

In our opinion all the above questions are generated by a dogmatic vision of mathematics. Very useful for the study of the logical structure of formal systems, for pondering on the way in which mathematics is used in the investigation of nature and is begining to be used in the study of man and society. The definition of mathematics should take into account the fact that combinatorics *was* and *is* also a branch of mathematics.

Therefore both the one who classifies and defines mathematical activity from a philosophical ground and the one who organises the programmes of high schools and universities, ought not to do so disregarding combinatorics.

I am not sure whether we do know what mathematics is, but if we do, we have to do it in such a way that the statement

Combinatorics is a part of mathematics

be true.

That is why this is a welcome volume; it makes us think.

It makes us think mathematically.

It makes us think of mathematics.

Undoubtedly, the reading of each chapter of this book makes us feel an intense mathematical joy.

Many people find its essence in counting problems. What such a problem is the reader will learn. Is the curious example 2 from page 2 one of this kind? Many problems of combinatorics refer to the partition of a set into subsets or to the arrangement of the elements of a set in a certain way. But in order to characterize the standpoint from which this type of problem is viewed in this book, the reader is urged to compare the things stated in chapters 7 and 8 with what is usually said about permutations in books on group theory.

Each problem of combinatorics requires an unrestrained use of the human mind, of the mathematical human mind.

It requires less information than intellectual power.

The methods used differ from problem to problem.

That is contrary to what happens in certain great mathematical theories in which outstanding results are obtained with great economy of means. Here there is a variety of means, therefore some people have doubts as to the importance of the results.

Hence the pleasure felt in solving problems of combinatorics, an opportunity for ever renewing this joy.

This book, this science, allows us to define mathematics as follows: a great delight.

Gr. C. MOISIL

CHAPTER 1

Sets and Functions

In the following we shall work with finite sets, i.e. sets which contain a finite number of elements. The number of elements of a finite set X will be denoted by $|X|$.

Given two sets X and Y, we say that a *function* f from X to Y has been defined, and we write: $f: X \to Y$, if with every element $x \in X$ is associated an element $y \in Y$, denoted by $f(x)$ and called the *image* of x under the function f. By $f(X)$ we denote the set $\{f(x) | x \in X\}$ (that is, the set of all images of the elements of X by the function f). We also say that the set X is the *domain* of the function f, and the set Y the *codomain* of the function f.

If $f(x) = y$, this correspondence of elements is denoted by $x \mapsto y$.

A function f is said to be injective or an *injection*, if it associates different elements of Y with different elements from X, i.e. $x_1 \neq x_2$ implies $f(x_1) \neq f(x_2)$. If the sets X and Y are finite and there exists an injection $f: X \to Y$, then $|X| \leqslant |Y|$; an illustration of this property is given in Fig. 1.1.

The function f is said to be surjective or a *surjection*, if $f(X) = Y$ or, equivalently, if for every $y \in Y$ there exists an element $x \in X$ (not necessarily unique) such that $f(x) = y$. In this case we say that f is a function from X *onto* Y. If the sets X and Y are finite and there exists a surjection $f: X \to Y$, then $|X| \geqslant |Y|$ (Fig. 1.2).

A function f which is both injective and surjective is said to be bijective or a *bijection*. If there exists a bijection between two finite sets X and Y, then from the above two inequalities we deduce that $|X| = |Y|$ (Fig. 1.3).

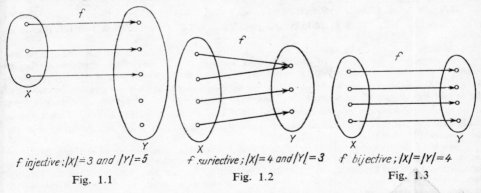

f injective; $|X|=3$ and $|Y|=5$

Fig. 1.1

f surjective; $|X|=4$ and $|Y|=3$

Fig. 1.2

f bijective; $|X|=|Y|=4$

Fig. 1.3

In the counting problems with which we deal in this book, a frequent procedure will be the following: instead of finding a formula for the number of elements of a finite set X which is given by certain properties of its elements, we determine the number of elements of another set Y, for which there is a bijection between X and Y.

Example 1. Given a finite set X, whose elements are denoted by $1, 2, \ldots, n$, let us determine the number of subsets of X, the set of which is denoted by $\mathscr{P}(X)$.

To do this, we shall establish a bijection between the subsets of X and the n-digit binary words, $x_1 x_2 \ldots x_n$ with $x_i \in \{0, 1\}$ where the concatenation of the letters $x_1, x_2, \ldots, x_n$ does not mean the product of their values. The bijection f is defined as follows: if $X_1 \subset X$, then $f(X_1) = x_1 x_2 \ldots x_n$ where $x_i = 0$ if $x_i \notin X_1$ and $x_i = 1$ if $x_i \in X_1$. For instance, if $n = 5$ and $X_1 = \{1, 3, 5\}$, then $f(X_1) = 10101$. By this function, the set X is associated with the word $\underbrace{11 \ldots 1}_{n \text{ times}}$, while the empty set $\varnothing$ is associated with the word $\underbrace{00 \ldots 0}_{n \text{ times}}$. It is obvious that f is a bijection between $\mathscr{F}(X)$ and the binary words of length n, for to different subsets correspond different words and every binary word $x_1 x_2 \ldots x_n$ is the image of a subset X_1 defined as follows: $X_1 = \{i \mid x_i = 1\}$, which implies that $f(X_1) = x_1 x_2 \ldots x_n$. But the number of n-digit binary words is 2^n, because the first binary digit can be chosen in two ways, the second in two ways, etc., resulting in the number $\underbrace{2 \cdot 2 \ldots 2}_{n \text{ factors}} = 2^n$ of binary words. We have thus proved that the number of elements of $\mathscr{P}(X)$ equals $2^{|X|}$.

Similarly, the construction of certain injective functions yields a general method for establishing inequalities. We give below two illustrative examples.

Example 2. *(Erdős, Szekeres).* In a sequence of $mn + 1$ distinct integers, $u_1, u_2, \ldots, u_{mn+1}$, either there exists a decreasing subsequence of length greater than m, or there is an increasing subsequence of length greater than n.

By definition a subsequence of the given sequence is a sequence with elements from the original sequence written in the same order as in the original sequence. For instance, let $m = 3$, $n = 4$ and let the sequence be $1, -2, 7, 3, 2, 4, 5, 6, 14, 10, 8, 9, 13$. This sequence does not contain decreasing subsequences of length greater than 3, a decreasing subsequence of length 3 being $7, 3, 2$, but it contains increasing subsequences of length greater than 4, an increasing subsequence of length 8 being $1, 2, 4, 5, 6, 8, 9, 13$.

In order to prove this property, let us denote by l_i^- the length of the longest decreasing subsequence which begins with u_i and by l_i^+ the length of the longest increasing subsequence which begins with u_i. Let us assume that our statement is false: this means that every decreasing subsequence has a length less than or equal to m and every increasing subsequence has a length less than or equal to n.

We now define a function f on the set $\{u_1, u_2, \ldots, u_{mn+1}\}$ with values in the Cartesian product $\{1, 2, \ldots, m\} \times \{1, 2, \ldots, n\}$ as follows: $f(u_i) = (l_i^-, l_i^+)$. Let us prove that this function is an injection.

Assume that $i < j$. In this case $u_i > u_j$ implies $l_i^- > l_j^-$, because we can add at least one element, namely u_i, to the left of the decreasing subsequence of maximum length which begins with u_j, thus obtaining a decreasing subsequence of greater length which begins with u_i; therefore, $l_i^- > l_j^-$. Similarly $u_i < u_j$ implies $l_i^+ > l_j^+$, because we can add at least one element, namely u_i, to the left of the increasing subsequence of maximum length which begins with u_j, thus obtaining an increasing subsequence of greater length which begins with u_i; therefore $l_i^+ > l_j^+$. Hence $u_i \neq u_j$ implies $(l_i^-, l_i^+) \neq (l_j^-, l_j^+)$ because these ordered pairs differ in at least one position; but the latter inequality means $f(u_i) \neq f(u_j)$.

We have thus proved that the function f is injective, hence the number of elements of the set $\{u_1, u_2, \ldots, u_{mn+1}\}$ is less than or equal to the number of elements of the codomain, which is the Cartesian product $\{1, 2, \ldots, m\} \times \{1, 2, \ldots, n\}$ and so contains mn elements. Thus our assumption has led to the conclusion $mn + 1 \leqslant mn$, which is false, and the property is established by *reductio ad absurdum*.

The fact that the Cartesian product $X \times Y$ with $|X| = m$ and $|Y| = n$ contains mn elements can be established by taking into account the fact that $X \times Y = \{(x, y) \mid x \in X \text{ and } y \in Y\}$, hence x can be chosen in m ways, y in n ways, resulting in mn distinct possibilities for forming the pairs (x, y).

Example 3. (*Erdös, Hajnal*). Given a set X with n elements and a family $\mathscr{F}$ of three-element subsets of X, such that every two of them have at most one common element, then there exists a subset of X which has at least $[\sqrt{2n}]$ elements and does not include any subset belonging to $\mathscr{F}$.

(By $[x]$ we denote the integer part of x, that is, the greatest integer less than or equal to x).

For example, let $n = 9$, $X = \{1, 2, 3, 4, 5, 6, 7, 8, 9\}$ and $\mathscr{F}$ be the family consisting of the following subsets of X: $\{1, 2, 3\}$, $\{3, 4, 5\}$, $\{5, 6, 7\}$, $\{1, 6, 8\}$, $\{1, 4, 7\}$, $\{2, 5, 8\}$, $\{7, 8, 9\}$, $\{1, 5, 9\}$. In this case there exists the set $\{1, 2, 7, 8\}$ which has $[\sqrt{18}] = 4$ elements and does not include any subset from the family $\mathscr{F}$.

In order to prove this property, let us take a subset $M \subset X$ which does not include any subset from the family $\mathscr{F}$ and is maximal with this property, that is, for every $x \in X \setminus M$ there exists a set $B \in \mathscr{F}$ such that $B \subset M \cup \{x\}$. Here $X \setminus M$ denotes the difference of the sets X and M, i.e. the set of elements which belong to X and do not belong to M. The existence of a maximal set with the above mentioned property is obvious: starting from a non-maximal set, say, from an element of X, we add new elements until we get a maximal set. Thus, the set $\{1, 2, 7, 8\}$ is maximal with the property that it does not include any subset from the above given family $\mathscr{F}$.

Let us put $|M| = r$, then $|X \setminus M| = n - r$.

According to maximality, for every $x \in X \setminus M$ there exists a subset which we denote by $A_x \subset M$, with $|A_x| = 2$, such that $\{x\} \cup A_x \in \mathscr{F}$.

In this way we have defined a function $f: X \setminus M \to \mathscr{P}_2(M)$, where $\mathscr{P}_2(M)$ stands for the set of all two-element subsets of M.

Let us prove that f is an injection, using the fact that every two sets from the family $\mathscr{F}$ have at most one common element.

Let $x, y \in X \setminus M$, with $x \neq y$. If $A_x = A_y$ then $\{x\} \cup A_x$, $\{y\} \cup A_y$ belong to the family $\mathscr{F}$ and contain two common elements, thus contradicting the hypothesis upon $\mathscr{F}$.

Therefore $x \neq y$ implies $A_x \neq A_y$, that is, $f(x) \neq f(y)$; hence f is an injection, therefore $|X \setminus M| \leqslant |\mathscr{P}_2(M)|$. But $|\mathscr{P}_2(M)| = \dfrac{r(r-1)}{2}$, where r is the number of elements in M, for the number of unordered pairs $\{x, y\}$ with $x, y \in M$, $x \neq y$, can be obtained in the following way: x can be chosen from M in r different ways and y from the remaining elements in $r - 1$ different ways. But the number $r(r-1)$ must be divided by 2, because each pair is obtained twice: once as $\{x, y\}$, then as $\{y, x\}$, and we have assumed that the order of the elements x and y is immaterial.

We have thus obtained $n - r \leqslant \dfrac{r(r-1)}{2}$ or $r^2 + r - 2n \geqslant 0$, hence $r \geqslant \dfrac{-1 + \sqrt{8n+1}}{2}$ by the rule for the sign of a binomial and also taking into account that r is a positive integer.

Setting $\dfrac{-1 + \sqrt{8n+1}}{2} = k$, we deduce that $2n = k^2 + k$, hence $[\sqrt{2n}] = [\sqrt{k(k+1)}]$. If k is a positive integer then $[\sqrt{k(k+1)}] = k$ (since $k < \sqrt{k(k+1)} < k + 1$) which implies $r \geqslant [\sqrt{2n}]$. Otherwise $r \geqslant [k] + 1$ and since $[\sqrt{k(k+1)}] \leqslant [k+1] = [k] + 1 \leqslant r$, we also obtain $r \geqslant [\sqrt{2n}]$.

We have proved that every subset M of X, maximal with respect to the property that it does not include any subset from $\mathscr{F}$, contains at least $[\sqrt{2n}]$ elements, thus concluding the proof.

Given two functions $f: X \to Y$ and $g: Y \to Z$, these can be composed to yield a new function denoted by $g \circ f: X \to Z$ and defined by $(g \circ f)(x) = g(f(x))$ for every $x \in X$. If $f: X \to Y$ is a bijection, we can define the inverse of the function f, denoted by $f^{-1}: Y \to X$, as follows: $f^{-1}(y) = x$ if $f(x) = y$. The function f^{-1} is well defined for every $y \in Y$, due to the fact that f is a bijection; f^{-1} is also a bijection and satisfies the relations $f^{-1} \circ f = I_X$ and $f \circ f^{-1} = I_Y$, where I_X and I_Y stand for the identity functions of X and Y, respectively. (I.e. $I_X: X \to X$ with $I_X(x) = x$ for every $x \in X$.)

Given $f: X \to Y$, no matter whether or not f is a bijection, for every $y \in Y$ the set $f^{-1}(y)$ is defined as $\{x | x \in X, f(x) = y\}$, i.e. the set of all elements of X such that their image by f is y. If f is bijective, then $f^{-1}(y)$ reduces to the single element x which is the image of y by the inverse of the function f; therefore the latter definition for $f^{-1}(y)$ includes, as a particular case, the former which applies only when f is a bijection, so having an inverse which is also denoted by f^{-1}.

The set of all functions $f: X \to Y$ is denoted by Y^X.

Problems

1.   Let X be a set with n elements and let $\mathscr{P}_h(X)$ denote the set of h-sets (h-element subsets) of X. If $M(n, k, h)$ stands for the minimum number of h-sets of a family $\mathscr{F} \subset \mathscr{P}_h(X)$, having the property that any k-set of X includes at least an h-set of $\mathscr{F}$ ($n \geqslant k \geqslant h \geqslant 1$), prove that

i) $M(n, k, h) \leqslant \dfrac{n}{h} M(n - 1, k - 1, h - 1)$;

ii) $M(n, k, h) \geqslant \dfrac{n}{n - h} M(n - 1, k, h)$;

(Katona, Nemetz, Simonovits, 1964)

iii) $M(n, k, h) \leqslant M(n - 1, k - 1, h - 1) + M(n - 1, k, h)$.

(Stanton, Kalbfleisch, 1970)

H i n t: If $\mathscr{F}$ is a minimal family of h-sets, prove that there exists $x \in X$ such that

$$|\{E | E \in \mathscr{F}, x \in E\}| \geqslant \frac{h}{n} M(n, k, h) \text{ and}$$

$$|\{E | E \in \mathscr{F}, x \notin E\}| \leqslant \frac{n - h}{n} M(n, k, h).$$

2.   Show that

$$M(n, k, h) \geqslant \left\{ \frac{n}{n - h} \left\{ \frac{n - 1}{n - h - 1} \left\{ \dots \left\{ \frac{k + 1}{k - h + 1} \right\} \dots \right\} \right\} \right\}$$

where $\{x\}$ is the smallest integer greater than or equal to x.

(Schönheim, 1964, Chvátal, 1971)

3.   If $n \geqslant k(h + 1)$ and $k \geqslant 1$, prove that

$$M(n, n - h, k) = h + 1.$$

(Tomescu, 1973)

4. If $f \circ g$ is an injective mapping, prove that g is injective too; if $f \circ g$ is surjective, prove that so is f.

5. If A is a finite set and $f: A \to A$ is injective (surjective), prove that f is bijective.

6. Let $f: A \to A$ be a function such that $\underbrace{f \circ f \circ \ldots \circ f}_{2k} = I_A$. Prove that f is a bijection.

7. Show that the minimum number of 3-subsets of an n-set X such that any pair of elements of X is contained in at least a 3-subset of X, equals $\dfrac{n^2}{6}$ if $n \equiv 0 \pmod 6$.

For a generalization see Schönheim [13].

8. There are n gossips each of whom knows some gossip not known to the others. They communicate by telephone, and whenever one gossip calls another, they tell each other all they know at that time. Prove that the minimum number of calls required before each gossip knows everything is equal to $2n - 4$ for $n \geqslant 4$.

(Baker, Shostak and Hajnal, Milner, Szemerédi, 1972)

N o t e: The simple "proof" in [2] is incorrect (see Corrigendum). For a proof see [1].

9. Let X be a set of n elements and let A be a subset of X with m elements $(0 \leqslant m \leqslant n)$.

Prove that the number of ways of writing A as an intersection of k sets is equal to $(2^k - 1)^{n-m}$ and the number of ways of writing A as a union of k sets is given by $(2^k - 1)^m$.

(Tainiter, 1968)

10. Let $X = (x_1, x_2, \ldots, x_n)$ and $Y = (y_1, y_2, \ldots, y_n)$ be two vectors. We say that X covers Y if $x_i = y_i$ for r values of i and if $r \geqslant n - 1$. Let G denote the set of p^n vectors $(y_1, y_2, \ldots, y_n)$, where $y_i \in \{1, 2, \ldots, p\}$ $(i=1, 2, \ldots, n)$. A set H of vectors $H_1, H_2, \ldots$ is called a covering set if every vector Y in G is covered by at least one vector H_i in H. Let $\sigma(n, p)$ be the minimum number of vectors which such a covering set H can contain.
Prove that:

i) $\sigma(2, p) = p$;

(Kalbfleisch, Stanton, 1969)

ii) $\sigma(n + 1, p) \leqslant p\,\sigma(n, p)$;

iii) $\dfrac{p^n}{n(p - 1) + 1} \leqslant \sigma(n, p) \leqslant p^n$.

(Taussky, Todd, 1948)

J. G. Kalbfleisch and R. G. Stanton [9] showed that $\sigma(3, p) = \left[\dfrac{p^2+1}{2}\right]$ and Zaremba [18] proved that if p is a prime or a prime power, and $n(p-1)+1$ is a power of p, then $\sigma(n, p) = \dfrac{p^n}{n(p-1)+1}$.

11.   If $|X| = n$ show that $\sum\limits_{A, B \subset X} |A \cap B| = n4^{n-1}$.

12.   Let $S = (X_i)_{1 \leqslant i \leqslant r}$ be a family of distinct subsets of X having the property $X_i \cap X_j \neq \varnothing$ for every $i, j = 1, ..., r$.
    Prove that the maximum value of r is equal to 2^{n-1}, where $n = |X|$.

(Katona, 1964)

H i n t: If we define the family $\bar{S} = (X \setminus X_i)_{1 \leqslant i \leqslant r}$, then $\bar{S}$ includes no set of S, hence max $r \leqslant 2^{n-1}$. This upper bound is attained by considering $Y = X \setminus \{x\}$ and the family of sets $\{Y_i \cup \{x\} \mid Y_i \subset Y\}$.

13.   Let $A_1, ..., A_n$ be any n sets. Take the largest subfamily $A_{i_1}, ..., A_{i_r}$ which is union-free, i.e. $A_{ij_1} \cup A_{ij_2} \neq A_{ij_3}$ ($1 \leqslant j_1, j_2, j_3 \leqslant r$), for every triple of distinct sets $A_{ij_1}, A_{ij_2}, A_{ij_3}$. Put $f(n) = \min r$, where the minimum is taken over all families of n distinct sets. Show that $\sqrt{2n} - 1 < f(n) < 2\sqrt{n} + 1$.

(Erdös, Shelah, 1972)

14.   Let X be a set having n elements and let $g(n)$ denote the smallest integer such that the subsets of X can be split into $g(n)$ classes where each of the classes is union free. Prove that $g(n) > \dfrac{n}{4}$.

(Erdös, Shelah, 1972)

BIBLIOGRAPHY

1. Baker, B., Shostak, R., *Gossips and telephones*, Discrete Math., **2**, 1972, 191–193.
2. Berman, G., *The gossip problem*, Discrete Math., **4**, 1973, 91; Corrigendum **4**, 1973, 397.
3. Chvátal, V., *Hypergraphs and Ramseyian theorems*, Proc. Amer. Math. Soc.,**27**, 1971, 434–440.
4. Erdös, P., Szekeres, G., *A combinatorial problem in geometry*, Compositio Math., **2**, 1935, 463–470.
5. Erdös, P., Hajnal, A., *On chromatic number of graphs and set-systems*, Acta Math. Acad. Sci. Hungar., **17**, 1966, 61–99.
6. Erdös, P., Shelah, S., *On problems of Moser and Hanson*, Graph Theory and Applications, edited by Y. Alavi, D. R. Lick and A. T. White, Proceedings of the Conference at Western Michigan University, 1972, Lecture Notes in Mathematics, Springer-Verlag, 1972.
7. Hajnal, A., Milner, E., Szemerédi, E., *A cure for the telephone disease*, Canad. Math. Bull., **15**, 1972, 447–450.
8. Kalbfleisch, J. G., Stanton, R. G., *Maximal and minimal coverings of (k − 1)-tuples by k-tuples*, Pacific J. Math., **26**, 1968, 131–140.
9. Kalbfleisch, J. G., Stanton, R. G., *A combinatorial problem in matching*, J. London Math. Soc., **44**, 1969, 60–64.

10. Katona, G., *Intersection theorems for systems of finite sets*, Acta Math. Acad. Sci. Hungar., **15**, 1964, 329—337.
11. Katona, G., Nemetz, T., Simonovits, M., *On a problem of Turán in the theory of graphs*, Mat. Lapok, **15**, 1964, 228—238.
12. Mills, W. H., *On the covering of pairs by quadruples*, I, II, J. Combinatorial Theory (A), **13**, 1972, 55—78 and **15**, 1973, 138—166.
13. Schönheim, J., *On coverings*, Pacific J. Math., **14**, 1964, 1405—1411.
14. Stanton, R., Kalbfleisch, J., *Coverings of pairs by k-sets*, International Conference on Combinatorial Mathematics, 1970, Ann. New York Acad. Sci., **175**, 1970, 366—369.
15. Tainiter, M., *Generating functions on idempotent semigroups with application to combinatorial analysis*, J. Combinatorial Theory, **5**, 1968, 273—288.
16. Taussky, O., Todd, J., *Covering theorems for groups*, Ann. Soc. Polonaise de Math., **21**, 1948, 303—305.
17. Tomescu, I., *Inégalités concernant les hypergraphes uniformes*, Cahiers Centre Études Recherche Opér., **15**, 1973, 355—362. (Colloque sur la théorie des graphes, I.H.E. de Bélgique, Bruxelles, 1973).
18. Zaremba, S. K., *A covering theorem for Abelian groups*, J. London Math. Soc., **26**, 1950, 71—72.

CHAPTER 2

Arrangements, Permutations, Combinations

Let X be a finite set with $|X| = n$, the elements of which are called *objects*, and Y be a finite set with $|Y| = m$, the elements of which are called *flats*. With every function f from X to Y, we can associate an arrangement of the set X of objects in the set Y of flats such that flat y_i contains the objects of the set $\{x \mid x \in X, f(x) = y_i\}$. Here the word arrangement has the meaning of placing or distributing the objects in the flats without any restriction whatever. Conversely, given an arrangement of the set X of objects in the set Y of flats, a function $f : X \to Y$ corresponds to this arrangement with $f(x_i) = y_j$ if object x_i is placed in flat y_j. The function f is well defined, since any object appears but in a single flat. Each of the above defined

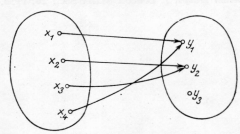

Fig. 2.1

correspondences constitutes a bijection between the two sets: the functions $f : X \to Y$ and the set of arrangements of the objects of the set X in the set Y of flats. The two functions are inverse to each other. For example, to the function in Fig. 2.1, corresponds the arrangement of the objects in the flats in Fig. 2.2.

A bijection can also be established, between the set of functions $f : X \to Y$ with $|X| = n$ and $|Y| = m$ and the set of n-tuples formed by elements of Y or of words of length n made up of letters from the set Y, as follows: to every function f, there corresponds the word $f(x_1) f(x_2) \ldots f(x_n)$, in which the order of the letters is essential.

Conversely, given a word formed by n letters from Y, namely $y_{i_1} y_{i_2} \ldots y_{i_n}$, one can associate with it the function f from X to Y defined by

$$f(x_i) = y_{i_1}, \quad f(x_2) = y_{i_2}, \ldots, f(x_n) = y_{i_n}.$$

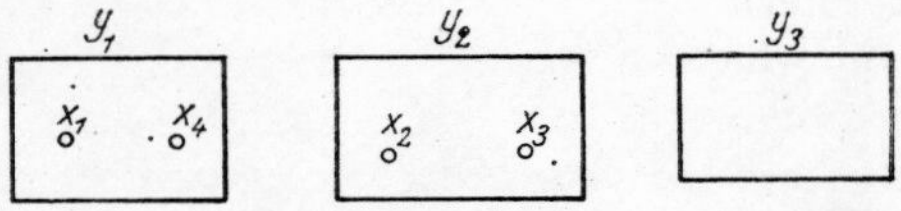

Fig. 2.2

PROPOSITION 1. *The number of functions* $f : X \to Y$ *is equal to* m^n *if the set* X *has* n *elements and the set* Y *has* m *elements.*

The number of functions $f: X \to Y$ is, from the above remark, equal to the number of words of length n built up from elements of the set Y. Since the first letter of the word can be choosen in m different ways, the second letter in m ways,..., the n-th one in m ways, the resulting number of words is equal to $\underbrace{m \cdot m \ldots m}_{n \text{ factors}} = m^n$.

It is obvious that this number also coincides with the number of arrangements of a set of n objects in a set of m flats. That is why the set of functions $f: X \to Y$ is always denoted by Y^X, whether the sets X and Y are finite or not.

PROPOSITION 2. *The number of injective functions* $f: X \to Y$, *where* $|X| = n$ *and* $|Y| = m$, *is equal to* $[m]_n = m(m - 1) \ldots (m - n + 1)$.

By definition, for every real x and every natural n we set:

$$[x]_n = \underbrace{x(x - 1) \ldots (x - n + 1)}_{n \text{ factors}}. \tag{2.1}$$

The injectivity of the function f is expressed by the fact that the letters of the word are pairwise distinct, while in the case of the arrangement of the objects in flats it is expressed by the fact that each flat contains at most one object. But the number of words formed by n different letters from the m-element set Y is equal to $m(m - 1) \ldots (m - n + 1) = [m]_n$, since the first letter of a word can be chosen in m different ways, the second letter in $m - 1$ ways, ..., and the n-th letter can be chosen from the remaining $m - n + 1$ letters in $m - n + 1$ different ways. A necessary and sufficient condition for the problem to be soluble, is that $m \geqslant n$. The numbers $[m]_n$ are also denoted by the symbol P_m^n and the name *permutations* of m objects taken n at a time from the set Y is given to the words formed by n different letters from the set Y, the number of which is equal to $[m]_n$. If $m = n$, the number $[n]_n$ is also denoted by P_n or by $n! = 1 \cdot 2 \cdot \ldots \cdot n$, and the arrangements of the objects are simply called *permutations* of n objects.

If $f: X \to Y$ is injective and $|X| = |Y| = n$, it follows that f is also a surjection and, therefore, a bijection. Hence the number of bijective mappings $f: X \to Y$, if $|X| = |Y| = n$, is equal to $n!$.

By an abuse of terminology, the name of permutations of the n-set Y is also given to the words of length n, each of them containing all the letters from Y. Obviously, in this case, every word consists of pairwise different letters.

The problem of determining the number of surjective functions $f: X \to Y$, when $|Y| \leqslant |X|$, is somewhat more intricate; it will be solved as an application of the *principle of inclusion and exclusion*.

Let us now suppose that the set $X = \{x_1, x_2, \ldots, x_n\}$ of objects is to be arranged in the flats cf the set $Y = \{y_1, y_2, \ldots, y_m\}$. Every flat may contain any number of objects from X, but if the order of the objects in a flat changes, the resulting arrangements will be different. Such an arrangement will be called an *ordered arrangement in flats*.

PROPOSITION 3. *The number of ordered arrangements of n objects in m flats is equal to*

$$[m]^n = m(m + 1) \dots (m + n - 1).$$

By definition, we shall use the notation

$$[x]^n = \underbrace{x(x + 1) \dots (x + n - 1)}_{n \text{ factors}} \tag{2.2}$$

for any real x and natural n.

To demonstrate this result, we shall denote by T_n the set of all ordered arrangements of n objects in m flats. Each ordered arrangement of $n - 1$ objects in m flats, namely $x_{i_1} x_{i_2} \dots x_{i_k} | \dots | x_{j_1} \dots x_{j_l}$, is characterized by a sequence of $(n - 1) + (m - 1)$ symbols, since there are $n - 1$ objects and $m - 1$ bars separating the m flats. The object x_n can be added to such an arrangement in $(n - 1) + (m - 1) + 1 = n + m - 1$ different ways. Indeed, if we give to each object and to each bar an ordering number, representing its place in the above sequence, the numbering starting from the left, the object x_n may be placed before the first element of the sequence (which can be a bar if the first flat is unoccupied) or after the element labelled by i, with $i = 1, 2, \dots, m + n - 2$, which gives a total of $m + n - 1$ possible placings. Since in this way, starting from the arrangements in T_{n-1}, all the arrangements in T_n are obtained without repetition, it follows that $|T_n| = (m + n - 1) |T_{n-1}|$. By iterating this relation, we obtain $|T_{n-1}| = (m + n - 2) |T_{n-2}|$, etc. Therefore $|T_n| = (m + n - 1)(m + n - 2) \dots (m + 1) |T_1| = [m]^n$, because $|T_1|$, the number of ordered arrangements of a single object in m flats, is equal to m.

If the set $Y = \{y_1, y_2, \dots, y_m\}$ is an ordered set, with $y_1 < y_2 < \dots < y_m$, then a *word* consisting of n letters from Y and denoted by $y_{i_1} y_{i_2} \dots y_{i_n}$ is called *increasing* provided that $y_{i_1} \leqslant y_{i_2} \leqslant \dots \leqslant y_{i_n}$.

For instance, if $Y = \{a, b, c, d\}$, with the order $a < b < c < d$, there will be 20 increasing words consisting of 3 letters from Y, namely:

aaa, abb, acc, add, aab, abc, acd, aac, abd, aad,
bbb, bbc, bbd, bcc, bcd, bdd, ccc, ccd, cdd, ddd.

The elements a, b, c, d called here *letters* or *symbols*, belong to a set endowed with a binary relation, denoted by $\leqslant$, which is reflexive, antisymmetric and transitive, and with $a \leqslant b \leqslant c \leqslant d$. It is called a *partial order relation*. By definition, we have $a < b$ if $a \leqslant b$ and $a \neq b$. It follows from this definition that in a partially ordered set, there may exist elements which are pairwise incomparable with respect to the partial order.

PROPOSITION 4. *The number of increasing words formed by n letters out of m symbols, is equal to* $\dfrac{[m]^n}{n!}$.

With each ordered arrangement of n objects $x_1, x_2, \dots, x_n$ in m flats $y_1, y_2, \dots, y_m$, we associate an increasing word as follows: if, in flat y_i, there are p_i objects from the set $\{x_1 x_2, ..., x_n\}$, the contribution of this flat to the increasing word which we are constructing will be $\underbrace{y_i ... y_i}_{p_i \text{ times}}$, while if y_i contains no object at all, the letter y_i will not appear in the increasing word. Thus we shall obtain the following increasing word

$$\underbrace{y_1 y_1 ... y_1}_{p_1 \text{ times}} \underbrace{y_2 ... y_2}_{p_2 \text{ times}} ... \underbrace{y_m ... y_m}_{p_m \text{ times}}$$

with $\sum\limits_{i=1}^{m} p_i = n$. For example

$$x_2 | x_3 x_6 x_1 | x_4 x_6 | \quad | x_5 ... \rightarrow y_1 y_2 y_2 y_2 y_3 y_3 y_5 ...$$

$$y_1 \quad y_2 \quad y_3 \quad y_4 \quad y_5$$

It is also to be remarked that the nature of the objects placed in the flats has no influence upon the increasing word so formed (only their number in each flat is significant); therefore, if the objects are permuted in $n!$ different ways, the result will be the same increasing word. By the above method, the $[m]^n$ ordered arrangements of n objects in m flats produce all the increasing words, formed by n letters out of m symbols, but an increasing word results from exactly $n!$ different ordered arrangements of the objects in the flats. It follows that the number of increasing words formed by n letters from the set Y with m elements is equal to $\dfrac{[m]^n}{n!}$.

For $n = 3$ and $m = 4$, we obtain 20 as before. The increasing words formed by n symbols from a set of m symbols are also called *combinations with repetition* of the m objects of the set Y taken n at a time.

Example 1 (*de Moivre*). Let us find the number of ways of writing the natural number m as a sum of n nonnegative integers: $m = u_1 + u_2 + ... + u_n$, two sums differing with respect to the integers or to their order.

We say that two sums differ with respect to the terms when the set of terms contained in the first sum differs from the set of terms contained in the second sum (this terminology is also used in Chap. 5, 6 and 9).

If the partial sums are denoted by $s_p = \sum\limits_{k=1}^{p} u_k$, each sum will be defined by an increasing word $s_1 s_2 ... s_{n-1}$ having the property $0 \leqslant s_1 \leqslant s_2 \leqslant ... \leqslant s_{n-1} \leqslant m$. Indeed, $u_k = s_k - s_{k-1}$ for $2 \leqslant k \leqslant n - 1$, while $u_1 = s_1$ and $u_n = m - s_{n-1}$. The number sought will be equal to the number of increasing words formed by $n - 1$ letters from the set $\{0, 1, ..., m\}$ with $m + 1$ elements, therefore to

$$\frac{[m+1]^{n-1}}{(n-1)!} = \frac{(m+n-1)!}{m!\,(n-1)!}.$$

For $m = 5$ and $n = 2$, we obtain $\dfrac{6!}{1!\,5!} = 6$ different ways, namely:

$$5 = 0 + 5 = 5 + 0 = 1 + 4 = 4 + 1 = 2 + 3 = 3 + 2.$$

If we assume further that the finite set X is endowed with an order relation so that $x_1 < x_2 < \ldots < x_n$, then a function $f: X \to Y$ is called *increasing* if for any $x_i, x_j \in X$ with $x_i < x_j$, we have $f(x_i) \leqslant f(x_j)$ and f is called *strictly increasing* if $f(x_i) < f(x_j)$.

The above defined bijection from the set of words formed by n letters from Y, onto the set of functions $f: X \to Y$, where $|Y| = m$, $|X| = n$, is in this case a bijection from the set of increasing words formed by n letters from Y, onto the set of increasing functions $f: X \to Y$. Thus if $|X| = n$, $|Y| = m$ and the elements of X and of Y, respectively, can be ordered like a chain, then the number $\dfrac{[m]^n}{n!}$ also represents the number of increasing functions $f: X \to Y$.

In this case it is also said that the sets X and Y are *totally ordered*.

To each strictly increasing function $f: X \to Y$, there corresponds a strictly increasing word, i.e. a word $y_{i_1} y_{i_2} \ldots y_{i_n}$ such that $y_{i_1} < y_{i_2} < \ldots < y_{i_n}$.

In order to find the number of strictly increasing words formed by n letters from an alphabet of m letters (which is equal to the number of strictly increasing functions defined on a set of n elements with values in a set with m elements, both sets being totally ordered), we shall proceed as in the proof of proposition 4. Indeed, the set (containing $[m]_n$ elements) of words with n different letters from an alphabet of m letters, can be obtained without repetitions from the set of strictly increasing words formed by n letters from the same m-letter alphabet by permuting in $n!$ different ways the letters of each word. We thus obtain

$$\frac{[m]_n}{n!} = \frac{m!}{n!(m-n)!}$$

strictly increasing words. Therefore we have

Proposition 5. *The number of strictly increasing words formed by n letters out of m symbols is equal to* $\dfrac{[m]_n}{n!}$. *This number is also equal to the number of n-element subsets of a set Y with m elements.*

Labelling each element of the set Y by an index, we may assume that $y_1 < y_2 < \ldots < y_m$. There exists a bijection f between the family of subsets of Y with n elements and the set of strictly increasing words formed by n letters from Y, defined as follows: $f(Y_1) = y_{i_1} y_{i_2} \ldots y_{i_n}$, where $Y_1 \subset Y, |Y_1| = n$, and the elements $y_{i_1}, y_{i_2}, \ldots, y_{i_n}$ are the very elements of the set Y_1, written in increasing order of the indices, i.e. we suppose that $i_1 < i_2 < \ldots < i_n$.

The numbers $\dfrac{[m]_n}{n!}$ are called *combinations* of m objects taken n at a time and are denoted by C_n^m or by $\dbinom{m}{n}$. In this book we make use of the latter notation, which for the sake of simplification of certain formulae, is defined as follows:

$$\binom{m}{n} = \frac{[m]_n}{n!} = \frac{m(m-1)\dots(m-n+1)}{n!} \text{ for } n \neq 0 \text{ and } m \geq n;$$

$$\binom{m}{0} = 1 \text{ for } m \geq 0 \text{ and } \binom{m}{n} = 0 \text{ for } m < n.$$

The strictly increasing words formed by n letters from Y, as well as the subsets with n elements of Y, are also called combinations of m elements of the set Y taken n at a time.

Example 2. Let us find the number of ways of writing the natural number m as a sum of n positive integers, $m = u_1 + u_2 + \dots + u_n$ with u_k an integer, $u_k > 0$ for $1 \leq k \leq n$; two such sums to be considered as distinct if they differ from each other in respect of the terms u_k or of their order.

By denoting, as in example 1, the partial sums by $s_p = \sum_{k=1}^{p} u_k$, each sum is completely determined by the strictly increasing word $s_1 s_2 \dots s_{n-1}$ with the property $1 \leq s_1 < s_2 < \dots < s_{n-1} \leq m-1$; therefore the number of ways of writing m as a sum of n positive integers is equal to the number of strictly increasing words formed from $n-1$ letters from an alphabet of $m-1$ letters: $\{1, 2, \dots, m-1\}$, i.e. to

$$\binom{m-1}{n-1}.$$

The numbers $\binom{m}{n}$ satisfy the recurrence relation $\binom{m}{n} = \binom{m-1}{n} + \binom{m-1}{n-1}$, which may be taken as a basis for computing these numbers with the aid of the *Pascal triangle*. They also satisfy $\binom{m}{n} = \binom{m}{m-n}$, which is called *the complementary combinations formula* and is useful for computing these numbers when $n > m-n$. These relations can be deduced by an elementary computation, taking into account the definition of combinations. We have $\binom{m}{0} = 1$, then these numbers increase as the lower index increases and reach a maximum for $n = \dfrac{m}{2}$ when m is even, or two equal maxima for $n = \dfrac{m-1}{2}, \dfrac{m+1}{2}$ when m is odd, and then they decrease to $\binom{m}{m} = 1$. The numbers $\binom{m}{n}$ are also called binomial numbers, because they appear as coefficients in the binomial formula:

$$(a + b)^n = \sum_{k=0}^{n} \binom{n}{k} a^k b^{n-k}. \tag{2.3}$$

To prove this formula, let us consider the product $(a_1 + b_1)(a_2 + b_2) \dots (a_n + b_n)$. For any $K \subset N = \{1, 2, \dots, n\}$, set

$$a^K = \prod_{i \in K} a_i \text{ and } b^K = \prod_{i \in K} b_i.$$

For every product of factors a_i and b_j, obtained by multiplying the parentheses, the set of indices of the factors a_i will be complementary to the set of indices of the factors b_j with respect to the set $\{1, 2, \dots, n\}$. Indeed, in such a product of factors a_i and b_j, all indices in the set N appear once only. Therefore,

$$\prod_{k=1}^{n} (a_k + b_k) = \sum_{K \subset N} a^K b^{N \setminus K}$$

where we define

$$a^{\varnothing} b^N = b^N \text{ and } a^N b^{\varnothing} = a^N.$$

Now taking $a_1 = a_2 = \dots = a_n = a$ and $b_1 = b_2 = \dots = b_n = b$, we obtain formula (2.3), since a set $K \subset N$ with k elements can be chosen in $\binom{n}{k}$ different ways and for any choice of $K \subset N$ with $|K| = k$, all products $a^K b^{N \setminus K}$ have the same value, namely $a^k b^{n-k}$.

Formula (2.3) is true for any elements a, b of a commutative ring and in particular for the real or complex numbers. For $a = b = 1$ in (2.3), we obtain the sum of the binomial numbers $\sum_{k=0}^{n} \binom{n}{k} = 2^n$.

Various generalizations can be given to the binomial formula, by making use of the concept of a derivative operator associated with a normal family of polynomials.

A *normal family of polynomials* is a set of polynomials, $P_0(x)$, $P_1(x)$, $P_2(x)$, ..., of a real variable x, having the property $P_0(x) = 1$ and such that for any $n \geqslant 1$, $P_n(x)$ is a polynomial of degree n in x which vanishes for $x = 0$.

A *derivative operator* D associated with the polynomials $P_n(x)$ is a function that associates with each polynomial $p(x)$, the polynomial denoted by $\mathrm{D}p(x)$, such that:

1) D is linear, i.e.:

$$\mathrm{D}(\lambda p(x)) = \lambda \mathrm{D}p(x)$$

for any real λ and

$$\mathrm{D}(p_1(x) + p_2(x)) = \mathrm{D}p_1(x) + \mathrm{D}p_2(x)$$

for any polynomials $p_1(x)$ and $p_2(x)$.

2) D is applied to the polynomials $P_n(x)$ of the normal family as follows:

$$DP_n(x) = nP_{n-1}(x), \text{ if } n \neq 0$$

and

$$DP_n(x) = 0, \text{ if } n = 0.$$

PROPOSITION 6. *For every normal family of polynomials $P_n(x)$, there exists one and only one derivative operator D.*

The proof follows from the fact that any polynomial $p_n(x)$ of degree n is uniquely expressible in the form

$$p_n(x) = a_n P_n(x) + a_{n-1}P_{n-1}(x) + \ldots + a_0 P_0(x),$$

where $a_n, a_{n-1}, \ldots, a_0$ are real numbers. For, let us denote by a_n the quotient of the coefficient of x^n in $p_n(x)$ by the coefficient of x^n in $P_n(x)$, then $p_{n-1}(x) = p_n(x) - a_n P_n(x)$ is a polynomial of degree at most $n - 1$. If we continue this procedure, denoting by a_{n-1} the quotient of the coefficient of x^{n-1} in $p_{n-1}(x)$ by the coefficient of x^{n-1} in $P_{n-1}(x)$, we shall find that $p_{n-2}(x) = p_{n-1}(x) - a_{n-1}P_{n-1}(x)$ is a polynomial of degree at most $n - 2$, etc. and finally we shall obtain the required expansion.

The uniqueness of this expression follows immediately, since the polynomials $P_n(x)$ being of different degrees, constitute a basis for the vector space of polynomials with real coefficients. Now the operator D is defined by relations 2) for the polynomials $P_n(x)$, while for an arbitrary polynomial $p_n(x)$ of degree n it must be defined by

$$Dp_n(x) = na_n P_{n-1}(x) + (n-1)a_{n-1}P_{n-2}(x) + \ldots + a_1,$$

in order to satisfy axiom 1). The operator D is unique since the expression of $p_n(x)$ as a sum of polynomials from the given normal family is unique.

PROPOSITION 7. *If $p_n(x)$ is a polynomial of degree n in x, then the unique expression of $p_n(x)$ as a sum of polynomials from the normal family has the form*

$$p_n(x) = p_n(0) + \frac{Dp_n(0)}{1!}P_1(x) + \frac{D^2p_n(0)}{2!}P_2(x) + \ldots + \frac{D^np_n(0)}{n!}P_n(x). \quad (2.4)$$

We first mention that $D^k p_n(x)$ represents the result of applying the operator D k times to the polynomial $p_n(x)$, while $D^k p_n(0)$ represents the value at $x = 0$ of this polynomial. We have seen from the proof of the previous proposition that there is a unique expansion

$$p_n(x) = a_0 P_0(x) + a_1 P_1(x) + \ldots + a_n P_n(x).$$

It still remains to be shown that $a_k = \dfrac{D^k p_n(0)}{k!}$. Taking $x = 0$, we obtain $p_n(0) = a_0$, since $P_0(x) = 1$. By applying the linear operator D to both sides, we obtain

$$Dp_n(x) = a_1 P_0(x) + 2a_2 P_1(x) + \dots + na_n P_{n-1}(x).$$

Taking $x = 0$, we obtain $Dp_n(0) = a_1$. By applying D again, $D^2 p_n(0) = 2a_2 P_0(x) + 3 \cdot 2 \, a_3 P_1(x) + \dots + n(n-1) \, a_n P_{n-2}(x)$, from which we deduce $D^2 p_n(0) = 2a_2$. After obtaining $D^k p_n(x) = k! \, a_k P_0(x) + \dots + n(n-1) \dots (n-k+1) \, a_n P_{n-k}(x)$, we deduce $D^k p_n(0) = k! a_k$ and by applying the operator D again, we obtain $D^{k+1} p_n(x) = (k+1)! a_{k+1} P_0(x) + \dots + n(n-1) \dots (n-k) a_n P_{n-k-1}(x)$. We have thus obtained by induction an expression for $D^k p_n(x)$, from which we deduce $D^k p_n(0) = k! a_k$ for any $k = 0, \dots, n$.

Taking $P_k(x) = x^k$ for any k, the derivative operator associated with this normal family is the usual derivative $\dfrac{dp_n(x)}{dx}$ and formula (2.4) is the classical *Taylor-Maclaurin formula* for polynomials. If, for example, we consider the polynomial $p_n(x) = (x + y)^n$, and expand it according to formula (2.4), with $Dp_n(x) = \dfrac{dp_n(x)}{dx}$, we obtain

$$Dp_n(x) = n(x + y)^{n-1}; \qquad D^2 p_n(x) = n(n-1)(x+y)^{n-2},$$

and generally

$$D^k p_n(x) = n(n-1) \dots (n-k+1)(x+y)^{n-k}$$

giving the Taylor formula

$$(x + y)^n = y^n + \binom{n}{1} xy^{n-1} + \binom{n}{2} x^2 y^{n-2} + \dots + \binom{n}{n} x^n,$$

i.e. the binomial formula.

Let us now consider the normal family of polynomials $P_n(x) = [x]_n = x(x - 1)\dots(x - n + 1)$ and $P_0(x) = 1$. The derivative operator associated with this normal family is the operator Δ which transforms the polynomial $p_n(x)$ as follows:

$$\Delta p_n(x) = p_n(x + 1) - p_n(x).$$

For

$$\Delta [x]_n = [x + 1]_n - [x]_n = (x + 1)[x]_{n-1} - (x - n + 1)[x]_{n-1} = n[x]_{n-1}.$$

Let us expand the polynomial $p_n(x) = [x + y]_n$ according to formula (2.4). We obtain

$$\Delta^k p_n(x) = n(n-1) \dots (n-k+1)[x+y]_{n-k},$$

therefore

$$[x + y]_n = \sum_{k=0}^{n} \binom{n}{k} [x]_k [y]_{n-k}. \tag{2.5}$$

Formula (2.5) is called the *Vandermonde formula*.

Let us consider the polynomials $P_n(x) = [x]^n = x(x + 1) \dots (x + n - 1)$ and $P_0(x) = 1$, which form a normal family. The derivative operator associated with this normal family is denoted by ∇ and, for any polynomial $p_n(x)$, it is defined as follows:

$$\nabla p_n(x) = p_n(x) - p_n(x - 1).$$

∇ is linear and therefore

$$\nabla [x]^n = [x]^n - [x - 1]^n = (x + n - 1) [x]^{n-1} + (x - 1) [x]^{n-1} = n[x]^{n-1}.$$

According to the uniqueness of the derivative operator associated with a normal family, ∇ is defined by the above equation for any polynomial. We thus have

$$\nabla^k [x + y]^n = n(n - 1) \dots (n - k + 1) [x + y]^{n-k}$$

hence formula (2.4) becomes

$$[x + y]^n = \sum_{k=0}^{n} \binom{n}{k} [x]^k [y]^{n-k}. \tag{2.6}$$

This expansion is called the *Nörlund formula*.

PROPOSITION 8. *The number of arrangements of a set of objects* $X = \{x_1, x_2, \dots, x_n\}$ *in a set* $Y = \{y_1, y_2, \dots, y_p\}$ *of flats such that flats* $y_1, y_2, \dots, y_p$ *contain* $n_1, n_2, \dots, n_p$ *objects respectively,* $(n_1 + n_2 + \dots + n_p = n)$ *is equal to*

$$\frac{n!}{n_1! n_2! \dots n_p!}.$$

The n_1 objects in flat y_1 can be chosen in $\binom{n}{n_1}$ different ways, n_2 of the $n - n_1$ remaining objects can be placed in flat y_2 in $\binom{n - n_1}{n_2}$ different ways, etc., resulting in a total number of possibilities equal to

$$\binom{n}{n_1} \binom{n - n_1}{n_2} \binom{n - n_1 - n_2}{n_3} \dots \binom{n_p}{n_p}$$

$$= \frac{n!}{n_1!(n - n_1)!} \cdot \frac{(n - n_1)!}{n_2!(n - n_1 - n_2)!} \dots \frac{n_p!}{n_p!} = \frac{n!}{n_1! n_2! \dots n_p!}.$$

The numbers $\dfrac{n!}{n_1!n_2!\ldots n_p!}$ are also denoted by $\begin{pmatrix} n \\ n_1, n_2, \ldots, n_p \end{pmatrix}$ and are called *multinomial numbers;* they generalize combinations (the case $p = 2$) and appear in the *multinomial formula* that extends relation (2.3):

$$(a_1 + a_2 + \ldots + a_p)^n = \sum_{\substack{n_1,\ldots,n_p \geqslant 0 \\ n_1+\ldots+n_p=n}} \begin{pmatrix} n \\ n_1, n_2, \ldots, n_p \end{pmatrix} \cdot a_1^{n_1} a_2^{n_2} \ldots a_p^{n_p}. \qquad (2.7)$$

To prove this formula, we proceed as in the demonstration of the binomial formula. We consider the product

$$(a_1^1 + a_2^1 + \ldots + a_p^1)(a_1^2 + a_2^2 + \ldots + a_p^2) \ldots (a_1^n + a_2^n + \ldots + a_p^n)$$

$$= \sum_{\substack{n_1,\ldots,n_p \geqslant 0 \\ n_1+\ldots+n_p=n}} (a_1^{i_1} a_1^{i_2} \ldots a_1^{i_{n_1}})(a_2^{j_1} a_2^{j_2} \ldots a_2^{j_{n_2}}) \ldots (a_p^{k_1} a_p^{k_2} \ldots a_p^{k_{n_p}}), \qquad (2.8)$$

where the upper indices in the sum, do not indicate powers and are pairwise different in each term. The latter sum indicates, by each parenthesis, an arrangement of the set of objects $\{1, 2, \ldots, n\}$ in the flats $y_1, y_2, \ldots, y_p$, with flat y_1 containing n_1 objects, ..., flat y_p containing n_p objects.

Indeed, by placing objects $i_1, i_2, \ldots, i_{n_1}$ in flat y_1, ..., objects $k_1, k_2, \ldots, k_{n_p}$ in flat y_p, we do obtain such an arrangement and by so doing, we obtain without repetitions all the arrangements in flats with the stated properties, their number being equal to

$$\frac{n!}{n_1!n_2!\ldots n_p!}.$$

Now if we take $a_i^1 = a_i^2 = \ldots = a_i^n = a_i$ for each $i = 1, 2, \ldots, p$, the sum in (2.8) becomes

$$\sum_{\substack{n_1,\ldots,n_p \geqslant 0 \\ n_1+\ldots+n_p=n}} \begin{pmatrix} n \\ n_1, n_2, \ldots, n_p \end{pmatrix} a_1^{n_1} a_2^{n_2} \ldots a_p^{n_p}$$

(where the upper indices do now indicate powers) since the products contained in (2.8) generate $\begin{pmatrix} n \\ n_1, n_2, \ldots, n_p \end{pmatrix}$ terms equal to $a_1^{n_1} a_2^{n_2} \ldots a_p^{n_p}$ for every decomposition of n into the form $n = n_1 + \ldots + n_p$.

Example 3. Let us show that

$$\begin{pmatrix} n \\ n_1, n_2, \ldots, n_p \end{pmatrix} = \sum_{\{i \,|\, n_i \neq 0\}} \begin{pmatrix} n-1 \\ n_1, n_2, \ldots, n_{i-1}, n_i - 1, n_{i+1}, \ldots, n_p \end{pmatrix}. \qquad (2.9)$$

To this end, we write

$$(a_1 + a_2 + \ldots + a_p)^n = (a_1 + \ldots + a_p)(a_1 + \ldots + a_p)^{n-1}.$$

By equating the coefficients of the term $a_1^{n_1} a_2^{n_2} \ldots a_p^{n_p}$ on both sides, we obtain (2.9), since

$$\begin{pmatrix} n \\ n_1, n_2, \ldots, n_p \end{pmatrix} a_1^{n_1} \ldots a_p^{n_p} = \sum_{\{i \mid n_i \neq 0\}} a_i \begin{pmatrix} -1 \\ n_1, \ldots, n_i-1, \ldots, n_p \end{pmatrix} a_1^{n_1} \ldots a_i^{n_i-1} \ldots a_p^{n_p}.$$

Example 4. The following relation holds:

$$\sum_{k} \sum_{\substack{s_i \geqslant 0; \ s_1+s_2+\ldots+s_k=h \\ s_1+2s_2+\ldots+ks_k=m}} \begin{pmatrix} h \\ s_1, s_2, \ldots, s_k \end{pmatrix} = \begin{pmatrix} m-1 \\ h-1 \end{pmatrix}. \tag{2.10}$$

Let us consider all the decompositions of m as a sum of h positive integers, the number of which, according to example 2, is equal to $\begin{pmatrix} m-1 \\ h-1 \end{pmatrix}$. If we denote by s_i the number of terms equal to i in such a sum, then $\sum_{i=1}^{k} s_i = h$ and $\sum_{i=1}^{k} is_i = m$, where k stands for the number of sets of terms such that the terms of one set are equal to one another, whereas any two terms belonging to different sets are different.

If the numbers s_i having these properties are fixed, the number of ways of writing m as a sum of h positive terms, so that s_i terms be equal to i, is equal to the number of arrangements of a set of h objects in a set of k flats, so that the first flat contains s_1 objects, the second s_2 objects,..., the last s_k objects, any two sums being different from each other only in the order of the terms, not with respect to their nature; therefore this number is $\begin{pmatrix} h \\ s_1, s_2, \ldots, s_k \end{pmatrix}$. Hence counting in two different ways the number of possible decompositions of m as a sum of h positive integers, we shall obtain the relation (2.10).

Example 5. Let us consider the graph in Fig. 2.3 representing the graph associated with a contact network, which realizes a symmetric Boolean function [23].

Each node may be characterized by two co-ordinates, one corresponding to the horizontal side and the other to the oblique side of the graph. By so doing, the co-ordinates of the input are (0, 0), while those of any point M are (p, q), provided that any path joining input I to node M contains p horizontal and q oblique arcs.

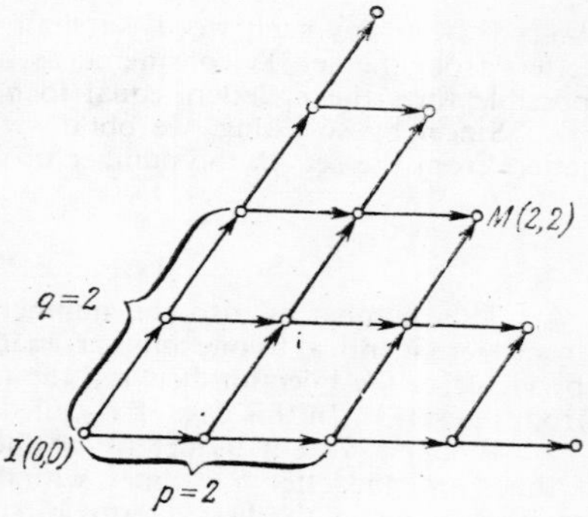

We recall that a path in a graph is a sequence of nodes pairwise joined by arcs so that, with the possible exception of the initial and the final nodes, the end of an arc reaching this node coincides with the origin of the following arc. Thus, $(x_1, x_2, \ldots, x_n)$ constitutes a path, provided that there is an arc going from x_1 to x_2, an arc from x_2 to $x_3, \ldots$, and an arc from x_{n-1} to x_n.

The problem now is to determine the number of paths between the input $I(0, 0)$ and an arbitrary node $M(p, q)$ of the graph. To this end, let us remark that any path joining input I to node $M(p, q)$ contains $p + q$ arcs, from which p have to be horizontal. Since the number of ways of choosing the p horizontal arcs out of the $p + q$ arcs is equal to $\begin{pmatrix} p+q \\ p \end{pmatrix}$, this is also the number of paths between input I and node $M(p, q)$.

Fig. 2.3.

More generally, given two nodes $M_1(p_1, q_1)$ and $M_2(p_2, q_2)$ of this graph, if $p_1 \leqslant p_2$ and $q_1 \leqslant q_2$, there exist paths between M_1 and M_2, the number of which is equal to

$$\binom{p_2 - p_1 + q_2 - q_1}{p_2 - p_1} = \binom{p_2 - p_1 + q_2 - q_1}{p_2 - p_1, \ q_2 - q_1}.$$

This problem can be generalized as follows:

Let us denote by N^p, the set of ordered systems of p non-negative integers $a = (a_1, a_2, \ldots, a_p)$. A partial order relation (reflexive, antisymmetric, transitive), can be defined on the set N^p as follows: we say that $a \leqslant b$ if $a_i \leqslant b_i$ for every $i = 1, 2, \ldots, p$. This set can be represented by an infinite graph whose nodes are the elements a of the Cartesian product N^p, with an arc joining two nodes a and b, oriented from a to b, if and only if there is an index i such that $b_i = a_i + 1$ and $b_k = a_k$ for any $k \neq i$.

If $a \leqslant b$, the number of paths from node a to node b is equal to the multinomial number

$$\binom{\sum_{i=1}^{p} (b_i - a_i)}{b_1 - a_1, \ldots, b_p - a_p}.$$

Indeed, to go from a to b, we must add $b_1 - a_1$ to the first component of a, $b_2 - a_2$ to the second component, ..., $b_p - a_p$ to the p-th component, the order of these operations being arbitrary; therefore, the number of possibilities is equal to the number of arrangements of $\sum_{i=1}^{p} (b_i - a_i)$ objects in p flats, so that the first flat contains $b_1 - a_1$ objects, the second $b_2 - a_2$ objects, ..., and the last $b_p - a_p$ objects.

Example 6. If we consider the words formed by n letters from the n-element set Y, such that not all of them be different from one another, but n_1 symbols of Y be equal to a_1, n_2 symbols to $a_2, \ldots$, and n_p symbols to a_p, then the number of these words will be equal to

$$\binom{n}{n_1, n_2, \ldots, n_p}.$$

Indeed, from any such word we shall be able to form $n_1! n_2! \ldots n_p!$ distinct words formed of n letters from the set Y, considered as having all its letters pairwise distinct, by permuting in all possible ways the n_1 letters equal to $a_1, \ldots$, and the n_p letters equal to a_p.

Since, by so doing, we obtain without repetition all words formed by n pairwise distinct letters from the set Y, the number of which is $n!$, it follows that the number sought is equal to

$$\frac{n!}{n_1! \, n_2! \, \ldots \, n_p!}.$$

This number is also the number of bijections $f: X \to Y$ with $|X| = |Y| = n$ if n_1 objects from Y are identical to one another, n_2 identical to one another, ..., and n_p identical to one another, by virtue of the bijection between the functions $f: X \to Y$ and the words formed by $|X|$ symbols from the set Y. In this case Y is called a multiset.

Example 7. Let us determine the number of sequences of letters containing the letter a k times and the letter b m times, with the following property (P): for any i, $1 \leqslant i \leqslant m + k$, the number of a's in the first i terms of the sequence is not less than the number of b's.

It is clear that for this number to be greater than zero, it is necessary that the conditi on $k \geqslant m \geqslant 0$ be met.

The number of sequences of letters containing the letter a k times and the letter b m times is equal to $P(m, k) = \dfrac{(m + k)!}{m! \, k!} = \dbinom{m + k}{m}$. If we find the number of sequences which do not satisfy (P), then, by subtracting it from the number $\binom{m+k}{m}$, we shall have the answer to our problem. Let us first prove the following assertion: the number of sequences of m b's and k a's which do not satisfy (P) is equal to $P(m-1,\ k+1) = \binom{m+k}{m-1}$, that is, to the total number of sequences of $m-1$ b's and $k+1$ a's. The proof runs as follows. Take any sequence of m b's and k a's which does not satisfy (P). There exists a position $2s+1$ where $s \geqslant 0$, such that the sequence contains the letter b in the position $2s+1$ and in front of this position there are s a's and s b's. We take the smallest such s and place the letter a in front of this sequence, thus obtaining a sequence of m b's and $k+1$ a's, the first letter of the sequence being a and such that there are as many a's as b's among the first $2s+2$ letters. We shall change the b's to a's and the a's to b's in the first $2s+2$ positions of the sequence. Since in the first $2s+2$ positions the number of a's was equal to the number of b's, there will be no change in the total number of letters of each kind, and we get a sequence of m b's and $k+1$ a's. Now, the first letter is b. Thus, we have associated with each sequence of m b's and k a's which does not satisfy (P), a sequence of m b's and $k+1$ a's beginning with the letter b. This mapping is obviously one-to-one. We shall now demonstrate that in ths way it is possible to obtain any sequence of m b's and $k+1$ a's beginning with the letter b, i.e. this mapping is also onto.

Indeed, take such a sequence. Since we assume that $m \leqslant k$, hence $m < k+1$, there will be a position at which the letters a and b even out. If, up to the first such position, we replace all a's by b's and all b's by a's and delete the first letter a, then we get a sequence of a's and b's which does not satisfy (P). Now if we apply to this sequence the described mapping we get the initial sequence. We have thus established that the number of sequences of m b's and k a's which do not satisfy (P) is equal to the number of those sequences of m b's and $k+1$ a's which begin with the letter b.

If we delete the first letter, we get all possible sequences of $m-1$ b's and $k+1$ a's. Now the number of such sequences is $P(m-1,\ k+1) = \binom{m+k}{m-1}$. Since the number of all sequences of m b's and k a's is $\binom{m+k}{m}$, the number of sequences which satisfy (P) is equal to $\binom{m+k}{m} - \binom{m+k}{m-1} = \frac{k-m+1}{k+1}\binom{m+k}{m}$. For $k=m=n-1$ we obtain the so-called Catalan number $T_n = \frac{1}{n}\binom{2n-2}{n-1}$.

Example 8. We have seen that every subset U of a set X with n elements $x_1, x_2, \ldots, x_n$ is associated with a binary word $c_U = c_1 c_2 \ldots c_n$, defined as follows: $c_i = 1$ if $x_i \in U$ and $c_i = 0$ otherwise, and that this correspondence is a bijection from the family of subsets of X onto the set of binary words of length n. We now define a partial order relation (reflexive, antisymmetric, transitive) on the set of binary words of length n, as follows: we say that $a = a_1 \ldots a_n$ is less than or equal to $b = b_1 \ldots b_n$ and we write $a \leqslant b$, if $a_i \leqslant b_i$ for $i = 1, \ldots, n$.

Since the inclusion relation between the subsets of X, namely $U \subset V$, is equivalent to the relation $c_U \leqslant c_V$ in the set of binary words of length n, the determination of the greatest number of subsets of X which are pairwise incomparable with respect to set inclusion, reduces to the determination of the greatest number of binary words which are pairwise incomparable with respect to the relation $\leqslant$ in the set of binary words.

PROPOSITION 9 (Sperner). *The greatest number of subsets of X, where $|X|=n$, which are pairwise incomparable with respect to set inclusion, is equal to* $\binom{n}{\left[\frac{n}{2}\right]}$.

The proof amounts to showing that the greatest number of pairwise incomparable words

of length n, equals $\left(\begin{array}{c} n \\ \left[\dfrac{n}{2}\right] \end{array}\right)$, where $\left[\dfrac{n}{2}\right]$ is the integer part of $\dfrac{n}{2}$.

The set of binary n-letter words can be identified with a hypercube in n dimensions, i.e. with a non-oriented graph whose vertices are the 2^n binary words with n positions, an edge joining two vertices if any only if the corresponding binary words differ in a single position. For $n = 1, 2, 3, 4$ these graphs are illustrated in Fig. 2.4.

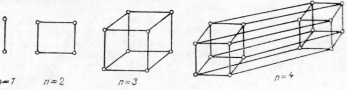

$n=1$ $n=2$ $n=3$ $n=4$

Fig. 2.4

The set of binary n-letter words, having $\left[\dfrac{n}{2}\right]$ positions equal to zero and $n - \left[\dfrac{n}{2}\right]$

positions equal to 1, contains $\left(\begin{array}{c} n \\ \left[\dfrac{n}{2}\right] \end{array}\right)$ elements, because the $\left[\dfrac{n}{2}\right]$ zero positions can be

chosen, from the set of n positions, in $\left(\begin{array}{c} n \\ \left[\dfrac{n}{2}\right] \end{array}\right)$ different ways, these binary words being

pairwise incomparable with respect to the order relation $\leqslant$ previously introduced. Therefore a

lower bound for the number sought is equal to $\left(\begin{array}{c} n \\ \left[\dfrac{n}{2}\right] \end{array}\right)$. To demonstrate the reverse inequality,

we shall use a construction of G. Hansel [11] which is realized by dividing the set of

n-dimensional hypercube vertices into $\left(\begin{array}{c} n \\ \left[\dfrac{n}{2}\right] \end{array}\right)$ chains with no common elements. A chain is,

by definition, a totally ordered set of elements, i.e. any two elements of it are comparable with respect to the order relation.

Since any two words from a set of binary words that are pairwise incomparable relative to the relation $\leqslant$ belong to different chains, it follows that the reverse inequality is also true.

We shall show that there exists a division of the vertex set of the n-dimensional hypercube

(denoted by B^n) into $\left(\begin{array}{c} n \\ \left[\dfrac{n}{2}\right] \end{array}\right)$ non-empty chains with no common elements, having the property

that the number of chains containing $n - 2p + 1$ vertices is equal to $\left(\begin{array}{c} n \\ p \end{array}\right) - \left(\begin{array}{c} n \\ p-1 \end{array}\right)$ for any

$0 \leqslant p \leqslant \left[\dfrac{n}{2}\right]$, where the notation $\left(\begin{array}{c} n \\ -1 \end{array}\right) = 0$ is adopted. The proof of this property is carried out by induction with respect to n. For $n = 1, 2$, the property is true, as can be seen from Fig. 2.4. Assuming that the property is true for $n - 1$, let us prove it for n.

Let us denote by B_0^n the set of words $c = x_1 \ldots x_n$ with $x_n = 0$ and by B_1^n the set of words c with $x_n = 1$. By hypothesis, there are partitions of B_0^n and B_1^n into chains having the property that any chain J_0 from the partition of B_0^n can be obtained from a chain J_1 from the partition of B_1^n, by changing the last component 1 into 0 for all the words in the chain.

We perform the following transformation: in every chain J_1 from the partition of B_1^n, we delete its greatest element c_1, and add it to the chain J_0, that corresponds to J_1 according to the correspondence previously described. The sets $J = J_0 \cup \{c_1\}$ and $J' = J_1 \backslash \{c_1\}$ are also chains of n-component binary words. We perform this transformation for all pairs of chains $J_0 \in B_0^n$ and $J_1 \in B_1^n$ and obtain a partition into chains of the set of binary n-letter words.

Each chain with k elements of the initial partition of B_0^n will yield a chain with $k + 1$ elements in the final partition of the vertex set of the n-dimensional hypercube, and each chain of B_1^n a chain with $k - 1$ elements. According to the inductive hypothesis, the number of chains with $n - 2p$ elements in the initial partition of B_0^n is equal to $\binom{n-1}{p} - \binom{n-1}{p-1}$, where $0 \leqslant p \leqslant \left[\dfrac{n-1}{2}\right]$, and the number of chains with $n - 2p + 2$ elements of B_1^n is equal to

$$\binom{n-1}{p-1} - \binom{n-1}{p-2}, \quad \text{with } 1 \leqslant p \leqslant \left[\frac{n}{2}\right].$$

The number of $(n - 2p + 1)$-element chains in the final partition of the set of vertices of the n-dimensional hypercube B^n is equal to

$$\binom{n-1}{p} - \binom{n-1}{p-1} + \binom{n-1}{p-1} - \binom{n-1}{p-2} = \binom{n}{p} - \binom{n}{p-1},$$

with $0 \leqslant p \leqslant \left[\dfrac{n}{2}\right]$, taking into account the recurrence relation for the binomial numbers.

If these numbers are summed over $p = 0, \ldots, \left[\dfrac{n}{2}\right]$, a total of $\left(\dfrac{n}{\left[\dfrac{n}{2}\right]}\right)$ chains formed by binary n-letter words is obtained.

For $n = 1, 2, 3$ and 4, these partitions into chains are:

1	11		111			1111	
0	10	01	110	101	011	1110 1101 1011 0111	
	00		100	001	010	1100 1001 1010 0110 0011 0101	
			100			1000 0001 0010 0100	
						0000	

$$n = 1 \qquad n = 2 \qquad n = 3 \qquad n = 4$$

An r-decomposition of a finite n-element set X is a mapping $\delta: \{1, 2, \ldots, r\} \to \mathscr{P}(X)$ such that $\bigcup_{i=1}^{r} \delta(i) = X$ and $\delta(i) \cap \delta(j) = \varnothing$ for $i \neq j$. An s-system of order r is a family σ of r-decompositions having the following property: for any i, the sets in the family $\{\delta(i) \mid \delta \in \sigma\}$

are pairwise incomparable with respect to set inclusion. Meshalkin [19] has proved the following generalization of Sperner's result: for any s-system σ, its cardinality satisfies

$$|\sigma| \leqslant \max_{\Sigma n_i = n} \frac{n!}{n_1! \ldots n_r!}.$$

A proof of the uniqueness of maximal s-systems is given in [13].

Other generalizations of Sperner's theorem were given by P. Erdös [7], E. Milner [20], G. Katona [14], [15], D. Kleitman [17] and J. Schönheim [22].

Problems

1. Prove that

$$\sum_{\alpha_1 + 2\alpha_2 + \ldots + n\alpha_n = p} \frac{k!}{\alpha_1! \ldots \alpha_n! \left(k - \sum_{i=1}^{n} \alpha_i\right)!} \binom{n}{1}^{\alpha_1} \ldots \binom{n}{n}^{\alpha_n} = \binom{nk}{p}$$

for any $p \leqslant k$.

(For $p = k - 1$ see Voloshin, 1972)

H i n t : Let X be a finite set with $|X| = nk$ elements and $X = A_1 \cup \ldots \cup A_k$ be a partition of X such that $|A_1| = |A_2| = \ldots = |A_k| = n$.

Count in two different ways the subsets of X having p elements.

2. Setting

$$\alpha_n = \binom{n}{0} + \binom{n}{3} + \binom{n}{6} + \ldots$$

$$\beta_n = \binom{n}{1} + \binom{n}{4} + \binom{n}{7} + \ldots$$

$$\gamma_n = \binom{n}{2} + \binom{n}{5} + \binom{n}{8} + \ldots,$$

prove that two of these numbers are equal and differ from the third by one.

H i n t : Apply Newton's binomial formula to the expression $(1 + \omega)^n$, where $\omega = \dfrac{-1 - i\sqrt{3}}{2}$.

3. Prove that $M(n, k, h) \leqslant \binom{n - k + h}{h}$.

(Tomescu, 1973)

H i n t : See problem 1, Chap. 1.

4. A graph with n labelled vertices consists of a set of vertices, denoted by $x_1, x_2, ..., x_n$, and a set of edges, joining some pairs of these vertices. Prove that the number of graphs with n labelled vertices and m edges, m_1 being coloured with the colour $a_1, ..., m_q$ being coloured with the colour a_q $(m_1 + m_2 + ... + m_q = m)$, is equal to

$$\binom{\binom{n}{2}}{\binom{n}{2} - m, \ m_1, \ m_2, \ ..., \ m_q}.$$

5. Let $A_1, A_2, ..., A_n$ be finite sets such that

$$|A_1| = |A_2| = ... = |A_n| \text{ and } \bigcup_{i=1}^{n} A_i = S.$$

For a fixed k $(1 \leqslant k \leqslant n)$, any union of k sets of this family equals S and any union of less than k sets is a proper subset of S. Show that:

i) $|S| \geqslant \binom{n}{k-1}$;

ii) If S is minimal, then $|A_i| = \binom{n-1}{k-1}$ for any $i = 1, 2, ..., n$ and the

number of elements common to any j sets, when S is minimal, equals $\binom{n-j}{k-1}$.

H i n t: Setting $M(x) = \{i \mid i = 1, 2, ..., n \text{ and } x \in A_i\}$ show that for any subset $T \subset \{1, 2, ..., n\}$ with $|T| = n - k + 1$, there exists $x \in S$ such that $M(x) = T$.

6. Prove that the number of sequences $(a_1, a_2, ..., a_{k+1})$ of nonnegative integers with the properties: $a_1 = 0$ and $|a_i - a_{i+1}| = 1$ is equal to $\left(\left[\dfrac{k}{2} \right] \right)$.

(Carlitz, 1971)

7. Let $g_0(n + 1)$ be the number of sequences $(a_1, a_2, ..., a_{n+1})$ of nonnegative integers such that $a_1 = 0$ and

$$|a_i - a_{i+1}| \leqslant 1 \text{ for } i = 1, 2, ..., n.$$

Prove that $g_0(n + 1) = c(n, n) + c(n, n + 1)$, where

$$(1 + x + x^2)^m = \sum_{k \geqslant 0} c(m, k)x^k.$$

(Carlitz, 1971)

8. Show that the number of sequences $(a_1, a_2, ..., a_{n+1})$ with nonnegative integers such that $a_1 = a_{n+1} = 0$ and $|a_i - a_{i+1}| = 1$ for $i = 1, 2, ..., n$ equals the Catalan number $\dfrac{1}{n+1}\dbinom{2n}{n}$.

(Carlitz, 1971)

H i n t: Use the sequence of partial sums for the sequence of exercise 26.

9. Prove the Erdös — Chao Ko — Rado Theorem:

If $X = \{1, 2, ..., n\}$ and $A_1, A_2, ..., A_m$ are different subsets of X such that: $|A_i| = k$ for $1 \leqslant i \leqslant m$; $k \leqslant \dfrac{n}{2}$ fixed and $A_i \cap A_j = \varnothing$ for $1 \leqslant i < j \leqslant m$, then

$$m \leqslant \binom{n-1}{k-1}.$$

H i n t: See a simple proof due to Katona in [16].

10. Prove that the number, denoted by $\begin{bmatrix} n \\ k \end{bmatrix}_q$ and called a Gaussian coefficient, of subspaces of dimension k of an n-dimensional vector space V over a finite field F with q elements, where q is a power of a prime, is equal to

$$\begin{bmatrix} n \\ k \end{bmatrix}_q = \frac{(q^n - 1)(q^{n-1} - 1)...(q^{n-k+1} - 1)}{(q^k - 1)(q^{k-1} - 1)...(q - 1)}.$$

11. Show that $\lim\limits_{q \to 1} \begin{bmatrix} n \\ k \end{bmatrix}_q = \binom{n}{k}$.

12. Derive the following properties of the Gaussian coefficients:

$$\begin{bmatrix} n \\ k \end{bmatrix}_q = \begin{bmatrix} n \\ n-k \end{bmatrix}_q,$$

$$\begin{bmatrix} n \\ k \end{bmatrix}_q = \begin{bmatrix} n-1 \\ k-1 \end{bmatrix}_q + q^k \begin{bmatrix} n-1 \\ k \end{bmatrix}_q. \qquad \text{(q-Pascal triangle).}$$

13. Prove that the number $N_{k, 1}$ of k-subspaces of an n-dimensional vector space over a finite field with q elements is given by

$$N_{k, 1} = \begin{bmatrix} n-1 \\ n-k \end{bmatrix}_q.$$

(Goldman, Rota, 1970)

14. Let $S_k(n) = 1^k + 2^k + \ldots + n^k$, where k is a nonnegative integer. Prove that $1 + \sum_{k=0}^{r-1} \binom{r}{k} S_k(n) = (n+1)^r$.

15. Prove that $\sum_{k=0}^{n} \binom{n}{k}^2 = \binom{2n}{n}$; $\sum_{k=0}^{m} \binom{n}{k} \cdot \binom{n}{m-k} = \binom{2n}{m}$ for $m \leqslant n$;

$\sum_{k=0}^{m} (-1)^k \binom{n-k}{n-m} \binom{n}{k} = 0$ for $0 < m \leqslant n$.

16. Verify the following identity:

$$(1 + x + x^2 + \ldots)^n = \sum_{r \geqslant 0} \binom{n+r-1}{r} x^r.$$

17. If q is a power of a prime, prove that

$$y^n = 1 + \sum_{k=0}^{n-1} \begin{bmatrix} n \\ k \end{bmatrix}_q (y-1)(y-q) \ldots (y-q^{n-k-1}).$$

(Cauchy)

H i n t: This equality counts, in two ways, all linear transformations of V_n (an n-dimensional vector space over the finite field $GF(q)$) into a space Y with y vectors [10] over $GF(q)$.

18. Prove that $\sum_{k=1}^{n} \binom{n}{k} k^{k-1} (n-k)^{n-k} = n^n$, where $0^0 = 1$.

H i n t: Use Abel's generalization of the binomial theorem [21] which states

$$x^{-1}(x + y + ma)^m = \sum_{k=0}^{m} \binom{m}{k} (x + ka)^{k-1} (y + (m-k)a)^{m-k}$$

for $y = 0$, $a = 1$ and compare the coefficients of x.

19. A Boolean function is a mapping $f : B^n \to B$ where $B = \{0, 1\}$. A Boolean function is called symmetric if it is invariant under the permutations of its variables, i.e. $f(x_1, \ldots, x_n) = f(x_{\sigma(1)}, \ldots, x_{\sigma(n)})$ where σ is any bijection of the set $\{1, 2, \ldots, n\}$ onto itself. Prove that the number of symmetric Boolean functions (of n variables) is equal to 2^{n+1}.

20. Let r denote a natural number, and let M denote a set. Then an r-partition of M is a set of r disjoint subsets of M whose union is M (these subsets may be empty). We shall consider the smallest integer $m_r(p)$ for which there exists a

finite set M and a family $F = \{A_1, ..., A_{m_r(p)}\}$ of subsets of M, each containing $p \geqslant 2$ elements, such that F has the property B_r: for any r-partition Π of M there exist an index i, $1 \leqslant i \leqslant m_r(p)$ and a class M_j of Π $(1 \leqslant j \leqslant r)$ such that $A_i \subset M_j$.
Prove that

$$r^{p-1} < m_r(p) \leqslant \binom{rp - r + 1}{p}.$$

(Herzog, Schönheim, 1972)

21. For the numbers previously defined prove that $m_r(2) = \binom{r+1}{2}$. More-over, if F has property B_r and $|F| = \binom{r+1}{2}$ then F consists of a complete graph K_{r+1}.

(Herzog, Schönheim, 1972)

22. Let n be a positive integer. Let $[j] = (1^{j_1}, 2^{j_2}, ..., n^{j_n})$ denote a partition of n, where $j_1 + 2j_2 + ... + nj_n = n$. Prove that

$$\sum_{[j]} (-1)^{j_1+j_2+...+j_n+1} \frac{(j_1 + j_2 + ... + j_n - 1)!}{j_1! j_2! ... j_n!} = \frac{1}{n}.$$

(Sheehan, 1970)

H i n t: $\dfrac{1}{n}$ is the coefficient of x^n in $\log \dfrac{1}{1-x} = x + \dfrac{x^2}{2} + \cdots + \dfrac{x^n}{n} + ...$, and the coefficient of x^n in

$$\log (1+(x+x^2+ ...+x^n)) = (x+x^2+...+x^n) - \frac{(x + ... + x^n)^2}{2} + \cdots$$
$$+ (-1)^{n-1} \frac{(x + ... + x^n)^n}{n} + \cdots .$$

23. If the associative law for multiplication is not satisfied, then the product of n factors may be written in different ways (preserving the order of the factors). For example, four elements a, b, c, d can be multiplied together in five ways: $(ab)(cd)$; $a(b(cd))$; $((ab)c)d$; $(a(bc))d$; $a((bc)d)$.

Prove that the number of ways of multiplying n numbers arranged in a given order equals the Catalan number $T_n = \dfrac{1}{n}\binom{2n - 2}{n - 1}$.

H i n t: If we denote the number of ways of multiplying k numbers by $E(k)$, we get $E(n) = \sum\limits_{k=1}^{n-1} E(k) E(n - k)$ where $E(1) = 1$. Consider the generating function $f(x) = E(1)x + E(2)x^2 + ... + E(n)x^n + ...$ and obtain $f^2(x) = f(x) - x$. Hence

$$f(x) = \frac{1 - \sqrt{1-4x}}{2} = \frac{1}{2}\left[1 - \left(1 - 2x - \ldots - \frac{2}{n}\binom{2n-2}{n-1}x^n - \ldots\right)\right] =$$

$$= x + \binom{2}{1}x^2 + \ldots + \frac{1}{n}\binom{2n-2}{n-1}x^n + \ldots.$$

24. For a convex $(2n-2)$-gon $A_1 A_2 \ldots A_{2n-2}$ prove that the number of ways of joining the vertices in pairs so that the resulting line segments do not intersect inside the $(2n-2)$-gon is equal to T_n.

25. Show that the number of ways of subdividing a convex $(n+1)$-gon $A_1 A_2 \ldots A_{n+1}$ into non-overlapping triangles by means of $n-2$ diagonals equals T_n. (This result has been rediscovered many times, but we may trace it back to Euler [9]).

H i n t: Establish a bijective mapping from the set of divisions of a convex $(n+1)$-gon into non-overlapping triangles by means of $n-2$ diagonals onto the set of nonassociative products with n factors in the order $x_1, x_2, \ldots, x_n$.

26. Prove that the number of sequences $(x_1, x_2, \ldots, x_{2n-2})$ with $x_i = +1$ or -1 for $i = 1, 2, \ldots, 2n-2$ which satisfy the conditions:

i) $x_1 + x_2 + \ldots + x_k \geqslant 0$ for every $k = 1, 2, \ldots, 2n-2$;

ii) $x_1 + x_2 + \ldots + x_{2n-2} = 0$

is equal to T_n.

(Whitworth, 1878)

27. Show that the number of paths from A to B in the graph illustrated in Fig. 2.5 is equal to the Catalan number T_n.

(Chung, Feller, 1949)

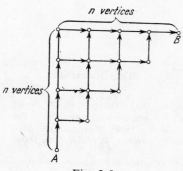

Fig. 2.5

H i n t: Assign $+1$ to each vertical arrow and -1 to each horizontal arrow and use exercise 26. For a more general result see [5].

28.   Let $F = (X_i)_{1 \leqslant i \leqslant k}$ be a family of h-subsets of a set X $(X_i \subset X, |X_i| = h$ for $i = 1, 2, ..., k)$. Show that $\min \left| \bigcup_{i=1}^{k} X_i \right|$ is equal to the smallest integer m having the property: $k \leqslant \binom{m}{h}$.

29.   A system S of non-empty subsets of an n-element set X is called a filter basis if for every $A, B \in S$ there exists $C \in S$ such that $C \subset A \cap B$.

Show that the number of filter bases of the set X equals $\sum_{k=0}^{n-1} \binom{n}{k} 2^{2^k - 1}$.

30.   A mapping $f : X \to X$ is said to be idempotent if $f(f(x)) = f(x)$ for every $x \in X$.

If $|X| = n$ prove that:

i) the number $i(n)$ of idempotent mappings $f : X \to X$ is equal to

$$\sum_{k=1}^{n} \binom{n}{k} k^{n-k};$$

ii) $1 + \sum_{n=1}^{\infty} i(n) \frac{x^n}{n!} = \exp(xe^x)$.

H i n t: Show that f is idempotent if and only if the mapping $\varphi : Y \to Y$, where $Y = f(X)$ and $\varphi(x) = f(x)$ for any $x \in Y$, is the identical mapping.

31.   A Boolean function of n arguments is said to be monotone if for any $(a_1, a_2, ..., a_n) \leqslant (b_1, b_2, ..., b_n)$ we have $f(a_1, a_2, ..., a_n) \leqslant f(b_1, b_2, ..., b_n)$.

If $\psi(n)$ denotes the number of monotone Boolean functions of n arguments, prove that:

$$2^{\binom{n}{\lceil \frac{n}{2} \rceil}} \leqslant \psi(n) \leqslant 3^{\binom{n}{\lceil \frac{n}{2} \rceil}}$$

(Hansel, 1966)

For sharper bounds see [18].

BIBLIOGRAPHY

1. Berge, C., *Principes de combinatoire*, Dunod, Paris, 1968.
   English edition: *Principles of combinatorics*, Academic Press, New York, 1971.
2. Carlitz, L., *Enumeration of certain types of sequences*, Math. Nachr., **49**, 1971, 125−147.
3. Carlitz, L., *A binomial identity related to ballots and trees*, J. Combinatorial Theory (A), **14**, 1973, 261−263.
4. Catalan, E. C., *Note sur une équation aux différences finies*, J. Math. Pures Appl., **3**, 1838, 508−516.
5. Chung, K. L., Feller, W., *Fluctuations in coin tossing*, Proc. Nat. Acad. Sci. U.S.A., **35**, 1949, 605−608.
6. Comtet, L., *Analyse combinatoire*, I, II, Presses Univ. de France, Paris, 1970.
7. Erdös, P., *On a lemma of Littlewood and Offord*, Bull. Amer. Math. Soc., **51**, 1945, 898−902.

8. Erdös, P., Chao Ko, Rado, R., *Intersection theorems for systems of finite sets*, Quart. J. Math. Oxford Ser., **12**, 1961, 313—318.
9. Euler, L., *Novi Commentarii Academiae Scientiarum Imperialis Petropolitanae*, **7**, 1758—1759, 13—14.
10. Goldman, J., Rota, G.-C., *On the foundations of combinatorial theory, IV, Finite vector spaces and Eulerian generating functions*, Studies in Appl. Math., **49**, 1970, 239—258.
11. Hansel, G., *Sur le nombre des fonctions booléennes monotones de n variables*, C. R. Acad. Sci. Paris, Ser. A, 262, 1966, 1088—1090.
12. Herzog, M., Schönheim, J., *The B_r property and chromatic numbers of generalized graphs*, J. Combinatorial Theory (B), **12**, 1972, 41—49.
13. Hochberg, M., Hirsch, W. M., *Sperner families, s-systems and a theorem of Meshalkin*, International Conference on Combinatorial Mathematics, 1970, Ann. New York Acad. Sci., 175, 1970, 224—237.
14. Katona, G., *On a conjecture of Erdös and a stronger form of Sperner's theorem*, Studia Sci. Math. Hungar., **1**, 1966, 59—63.
15. Katona, G. O. H., *A generalization of some generalizations of Sperner's theorem*, J. Combinatorial Theory (B), **12**, 1972, 72—81.
16. Katona, G. O. H., *A simple proof of the Erdös-Chao Ko-Rado theorem*, J. Combinatorial Theory (B), **13**, 1972, 183—184.
17. Kleitman, D., *On a lemma of Littlewood and Offord on the distribution of certain sums*, Math. Z., **90**, 1965, 251—259.
18. Kleitman, D., *On Dedekind's problem: the number of monotone Boolean functions*, Proc. Amer. Math. Soc., **21**, 1969, 677—682.
19. Meshalkin, L. D., *Generalizations of Sperner's theorem about the number of subsets of a finite set* (Russian), Teor. Verojatnost. Primenen., **8**, 1963, 219—220.
20. Milner, E. C., *A combinatorial theorem on systems of sets*, J. London Math. Soc., **43**, 1968, 204—206.
21. Riordan, J., *Combinatorial identities*, Wiley, New York, 1968.
22. Schönheim, J., *A generalization of results of P. Erdös, G. Katona and D. J. Kleitman concerning Sperner's theorem*, J. Combinatorial Theory (A), **11**, 1971, 111—117.
23. Shannon, C. E., *A symbolic analysis of relay and switching circuits*, Trans. A.I.E.E., **57**, 1938, 713—723.
24. Sheehan, J., *An identity*, Amer. Math. Monthly, **77**, 1970, 168.
25. Sperner, E., *Ein Satz über Untermengen einer endlichen Menge*, Math. Z., **27**, 1928, 544—548.
26. Tomescu, I., *Inégalités concernant les hypergraphes uniformes*, Cahiers Centre Études Recherche Opér., **15**, 1973, 355—362. (Colloque sur la théorie des graphes, I.H.E. de Belgique, Bruxelles, 1973).
27. Voloshin, Yu. M., *Enumeration of function compositions*, J. Combinatorial Theory (A), **12**, 1972, 202—216.
28. Whitworth, W. A., *Arrangements of m things of one sort and m things of another sort under certain conditions of priority*, Messenger of Math., **8**, 1878, 105—114.

CHAPTER 3

The Principle of Inclusion and Exclusion and Applications

In the following we consider a finite set X and a numerical function $m(x) \geqslant 0$ defined for every $x \in X$ and called *the measure of the element* x. As a matter of fact, the proofs in this chapter will not use the positivity of the measure, but only the additivity, in the following sense.

For every subset $A \subset X$ we define *the measure of the set* A, which we denote by $m(A)$, by the equation:

$$m(A) = \sum_{x \in A} m(x), \qquad \text{if } A \neq \varnothing$$

and

$$m(A) = 0, \qquad \text{if } A = \varnothing. \tag{3.1}$$

It follows from this definition that if $A, B \subset X$ and $A \cap B = \varnothing$ then

$$m(A \cup B) = m(A) + m(B).$$

The number of elements in a finite set is an example of a measure. In fact, if we define $m(x) = 1$ for every $x \in X$, then $m(A) = |A|$, i.e. the number of elements in the set A. If $m(x)$ is a probability distribution on the independent events of a given field and $\overline{A}$ is the set of all events with a certain property, then $m(A)$ is the probability of obtaining that property.

If $A \subset X$ and $\overline{A} = X \setminus A$ represents the complement of A in X, then (3.1) implies that $m(\overline{A}) = m(X) - m(A)$.

By definition we set

$$m\left(\bigcup_{i \in K} A_i\right) = m\left(\bigcap_{i \in K} A_i\right) = 0, \qquad \text{if } K = \varnothing.$$

PROPOSITION 1. *Let A_i $(i \in Q = \{1, 2, ..., q\})$ be subsets of X. Then*

$$m\left(\bigcup_{i \in Q} A_i\right) = \sum_{K \subset Q} (-1)^{|K|+1} m\left(\bigcap_{i \in K} A_i\right). \tag{3.2}$$

We prove this property by induction on $q = |Q|$. For $q = 2$ the relation (3.2) becomes

$$m(A_1 \cup A_2) = m(A_1) + m(A_2) - m(A_1 \cap A_2),$$

which is true, because from the sum of the measures of the sets A_1 and A_2 we must subtract the measure of the elements common to A_1 and A_2, i.e. the measure of $A_1 \cap A_2$, for this measure has been added twice: both in $m(A_1)$ and $m(A_2)$.

We now assume formula (3.2) to be valid for $|Q| \leqslant q - 1$ and we prove it for $|Q| = q$:

$$m(A_1 \cup A_2 \cup \ldots \cup A_q)$$

$$= m(A_1 \cup \ldots \cup A_{q-1}) + m(A_q) - m((A_1 \cup \ldots \cup A_{q-1}) \cap A_q).$$

But

$$(A_1 \cup \ldots \cup A_{q-1}) \cap A_q = \bigcup_{i<q} (A_i \cap A_q),$$

hence, by the induction hypothesis, we get

$$m(\bigcup_{i \in Q} A_i)$$

$$= m\left(\bigcup_{i=1}^{q-1} A_i\right) + m(A_q) - m(\bigcup_{i<q}(A_i \cap A_q))$$

$$= \sum_{i<q} m(A_i) - \sum_{i<j<q} m(A_i \cap A_j) + \sum_{i<j<k<q} m(A_i \cap A_j \cap A_k) - \ldots$$

$$+ m(A_q) - \sum_{i<q} m(A_i \cap A_q) + \sum_{i<j<q} m(A_i \cap A_j \cap A_q) -- \ldots .$$

Whence, rearranging the terms, we obtain

$$m(\bigcup_{i \in Q} A_i) = \sum_{i=1}^{q} m(A_i) - \sum_{1 \leqslant i<j \leqslant q} m(A_i \cap A_j) + \ldots + (-1)^{q+1} m\left(\bigcap_{i=1}^{q} A_i\right).$$

In the case when $m(A_i) = |A_i|$, formula (3.2) becomes

$$\left|\bigcup_{i=1}^{q} A_i\right| = \sum_{i=1}^{q} |A_i| - \sum_{1 \leqslant i<j \leqslant q} |A_i \cap A_j| + \ldots + (-1)^{q+1} \left|\bigcap_{i=1}^{q} A_i\right|, \tag{3.3}$$

which is known as *the principle of inclusion and exclusion*.

We shall see that formula (3.3) is very useful whenever we do not know the number of elements in the union $\bigcup\limits_{i=1}^{q} A_i$, but we do know the number of elements in each of the subsets $\bigcap\limits_{i \in K} A_i$ for every subset $K \subset Q = \{1, 2, ..., q\}$.

The next proposition gives a formula dual to (3.2):

PROPOSITION 2. *If A_i, with $i \in Q = \{1, 2, ..., q\}$, are subsets of X, then*

$$m(\bigcap_{i \in Q} A_i) = \sum_{K \subset Q} (-1)^{|K|+1} m(\bigcup_{i \in K} A_i). \tag{3.4}$$

The proof of this equation can be obtained from the proof of proposition 1 by interchange of unions and intersections.

PROPOSITION 3 *(Sylvester). If $A_1, A_2, ..., A_q \subset X$, the measure of the set of all elements of X which do not belong to any of the sets A_i, equals*

$$M_q^0 = m(X) + \sum_{K \subset Q} (-1)^{|K|} m(\bigcap_{i \in K} A_i). \tag{3.5}$$

This formula is deduced from (3.2), for

$$M_q^0 = m(\bigcap_{i \in Q} \overline{A}_i) = m(\overline{\bigcup_{i \in Q} A_i}) = m(X) - m(\bigcup_{i \in Q} A_i)$$

$$= m(X) - \sum_{K \subset Q} (-1)^{|K|+1} m(\bigcap_{i \in K} A_i)$$

$$= m(X) + \sum_{K \subset Q} (-1)^{|K|} m(\bigcap_{i \in K} A_i).$$

Summing first with respect to $K \subset Q$ with $|K| = k$ and then with respect to $k = 1, ..., q$, the last sum can be written

$$m(X) + \sum_{\substack{K \subset Q \\ K \neq \varnothing}} (-1)^{|K|} m(\bigcap_{i \in K} A_i) = m(X) + \sum_{k=1}^{q} (-1)^k \sum_{\substack{K \subset Q \\ |K|=k}} m(\bigcap_{i \in K} A_i).$$

PROPOSITION 4 *(the sieve formula of Jordan). If $A_1, A_2, ..., A_q \subset X$, the measure of the set of elements in X which belong to p sets A_i equals*

$$M_q^p = \sum_{k=p}^{q} (-1)^{k-p} \binom{k}{p} \sum_{\substack{K \subset Q \\ |K|=k}} m(\bigcap_{i \in K} A_i). \tag{3.6}$$

Formula (3.6) is a generalization of Sylvester's formula (for $p = 0$), if we take $m(\bigcap_{i \in \varnothing} A_i) = m(X)$.

Let $P \subset Q = \{1, 2, ..., q\}$ be a set such that $|P| = p$. The measure of elements which belong to all the sets A_i with $i \in P$ and do not belong to any of the sets A_j with $j \in Q \setminus P$ is

$$m(\bigcap_{i \in P} A_i \cap \bigcap_{j \in Q \setminus P} \overline{A_j}).$$

But the set of these elements coincides with the set of elements which belong to the set $\bigcap_{i \in P} A_i$ and do not belong to any set

$$A_j \cap \bigcap_{i \in P} A_i \subset \bigcap_{i \in P} A_i \text{ for } j \in Q \setminus P.$$

Hence

$$m(\bigcap_{i \in P} A_i \cap \bigcap_{j \in Q \setminus P} \overline{A_j})$$

$$= m(\bigcap_{i \in P} A_i) - \sum_{\substack{K \supset P \\ |K| = p+1}} m(\bigcap_{i \in K} A_i) + \sum_{\substack{K \supset P \\ |K| = p+2}} m(\bigcap_{i \in K} A_i) - \ldots$$

$$= \sum_{K \supset P} (-1)^{|K| - |P|} m(\bigcap_{i \in K} A_i).$$

In order to obtain an expression for M_q^p we must sum with respect to all the subsets $P \subset Q$ with $|P| = p$ elements:

$$M_q^p = \sum_{|P| = p} m(\bigcap_{i \in P} A_i \cap \bigcap_{j \in Q \setminus P} \overline{A_j})$$

$$= \sum_{|P| = p} \sum_{K \supset P} (-1)^{|K| - |P|} m(\bigcap_{i \in K} A_i)$$

$$= \sum_{\substack{K \subset Q \\ |K| \geqslant p}} \sum_{\substack{P \subset L \\ |P| = p}} (-1)^{|K| - |P|} m(\bigcap_{i \in K} A_i)$$

by changing the summation order. But the index set $P \subset K$ with $|P| = p$ and $|K| = k$ can be chosen in $\binom{k}{p}$ ways for each choice of K and since $m(\bigcap_{i \in K} A_i)$ does not depend on P, it follows that

$$M_q^p = \sum_{k = p}^{q} (-1)^{k - p} \binom{k}{p} \sum_{\substack{K \subset Q \\ |K| = k}} m(\bigcap_{i \in K} A_i),$$

which completes the proof.

If we want to count the elements which belong to none of the sets $A_1, A_2, ..., A_q \subset X$, we start from the table of elements of X and we eliminate the elements of A_1 then from the remaining elements we eliminate the elements of A_2 etc., thus obtaining finally $|X \setminus \bigcup_{i=1}^{q} A_i|$. This procedure is identical with the *sieve of Eratosthenes*, which constructs all prime numbers up to any given positive integer.

The formulae obtained above are known as *sieve-like formulae* because of their analogy with the sieve of Eratosthenes and they apply in a broad class of counting problems, as will be seen in the sequel.

Example 1. *Determination of Euler's totient function*, $\varphi(n)$. Given a positive integer n, let us determine the number $\varphi(n)$ of positive integers smaller than n and relatively prime to it.

Two integers a and b are said to be *relatively prime* if their greatest common divisor, denoted by (a, b), equals 1.

If $n = p_1^{i_1} p_2^{i_2} \ldots p_q^{i_q}$ is the decomposition of n into q distinct prime factors and A_i stands for the set of natural numbers not greater than n which are multiples of p_i, then

$$|A_i| = \frac{n}{p_i} , \quad |A_i \cap A_j| = \frac{n}{p_i p_j} , \ldots$$

sor the distinct prime numbers p_i and p_j are a fortiori relatively prime, etc. The natural numbers fmaller than n and relatively prime to n are the numbers from the set $X = \{1, 2, \ldots, n\}$ which belong to none of the sets A_i ($i = 1, 2, \ldots, q$), their number being given by Sylvester's formula (3.5) with $m(A) = |A|$ for every $A \subset X$:

$$\varphi(n) = n - \sum_{i=1}^{q} |A_i| + \sum_{1 \leqslant i < j \leqslant q} |A_i \cap A_j| - \sum_{1 \leqslant i < j < k \leqslant n} |A_i \cap A_j \cap A_k| + \ldots$$

$$= n - \sum_{i=1}^{q} \frac{n}{p_i} + \sum_{\leqslant i < j \leqslant q} \frac{n}{p_i p_j} - \sum_{1 \leqslant i < j < k \leqslant q} \frac{n}{p_i p_j p_k} + \ldots + (-1)^q \frac{n}{p_1 p_2 \ldots p_q} .$$

We notice that this sum is precisely the development of the product

$$\varphi(n) = n \left(1 - \frac{1}{p_1}\right) \left(1 - \frac{1}{p_2}\right) \ldots \left(1 - \frac{1}{p_q}\right). \tag{3.7}$$

We mention that Euler's totient function $\varphi(n)$ is used in number theory. For instance, if $n = 12 = 2^2 \cdot 3$, $\varphi(12) = 12 \left(1 - \frac{1}{2}\right) \left(1 - \frac{1}{3}\right) = 4$ and the positive integers smaller than 12 and relatively prime to 12 are 1, 5, 7 and 11.

The same principle enables us to determine how many natural numbers not greater than a given natural number are divisible by certain integers. Thus, for example, let us find out how many natural numbers not greater than 500 are divisible by 2, by 3, by 5 or by 7. Let A_1

be the set of natural numbers not greater than 500 which are multiples of 2, ..., A_4 the set of natural numbers not greater than 500 which are multiples of 7; then formula (3.3) yields:

$$|A_1 \cup A_2 \cup A_3 \cup A_4| = |A_1| + |A_2| + |A_3| + |A_4| - |A_1 \cap A_2| - |A_1 \cap A_3|$$

$$- |A_1 \cap A_4| - |A_2 \cap A_3| - |A_2 \cap A_4| - |A_3 \cap A_4| + |A_1 \cap A_2 \cap A_3|$$

$$+ |A_1 \cap A_2 \cap A_4| + |A_2 \cap A_3 \cap A_4| + |A_1 \cap A_3 \cap A_4| - |A_1 \cap A_2 \cap A_3 \cap A_4|$$

$$= \left[\frac{500}{2}\right] + \left[\frac{500}{3}\right] + \left[\frac{500}{5}\right] + \left[\frac{500}{7}\right] - \left[\frac{500}{6}\right] - \left[\frac{500}{10}\right] - \left[\frac{500}{14}\right] - \left[\frac{500}{15}\right]$$

$$- \left[\frac{500}{21}\right] - \left[\frac{500}{35}\right] + \left[\frac{500}{30}\right] + \left[\frac{500}{42}\right] + \left[\frac{500}{105}\right] + \left[\frac{500}{70}\right] - \left[\frac{500}{210}\right] = 385.$$

It also follows that there exist $500 - 385 = 115$ natural numbers smaller than 500 which are not divisible by any of the numbers 2, 3, 5 and 7.

Example 2. The problem of fixed points. Let p be a permutation of the set $X = \{1, 2, ..., n\}$, that is, a word $p(1) p(2)... p(n)$ of length n, where $p(1), p(2), ..., p(n)$ are the n distinct elements of the set X. We have seen in Chap. 2 that to this word corresponds a bijection of the set X onto itself, which we denote by

$$p = \begin{pmatrix} 1 & 2 & ... & n \\ p(1) & p(2) & ... & p(n) \end{pmatrix}.$$

We say that the permutation p has a coincidence at i if $p(i) = i$. A number i with this property is also called a *fixed point* of the permutation. For example, the permutation

$$p = \begin{pmatrix} 1 & 2 & 3 & 4 & 5 & 6 & 7 & 8 & 9 \\ 3 & 2 & 1 & 4 & 7 & 6 & 5 & 8 & 9 \end{pmatrix}$$

has five fixed points: 2, 4, 6, 8 and 9.

Let us find first the number $D(n)$ of permutations of n objects without fixed points. To this end, denote by A_i the set of $(n-1)!$ permutations which have a fixed point in i and apply the principle of inclusion and exclusion (3.3) to find the number of permutations which have at least one fixed point. This number is

$$|A_1 \cup A_2 \cup ... \cup A_n| = \sum_{i=1}^{n} |A_i| - \sum_{1 \leqslant i < j \leqslant n} |A_i \cap A_j| + ... + (-1)^{n-1} \left|\bigcap_{i=1}^{n} A_i\right|.$$

But $|A_{i_1} \cap A_{i_2} \cap ... \cap A_{i_k}| = (n-k)!$, because a permutation from the set $A_{i_1} \cap ... \cap A_{i_k}$ has fixed points at $i_1, i_2, ..., i_k$, the other positions being chosen in all $(n-k)!$ possible ways. On the other hand, k positions $i_1, i_2, ..., i_k$ can be chosen from the set of n positions in $\binom{n}{k}$ ways, hence

$$\left|\bigcup_{i=1}^{n} A_i\right| = \binom{n}{1}(n-1)! - \binom{n}{2}(n-2)! + ... + (-1)^{n-1}\binom{n}{n}.$$

The number of permutations of n objects without fixed points is obtained by subtracting the number of permutations with at least one fixed point from the total number of $n!$ permutations, therefore

$$D(n) = n! - \binom{n}{1}(n-1)! + \ldots + (-1)^k \binom{n}{k}(n-k)! + \ldots + (-1)^n \binom{n}{n},$$

which can also be written

$$D(n) = n!\left(1 - \frac{1}{1!} + \frac{1}{2!} - \frac{1}{3!} + \ldots + \frac{(-1)^k}{k!} + \ldots + \frac{(-1)^n}{n!}\right). \tag{3.8}$$

The number $D(n)$ can also be computed recursively, taking into account the relations $D(1) = 0$; $D(n) = nD(n-1) + (-1)^n$. The number of permutations of n objects with p fixed points is therefore equal to $\binom{n}{p}D(n-p)$, for the p fixed points can be chosen in $\binom{n}{p}$ ways while the other points are not fixed, hence for each choice of p fixed points there exist $D(n-p)$ permutations of the remaining objects without fixed points. Notice that in this way each permutation with p fixed points is obtained once and only once, which implies the expression obtained above.

Example 3. Determination of the number of functions which actually depend on all the variables. Let $g : E^k \to F$, where $E = \{e_1, \ldots, e_n\}$ and $F = \{f_1, \ldots, f_m\}$, that is, a function $g(x_1, \ldots, x_k)$ of k arguments taking values in the set F. The number of these functions g is $|F|^{|E^k|} = m^{n^k}$. We say that g does not actually (or essentially) depend on the variable x_i, if it is constant with respect to the component x_i, i.e. for every system of values $(x'_1, \ldots, x'_{i-1}, x'_{i+1}, \ldots, x'_k) \in E^{k-1}$ and for every $\alpha, \beta \in E$, the relation

$$g(x'_1, \ldots, x'_{i-1}, \alpha, x'_{i+1}, \ldots, x'_k) = g(x'_1, \ldots, x'_{i-1}, \beta, x'_{i+1}, \ldots, x'_k)$$

holds.

The number of functions which do not actually depend on q fixed variables ($q \le k$) equals the number of functions $g : E^{k-q} \to F$, which is $m^{n^{k-q}}$, because these functions are constant in q variables and can be identified with the F-valued functions with $k - q$ arguments in the set E.

Setting

$$A_i = \{g : E^k \to F \mid g \text{ does not actually depend on the variable } x_i\}$$

and denoting by $E(n, m, k)$ the number of functions $g : E^k \to F$ which essentially depend on all the variables, the Sylvester formula implies

$$E(n, m, k) = m^{n^k} - \sum_{i=1}^{k}|A_i| + \sum_{1 \le i < j \le k}|A_i \cap A_j| - \ldots + (-1)^{k-1}\left|\bigcap_{i=1}^{k} A_i\right|$$

Taking into account the definition of the sets A_i, we deduce that

$$E(n, m, k) = m^{n^k} - \binom{k}{1}m^{n^{k-1}} + \binom{k}{2}m^{n^{k-2}} - \ldots + (-1)^k m, \tag{3.9}$$

since the p variables on which the function g does not actually depend, can be chosen from the set of k variables in $\binom{k}{p}$ distinct ways.

If $E = F = \{0, 1\}$, then the functions $g: E^k \to F$ are Boolean functions of k variables and their number is 2^{2^k}. The number of Boolean functions of k variables which actually depend on all the variables can be obtained from formula (3.9) by taking $m = n = 2$:

$$E(2, 2, k) = 2^{2^k} - \binom{k}{1} 2^{2^{k-1}} + \binom{k}{2} 2^{2^{k-2}} - \ldots + (-1)^k 2. \tag{3.10}$$

If a Boolean function does not essentially depend on a variable x_i, there exists a disjunctive normal form which, using only disjunction, conjunction and negation, generates the function g and does not contain the variable x_i. For example, if $k = 2$ then $E(2, 2, 2) = 10$, therefore 10 functions out of the 16 Boolean functions of two variables, actually depend on both variables.

The six Boolean functions which do not essentially depend on both variables are given in Table 3.1.

TABLE 3.1

x_1	x_2	$g(x_1, x_2)$	x_1	x_2	$g(x_1, x_2)$	x_1	x_2	$g(x_1, x_2)$
0	0	0	0	0	1	0	0	0
0	1	0	0	1	1	0	1	0
1	0	0	1	0	1	1	0	1
1	1	0	1	1	1	1	1	1
	$a: g = 0$			$b: g = 1$			$c: g = x_1$	

x_1	x_2	$g(x_1, x_2)$	x_1	x_2	$g(x_1, x_2)$	x_1	x_2	$g(x_1, x_2)$
0	0	1	0	0	0	0	0	1
0	1	1	0	1	1	0	1	0
1	0	0	1	0	0	1	0	1
1	1	0	1	1	1	1	1	0
	$d: g = \bar{x}_1$			$e: g = x_2$			$f: g = \bar{x}_2$	

The functions a and b are constant functions and do not depend either on x_1 or on x_2, the functions c and d do not depend on x_2 but do depend on x_1, while e and f do not depend on x_1 but do depend on x_2.

Example 4 indicates an application to graph theory. Given a finite undirected graph $G = (X, U)$, where X is the set of vertices and U the set of edges, a complete subgraph C is a set of vertices of the graph G which are joined in all possible ways by edges from U. A complete subgraph with k vertices will be called a complete k-subgraph. We assume in the following that $2 \leqslant k \leqslant n$, where n is the number of vertices in G. The degree of a vertex $x \in X$, denoted by $d(x)$, is by definition the number of edges having the vertex x as an extremity. Obviously if the graph G does not include complete k-subgraphs, then there exist certain limitations on the degrees of its vertices and on the number of its edges; these are given in the following two classical results.

PROPOSITION 5 *(Zarankiewicz)*. *If the graph G does not include complete k-subgraphs, then the set of degrees of its vertices satisfies the inequality*

$$\min_{x \in X} d(x) \leqslant \left[\frac{(k-2)\,n}{k-1} \right]. \tag{3.11}$$

Let $(k-2)n = p(k-1) + r$ with $0 \leqslant r \leqslant k-2$ and assume the contrary hypothesis that for every $x \in X$ the degree $d(x) \geqslant p+1$.

Take a vertex $x_{i_1} \in X$ and denote by $A_{x_{i_1}}$ the set of those vertices in the graph G which are joined by edges to x_{i_1}. Now take another vertex $x_{i_2} \in A_{x_{i_1}}$ and the corresponding set $A_{x_{i_2}}$. The principle of inclusion and exclusion, in its dual form, yields

$$|A_{x_{i_1}} \cap A_{x_{i_2}}| = |A_{x_{i_1}}| + |A_{x_{i_2}}| - |A_{x_{i_1}} \cup A_{x_{i_2}}| \geqslant 2p + 2 - n,$$

because each of the sets $A_{x_{i_1}}$ and $A_{x_{i_2}}$ contains at least $p+1$ elements. Taking an arbitrary element $x_{i_3} \in A_{x_{i_1}} \cap A_{x_{i_2}}$ we obtain $|A_{x_{i_1}} \cap A_{x_{i_2}} \cap A_{x_{i_3}}| \geqslant 3(p+1) - 2n$, etc. We thus obtain by induction, for $x_{i_{k-1}} \in \bigcap_{j=1}^{k-2} A_{x_{i_j}}$:

$$\left| \bigcap_{j=1}^{k-1} A_{x_{i_j}} \right| = |A_{x_{i_{k-1}}}| + \left| \bigcap_{j=1}^{k-2} A_{x_{i_j}} \right| - \left| A_{x_{i_{k-1}}} \cup \bigcap_{j=1}^{k-2} A_{x_{i_j}} \right|$$

$$\geqslant p + 1 + (k-2)(p+1) - (k-3)n + n$$

$$= (k-1)(p+1) - (k-2)n = k - 1 - r > 0,$$

hence there is at least one vertex $x_{i_k} \in \bigcap_{j=1}^{k-1} A_{x_{i_j}}$. Therefore, as shown by the construction of the sets $A_{x_{i_j}}$, we have obtained a complete k-subgraph $\{x_{i_1}, x_{i_2}, \ldots, x_{i_k}\}$, which contradicts the hypothesis of the proposition.

PROPOSITION 6 *(Turán)*. *In the class of graphs G with n vertices and without complete k-subgraphs, the maximum number of edges in a graph is given by*

$$M(n, k) = \frac{k-2}{k-1} \cdot \frac{n^2 - r^2}{2} + \binom{r}{2}, \tag{3.12}$$

where r stands for the remainder in the division of n by $k-1$, i.e. $n = (k-1)t + r$ with $0 \leqslant r \leqslant k-2$.

The graph G for which this maximum is reached is unique up to an isomorphism and consists of $k-1$ classes of vertices, among which r classes contain $t+1$ vertices each, while each of the remaining classes contains t vertices and each vertex is joined by edges to all the vertices of the other classes.

Such a graph is called a complete $(k-1)$-partite graph. We give below a proof by induction of this theorem, which is simpler than the proof given by Turán [11].

For $n = 1, 2, \ldots, k-1$, the graph which has a maximum number of edges and does not include complete k-subgraphs is a complete n-graph and is of the form indicated above. Assume the theorem to be true for every $n' \leqslant n-1$. If the graph G has n vertices and does not include complete k-subgraphs, then from proposition 5 we infer that G has at least one vertex x of degree $d(x) \leqslant \left[\dfrac{(k-2)n}{k-1}\right]$. If we consider the subgraph $G_x = (X \setminus \{x\}, U_x)$, where U_x is obtained from the set U of edges by deleting all the edges which have x as an extremity, then this graph which does not include complete k-subgraphs may or may not be maximal with respect to the number of edges. If this subgraph is not maximal, we replace it by a k-maximal graph which, according to the inductive hypothesis, consists of $k-1$ classes of vertices, among which r' contain $t'+1$ vertices each, while each of the remaining classes contains t' vertices, where $n-1 = = (k-1)t' + r'$ with $0 \leqslant r' \leqslant k-2$, and each vertex is joined by edges to all the vertices which do not belong to its class. Now add vertex x to a class which contains t' vertices and connect x to all the vertices which do not belong to the class of x, thus obtaining a graph without complete k-subgraphs and which is unique up to isomorphism. The degree of vertex x in the graph obtained in this way

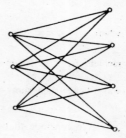

is $n - 1 - t' = n - 1 - \left[\dfrac{n-1}{k-1}\right]$. But a simple computation

Fig. 3.1

shows that $n - 1 - \left[\dfrac{n-1}{k-1}\right] = \left[\dfrac{(k-2)n}{k-1}\right]$, hence the n-vertex k-maximal graph must have the structure described above.

The maximum number of edges $M(n, k)$ given by (3.12) is half of the sum of the degrees of the vertices from the complete $(k-1)$-partite graph constructed above. We also see, by a simple computation, that the maximum number of edges of a graph without complete k-subgraphs can also be written $M(n, k) = \left[\dfrac{k-2}{k-1} \cdot \dfrac{n^2}{2}\right]$ for $2 \leqslant k \leqslant 8$ and for $k = n-1$, n. For $k = 3$ we conclude that the maximum number of edges of a graph G with n vertices and without triangles (complete 3-subgraphs) equals $M(n, 3) = \left[\dfrac{n^2}{4}\right]$ and this number is attained only when

G is a complete bipartite graph consisting of two classes of vertices which contain $\left[\dfrac{n}{2}\right]$ and $n - \left[\dfrac{n}{2}\right]$ vertices, respectively. A classical notation for this complete bipartite graph is $K_{\left[\frac{n}{2}\right], \, n - \left[\frac{n}{2}\right]}$. For $n = 7$ this graph is drawn in Fig. 3.1.

Example 5. Determination of the number $s_{n,m}$ of surjective functions. Consider the finite sets $X = \{x_1, x_2, \ldots, x_n\}$ and $Y = \{y_1, y_2, \ldots, y_m\}$ with $n \geqslant m$. For each $i \in \{1, 2, \ldots, m\}$, denote by A_i the set of all functions from X to Y for which y_i is not the image of any element of X, i.e.

$$A_i = \{f: X \to Y \mid y_i \notin f(X)\}.$$

The set of surjective functions from X onto Y coincides with the set of those Y-valued functions defined on X which do not belong to any of the sets A_i. Hence $s_{n,m} = m^n - |A_1 \cup A_2 \cup \ldots \cup A_m|$, for the total number of functions from X to Y is m^n. Therefore

$$s_{n,m} = m^n - \sum_{i=1}^{m} |A_i| + \sum_{1 \leqslant i < j \leqslant m} |A_i \cap A_j| - \sum_{1 \leqslant i < j < k \leqslant m} |A_i \cap A_j \cap A_k| + \ldots + (-1)^m \left| \bigcap_{i=1}^{m} A_i \right|.$$

But A_i is in fact the number of functions defined on X with values in $Y \setminus \{y_i\}$, hence $|A_i| = (m-1)^n$; $A_i \cap A_j$ is the set of functions defined on X with values in $Y \setminus \{y_i, y_j\}$, hence $|A_i \cap A_j| = (m-2)^n$ and in general

$$|A_{i_1} \cap A_{i_2} \cap \ldots \cap A_{i_l}| = (m-l)^n,$$

where $1 \leqslant i_1 < i_2 < \ldots < i_l \leqslant m$. Notice that $\bigcap_{i=1}^{m} A_i = \varnothing$ because $f(X)$ contains at least one element from Y, for every function $f \colon X \to Y$. But we can eliminate l elements from Y in $\binom{m}{l}$ ways, hence each sum $\sum_{\substack{K \subset \{1,2,\ldots,m\} \\ |K| = l}} |\bigcap_{i \in K} A_i|$ contains $\binom{m}{l}$ terms equal to $(m-l)^n$.

We conclude that

$$s_{n,m} = m^n - \binom{m}{1}(m-1)^n + \binom{m}{2}(m-2)^n - \ldots + (-1)^{m-1}\binom{m}{m-1}. \qquad (3.13)$$

For $m = n$, $s_{n,n}$ represents the number of bijective functions $f \colon X \to Y$ with $|X| = |Y| = n$, therefore $s_{n,n} = n!$ and we obtain the identity

$$n! = n^n - \binom{n}{1}(n-1)^n + \binom{n}{2}(n-2)^n - \ldots + (-1)^{n-1}\binom{n}{n-1}$$

whereas for $n < m$ we get $s_{n,m} = 0$.

Problems

1.　Prove that the number of h-cliques of the Turán graph with n vertices and $k-1$ parts, denoted by $E(n, k, h)$, satisfies the following recurrence relation:

$$E(n, k, h) = \sum_{i=0}^{h} \binom{k-i-1}{h-i} E(n-k+1, k, i)$$

where $E(n, k, 0) = 1$.

(Sauer, 1971)

2.　Prove that $E(n, k, h) \leqslant \binom{k-1}{h}\left(\dfrac{n}{k-1}\right)^h$ and

$$\lim_{n \to \infty} E(n, k, h) \Big/ \binom{n}{h} = \prod_{r=1}^{h-1}\left(1 - \frac{r}{k-1}\right).$$

(Tomescu, 1973)

3. An h-hypergraph H is a pair $H = (X, \mathscr{E})$, where X is a set of vertices and $\mathscr{E} \subset \mathscr{P}_h(X)$ is a set of h-sets of X.

A strong clique C is a set $C \subset X$ having the property that for any $x_i, x_j \in C$ there exists $E \in \mathscr{E}$ such that $x_i, x_j \in E$. If we define the degree of x by the equality $d_H(x) = \max\{|F| \mid F \subset \mathscr{E}, \ \forall E_1, E_2 \in F: E_1 \cap E_2 = \{x\}\}$, prove the following generalization of Zarankiewicz's result:

If an h-hypergraph H has no strong clique with k vertices, then

$$\min_{x \in X} d_H(x) \leqslant \left[\frac{(k-2)n}{(k-1)(h-1)} \right].$$

(Tomescu, 1973)

4. Show that

$$E(n, k, h) = \sum_{i=0}^{h} \binom{r}{i} \binom{k-1-i}{h-i} q^{h-i}$$

where $n = q(k-1) + r$ with $0 \leqslant r < k - 1$ and $0^0 = 1$.

(Sauer, 1971)

5. Prove that the maximum number of h-cliques of a graph with n vertices without k-cliques is equal to $E(n, k, h)$ i.e. the number of h-cliques of Turán's graph with n vertices and $k - 1$ parts.

(Zykov, 1949, Erdös, 1962, Sauer, Tomescu, 1971)

6. Let $A_k = E(2, 2, k)$ enumerate the Boolean functions $f(x_1, x_2, ..., x_k)$ with k essential arguments. Prove that:

$$A_k = 2^{2^k} - \binom{k}{k-1} A_{k-1} - \binom{k}{k-2} A_{k-2} - ... - \binom{k}{1} A_1 - A_0.$$

7. Let P be a nonempty, finite set with p members, and Q be a nonempty, finite set with q members. Let $N_k(p, q)$ be the number of binary relations of cardinality k with domain P and range Q. (Equivalently, $N_k(p, q)$ is the number of $p \times q$ matrices of 0's and 1's with exactly k entries equal to 1 and no row or column identically 0).

Prove that $\sum_{k=1}^{pq} (-1)^{k-1} N_k(p, q) = (-1)^{p+q}$.

(Leader, 1971)

8. There are r distinct things distributed among $n + p$ persons so that at least n of them receive at least one thing. Applying the principle of inclusion and exclusion prove that the number of ways of dividing the things is

$$(n + p)^r - \binom{n}{1}(n + p - 1)^r + \binom{n}{2}(n + p - 2)^r - ... + (-1)^n p^r.$$

9. There are n pairs of identical letters, different pairs consisting of distinct letters.

These letters are ordered in all possible ways so that no two identical letters come in succession. Prove that the number of distinct orders is

$$\frac{1}{2^n}\left[(2n)! - \binom{n}{1}2\,(2n-1)! + \binom{n}{2}2^2\,(2n-2)! - \ldots + (-1)^n\,2^n\,n!\right].$$

H i n t: The letters may be ordered in $\dfrac{(2n)!}{2^n}$ ways without restrictions. Apply the principle of inclusion and exclusion.

10. Let $H_n(r)$ enumerate the Hamiltonian cycles having r edges in common with a fixed Hamiltonian cycle of a complete graph with n vertices K_n.

Prove that

$$H_n(r) = \sum_{s=0}^{n-r}(-1)^s\binom{r+s}{s}S_{r+s}^{(n)}$$

for $n \geqslant 3$, $0 \leqslant r \leqslant n$, where:

$$S_k^{(n)} = \frac{(n-k-1)!}{2}\sum_{j=1}^{k}\left[\binom{k}{j}\binom{n-k-1}{j-1} + \binom{k-1}{j-1}\binom{n-k}{j}\right]2^j$$

$$(0 < k < n) \text{ and } S_0^{(n)} = \frac{(n-1)!}{2}; \quad S_n^{(n)} = 1.$$

(Baróti, 1973)

H i n t: If $N(A_{i_1}, A_{i_2}, \ldots, A_{i_k})$ denotes the number of Hamiltonian cycles of K_n using the edges $e_{i_1}, e_{i_2}, \ldots, e_{i_k}$ of K_n, then $S_k^{(n)} = \Sigma\, N(A_{i_1}, A_{i_2}, \ldots, A_{i_k})$. $H_n(r)$ is obtained by applying the principle of inclusion and exclusion.

11. Let $s_{n,m,r}$ denote the number of functions

$$f: X \to Y$$

where $|X| = n$, $|Y| = m$, with the property that $f(X) \supset Z$, where Z is an r-subset of Y.

Prove that the number of such functions is equal to $\displaystyle\sum_{i=0}^{n}(-1)^i\binom{r}{i}(m-i)^n$.

12. Show that $\displaystyle\sum_{d\,|\,n}\varphi(d) = n$, where φ is Euler's totient function.

(Gauss)

H i n t: Let $n = p_1^{i_1}\,p_2^{i_2}\,\ldots\,p_q^{i_q}$ where $p_1, \ldots, p_q$ are pairwise distinct prime numbers. Use induction on $i_1 + i_2 + \ldots + i_q$.

13.   Prove that the number of 3×3 arrays with row and column sums all equal to r is equal to

$$H_3(r) = \binom{r+2}{2}^2 - 3\binom{r+3}{4}.$$

(MacMahon, 1916)

14.   Let T be the set of all $a \times b$ (0, 1)-matrices M. Let R_0, C_0 be the subsets of all such M with at least one zero row and with at least one zero column, respectively. Let $R_1 = T \setminus R_0$. Prove that the number of M in the intersection $R_1 \cap C_0$ is

$$\sum_{j=1}^{b} (-1)^{j-1} \binom{b}{j}(2^{b-j} - 1)^a.$$

(Everett, Stein, 1973)

15.   Prove that: $\displaystyle\sum_{n=0}^{\infty} D(n) \frac{t^n}{n!} = \frac{e^{-t}}{1 - t}$;

$$D(n + 1) = (n + 1) D(n) + (-1)^{n+1};$$

$$D(n + 1) = n(D(n) + D(n - 1)).$$

BIBLIOGRAPHY

1. Baróti, G., *On the number of certain Hamilton circuits of a complete graph*, Period. Math. Hungar., **3**, 1973, 135—139.
2. Berge, C., *Principes de combinatoire*, Dunod, Paris, 1968.
   English edition: Principles of combinatorics, Academic Press, New York, 1971.
3. Biondi, E., Divieti, L., Guardabassi, G., *Counting paths, circuits, chains and cycles in graphs: a unified approach*, Canad. J. Math., **22**, 1970, 22—35.
4. Erdös, P., *On the number of complete subgraphs contained in certain graphs*, Magyar Tud. Akad. Mat. Kutató Int. Közl., **7** (1962), 459—464.
5. Everett, C. J., Stein, P. R., *The asymptotic number of (0, 1)-matrices with zero permanent*, Discrete Math., **6**, 1973, 29—34.
6. Jordan, C., *Sur la probabilité des épreuves répétées*, Bull. Soc. Math. France, **54**, 1926, 101—137.
7. Leader, S., *Combinatorics of matrices with 0's and 1's*, Amer. Math. Monthly, **80**, 1973, 84.
8. MacMahon, P. A., *Combinatory analysis*, vol. II, Cambridge Univ. Press, Cambridge, 1916.
9. Sauer, N., *A generalization of a theorem of Turán*, J. Combinatorial Theory (B), **10**, 1971, 109—112.
10. Tomescu, I., *Inégalités concernant les hypergraphes uniformes*, Cahiers Centre Études Recherche Opér., **15**, 1973, 355—362. (Colloque sur la théorie des graphes, I.H.E. de Belgique, Bruxelles, 1973).
11. Turán, P., *On the theory of graphs*, Colloq. Math., **3**, 1954, 19—30.
12. Zarankiewicz, K., *Sur les relations symétriques dans l'ensemble fini*, Colloq. Math. **1**, 1947, 10—14.
13. Zykov, A. A., *On certain properties of linear complexes* (in Russian), Mat. Sb., **24**, 1949, 163—188.

CHAPTER 4

The Numbers of Stirling, Bell and Fibonacci

Besides the binomial and multinomial numbers, the *numbers of Stirling, Bell and Fibonacci* play a special role in counting problems, as will be seen later.

In order to define the *Stirling numbers of the first kind*, which we denote by $s(n, k)$, we write the polynomial $[x]_n = x(x - 1) \ldots (x - n + 1)$ in increasing order of powers of x. The coefficients of this development are by definition the Stirling numbers of the first kind, i.e.

$$[x]_n = s(n, 0) + s(n, 1)x + s(n, 2)x^2 + \ldots + s(n, n)x^n. \tag{4.1}$$

The numbers $s(n, k)$ can be determined recursively, using the relations $s(n, 0)=0$, $s(n, n) = 1$ and $s(n + 1, k) = s(n, k - 1) - ns(n, k)$; the last relation is obtained by equating the coefficients of x^k on the two sides of the equation $[x]_{n+1} = =[x]_n(x-n)$. We thus obtain Table 4.1, where by definition we have set $s(n, k)=0$ for $n < k$.

TABLE 4.1

$s(n, k)$	$k = 0$	1	2	3	4	5	$\ldots$
$n = 1$	0	1	0	0	0	0	
2	0	-1	1	0	0	0	
3	0	2	-3	1	0	0	
4	0	-6	11	-6	1	0	
5	0	24	-50	35	-10	1	

In order to define the Stirling numbers of the second kind, we recall the notion of a *partition* of a set into classes, which we have already used in Chap. 2.

Given a set X, the subsets $A_1, A_2, \ldots, A_p$ form a partition of X, if $A_i \neq \varnothing$ for every i, the sets A_i and A_j are disjoint for every $i \neq j$, i.e. $A_i \cap A_j = \varnothing$, and $\bigcup_{i=1}^{p} A_i = X$. The sets A_i are said to be the *classes of the partition*.

The sets $A_i \subset X$ determine a partition of the set X, if the binary relation denoted $a \sim b$ and defined by "a and b are contained in the same subset A_i," is reflexive ($a \sim a$), symmetric ($a \sim b$ implies $b \sim a$) and transitive ($a \sim b$ and $b \sim c$ imply $a \sim c$). In this case we say that $\sim$ is an equivalence relation and the A_i's are the classes of this equivalence.

There exists a bijection from the set of all partitions of the set X onto the set of all equivalence relations on the set X. This bijection is established by associating with every partition Δ of X, the equivalence relation $\sim_\Delta$ between the elements of X, defined as follows: $a \sim_\Delta b$ if a and b are contained in the same class of Δ.

The *Stirling numbers of the second kind* are the numbers $S(n, m)$ of partitions of n-element sets into m classes. As the classes of the partition are non-empty, each of them contains at least one element, hence the problem has a solution only if $n \geqslant m$, i.e. if the number of elements is at least equal to the number of classes. For example, if we have four elements a, b, c, d, there are six partitions of the set $\{a, b, c, d\}$ into three classes, namely $\{(a), (b), (c, d)\}$, $\{(a), (c), (b, d)\}$, $\{(a), (d), (b, c)\}$, $\{(b), (c), (a, d)\}$, $\{(b), (d), (a, c)\}$, $\{(c), (d), (a, b)\}$, therefore $S(4, 3) = 6$.

Remark: Both the order of classes and the order of elements in a class are immaterial.

The Stirling numbers of the second kind can be computed recursively as follows. Notice first that $S(n, 1) = S(n, n) = 1$. If we consider the set of the $S(n, k - 1)$ partitions of n elements into $k - 1$ classes, we can obtain $S(n, k - 1)$ partitions of $n + 1$ elements into k classes, by adding to each partition a new class consisting of a single element, namely the $(n + 1)$-st. Now consider a partition of n elements into k classes: we can add the $(n + 1)$-st element to the classes already existing in k different ways. All the partitions of $n + 1$ elements into k classes are obtained without repetitions by one of the above two procedures, so we infer that $S(n + 1, k) = S(n, k - 1) + kS(n, k)$ for $1 < k < n$ and $S(n, 1) = S(n, n) = 1$. These relations enable a recursive computation of the numbers $S(n, k)$, yielding Table 4.2, where we have set $S(n, m) = 0$ for $m > n$.

TABLE 4.2

$S(n, m)$	$m = 1$	2	3	4	5	...
$n = 1$	1	0	0	0	0	
2	1	1	0	0	0	
3	1	3	1	0	0	
4	1	7	6	1	0	
5	1	15	25	10	1	

For a direct computation of the Stirling numbers of the second kind, we shall prove that $m! S(n, m) = s_{n,m}$. In fact, to every surjection f of the set $X = \{x_1, x_2, ..., x_n\}$

onto the set $Y = \{y_1, y_2, ..., y_m\}$ corresponds a partition of the set X into m classes, namely

$$f^{-1}(y_1), f^{-1}(y_2), ..., f^{-1}(y_m).$$

As the order of classes in a partition is immaterial, it follows that $m!$ surjective functions from X onto Y will generate the same partition of X. In other words, if we permute the elements of Y in $m!$ different ways, we get $m!$ surjections starting from a surjection f, but all of them will generate, in the way indicated above, the same partition of X into m classes. Since two different partitions cannot be obtained from two surjections which differ only by a permutation of the elements of Y, it follows from (3.13) that

$$S(n, m) = \frac{1}{m!} s_{n,m} = \frac{1}{m!} \sum_{k=0}^{m-1} (-1)^k \binom{m}{k} (m-k)^n. \qquad (4.2)$$

PROPOSITION 1. *The polynomials x^n are sums of polynomials $[x]_k$ with coefficients $S(n, k)$, that is*

$$x^n = \sum_{k=1}^{n} S(n, k)[x]_k. \qquad (4.3)$$

Indeed, take two sets X and Y, having n elements and m elements, respectively. Every function $f: X \to Y$ can be viewed as a surjective function if we suitably change its codomain, that is if we consider $f: X \to f(X)$, where $f(X) = \{f(x) | x \in X\} \subset Y$. Therefore the total number of functions from X to Y, equal to m^n, is also equal to the number of functions in the sets

$$\{f: X \to Y \mid |f(X)| = k\} \qquad \text{for } k = 1, 2, ..., m,$$

these function sets being pairwise disjoint. But the number of surjective functions defined on a set with n elements and with values in a k-element set $(n \geqslant k)$ equals $s_{n,k} = k! \, S(n, k)$. Since k elements can be chosen out of the m elements of Y in $\binom{m}{k}$ ways, it follows that

$$m^n = \sum_{k=1}^{n} \binom{m}{k} k! \, S(n, k) = \sum_{k=1}^{m} m(m-1) \ldots (m-k+1) \, S(n, k).$$

Notice that in this sum the index k can take values from 1 to n, because $[m]_k$ vanishes for $k = m+1, m+2, ..., n$. We have thus proved that

$$m^n = \sum_{k=1}^{n} m(m-1) \ldots (m-k+1) \, S(n, k) = \sum_{k=1}^{n} [m]_k S(n, k). \qquad (4.4)$$

Now we must prove that $x^n - \sum_{k=1}^{n} [x]_k S(n, k)$ is identically null. But this polynomial is of degree at most $n-1$, for $[x]_n S(n, n) = x(x-1) \ldots (x-n+1)$ contains the term x^n, the other terms of the sum do not contain x^n, hence x^n has coefficient zero in $x^n - \sum_{k=1}^{n} [x]_k S(n, k)$. The equality (4.4) being valid for $m = 1, 2, \ldots, n$, it follows that our polynomial of degree at most $n - 1$, vanishes for n distinct values of the variable x, hence it is the null polynomial.

Comparing (4.1) with (4.3) we can notice the analogy between the Stirling numbers of the first and second kinds.

PROPOSITION 2. *The Stirling numbers of the second kind satisfy the following recurrence relation* ($m \geqslant 2$):

$$S(n + 1, m) = \sum_{k=m-1}^{n} \binom{n}{k} S(k, m - 1). \tag{4.5}$$

Consider the table of the $S(n + 1, m)$ different partitions of an $(n + 1)$-element set X into m classes. If we delete the class containing the element $n + 1$, we obtain a partition of a set K with k elements ($m - 1 \leqslant k \leqslant n$) into $m - 1$ classes. We also have $k \geqslant m - 1$, for each of the remaining $m - 1$ classes contains at least one element, hence $|K| \geqslant m - 1$. The partitions into $m - 1$ classes obtained in this way are pairwise distinct, for otherwise the corresponding partitions of the $(n + 1)$-element set X into m classes would be equal, contrary to the hypothesis. Notice also that in this way we obtain all the partitions of a set $K \subset \{1, 2, \ldots, n+1\}$ with $|K| \geqslant m - 1$ into m classes. In fact, the partition $\{K_i\}_{1 \leqslant i \leqslant m}$ of K has been obtained from the partition $\{K_i\}_{1 \leqslant i \leqslant m} \cup \{X \setminus K\}$ of X by deleting the class $X \setminus K$ which contains the element $n + 1$. But k elements can be chosen out of n elements in $\binom{n}{k}$ different ways, hence the number of partitions of a set with $n + 1$ elements into m classes is

$$S(n + 1, m) = \sum_{k=m-1}^{n} \binom{n}{k} S(k, m - 1).$$

The sequence $S(n, m)$ is first increasing, then decreasing with respect to m. In [11] it is proved that the value of m from which the numbers $S(n, m)$ decrease is at most equal to $\left[\dfrac{n + 1}{2}\right]$ and converges asymptotically to $\dfrac{n}{\log n}$.

Example 1. Find the number of possibilities of arranging n different objects into m different cells, so that p cells are filled with one or more objects, and the other $m - p$ cells are empty.

The number of possibilities of arranging the n different objects into p different cells so that none of the cells is empty, equals the number of surjective functions defined on a set with n elements and with values in a p-element set, that is, $s_{n,p}$. Since the p filled cells can be chosen out of the m cells in $\begin{pmatrix} m \\ p \end{pmatrix}$ different ways, it follows that the number of solutions to this problem is

$$\begin{pmatrix} m \\ p \end{pmatrix} s_{n,p} = \begin{pmatrix} m \\ p \end{pmatrix} p! \, S(n, p) = [m]_p \, S(n, p),$$

where $[m]_p$ stands for the product $m(m - 1) \ldots (m - p + 1)$.

Example 2. We are given n different objects, say $1, 2, \ldots, n$. What is the number of partitions of the n objects into $k + r$ classes $(0 \leqslant r \leqslant n - k)$, if the objects denoted by $1, 2, \ldots, k$ must belong to k different classes?

Assume first $r \geqslant 1$. Since the order of classes in a partition is immaterial, let us assign the k objects $1, 2, \ldots, k$ to k distinct classes. Then p objects out of the remaining $n - k$ objects can be assigned to the remaining r classes in $S(p, r)$ different ways. Finally the $n - k - p$ remaining objects can be assigned to the k classes which contain the objects $1, 2, \ldots, k$, respectively, in k^{n-k-p} ways, because the first remaining object can be assigned to any of the k classes, thus resulting in k possibilities, the second object can also be assigned in k ways, etc. As p objects can be chosen out of $n - k$ objects in $\begin{pmatrix} n - k \\ p \end{pmatrix}$ ways, it follows that we obtain

$$\sum_{p=r}^{n-k} \begin{pmatrix} n - k \\ p \end{pmatrix} S(p, r) \, k^{n-k-p}$$ partitions with the above formulated properties. Since these partitions are obtained without repetitions, it follows that the number we have obtained is precisely the required number. If $r = 0$, this number of partitions equals k^{n-k}.

For $n=5$, $k=3$ and $r=1$, we get $\sum_{p=1}^{2} \begin{pmatrix} 2 \\ p \end{pmatrix} S(p, 1) \, 3^{2-p} = 7$ partitions, because $S(p, 1)=1$. These partitions are the following: $(1, 4) (2)(3)(5)$; $(1)(2, 4)(3)(5)$; $(1)(2)(3, 4)(5)$; $(1)(2)(3)(4, 5)$; $(1, 5) (2) (3) (4)$; $(1) (2, 5) (3) (4)$ and $(1) (2) (3, 5) (4)$.

A partition of type $1^{\lambda_1} 2^{\lambda_2} \ldots k^{\lambda_k}$ is by definition a partition of a set X into λ_1 classes with one element, λ_2 classes with two elements, $\ldots$, λ_k classes with k elements. In this case, the number of elements of X equals the sum of the numbers of elements in each class, because the classes are disjoint; hence $\lambda_1 + 2\lambda_2 + \ldots + k\lambda_k = n$. Let us now determine the number of partitions of a given type.

PROPOSITION 3. *The number of partitions of type* $1^{\lambda_1} 2^{\lambda_2} \ldots k^{\lambda_k} (\lambda_1 + 2\lambda_2 + \ldots + k\lambda_k = n)$ *of a set X with n elements, equals*

$$N(1^{\lambda_1} 2^{\lambda_2} \ldots k^{\lambda_k}) = \frac{n!}{\lambda_1! \, \lambda_2! \ldots \lambda_k! \, (1!)^{\lambda_1} (2!)^{\lambda_2} \ldots (k!)^{\lambda_k}}. \tag{4.6}$$

If a partition of the set X is of type $1^{\lambda_1} 2^{\lambda_2} \ldots k^{\lambda_k}$, we can obtain a permutation of the set X by writing the elements of the set X in the order in which they appear in the classes and then suppressing the parentheses. We also agree to write first the classes with one element, then the classes with two elements, $\ldots$, finally the classes with k elements. Thus, for example, the partition $\{(1),(2, 4), (3, 5)\}$ of the set $X = \{1, 2, 3, 4, 5\}$ is associated with the permutation $(1, 2, 4, 3, 5)$ of the set X.

Since the order of the elements in a class and the order of classes in a partition are immaterial, we obtain $(1!)^{\lambda_1} (2!)^{\lambda_2} \dots (k!)^{\lambda_k}$ different permutations of the set X from the same partition, by permuting the elements in each class, and from each of these permutations we obtain $\lambda_1! \lambda_2! \dots \lambda_k!$ permutations by permuting the classes with the same number of elements. It follows that from a single partition of type $1^{\lambda_1} 2^{\lambda_2} \dots k^{\lambda_k}$ of the set X we obtain $\lambda_1! \lambda_2! \dots \lambda_k!(1!)^{\lambda_1}(2!)^{\lambda_2} \dots (k!)^{\lambda_k}$ different permutations of the set X, while from different partitions of the set X we obtain different permutations of X. For if two partitions of type $1^{\lambda_1} 2^{\lambda_2} \dots k^{\lambda_k}$ are distinct, then there exist at least two distinct classes belonging to the two partitions and having a non-empty intersection. Permuting the elements of a class in all possible ways, as well as the classes which have the same number of elements, we obtain distinct permutations of the set X. This procedure generates all $n!$ permutations of the set X. In order to prove this assertion, consider a permutation $x_1 x_2 \dots x_n$ of the set X. There exists a partition of type $1^{\lambda_1} 2^{\lambda_2} \dots k^{\lambda_k}$ of the set X which contains each of the elements $x_1, x_2, \dots, x_{\lambda_1}$ in a single-element class, the elements x_{λ_1+1} and x_{λ_1+2} in a two-element class, etc.

Therefore, if we separate the elements which appear in the sequence $x_1 x_2 \dots x_n$ in the order in which they appear, namely λ_1 elements taken one at a time, $2\lambda_2$ elements taken 2 at a time, ..., $k\lambda_k$ elements taken k at a time, we define a partition of X of type $1^{\lambda_1} 2^{\lambda_2} \dots k^{\lambda_k}$. By permuting the elements of a class as well as the classes which contain the same number of elements, we are sure to obtain in particular the permutation considered above. Hence

$$N(1^{\lambda_1} 2^{\lambda_2} \dots k^{\lambda_k})\lambda_1! \lambda_2! \dots \lambda_k! (1!)^{\lambda_1}(2!)^{\lambda_2} \dots (k!)^{\lambda_k} = n!.$$

Thus, for example, the number of partitions of type $1^1 2^2$ of the set $\{1, 2, 3, 4, 5\}$ equals $\dfrac{5!}{1! \, 2! \, 1! \, (2!)^2} = 15$, namely:

(1) (2, 3) (4, 5);   (1) (2, 4) (3, 5);   (1) (2, 5) (3, 4);   (2) (1, 3) (4, 5);

(2) (1, 4) (3, 5);   (2) (1, 5) (3, 4);   (3) (1, 2) (4, 5);   (3) (1, 4) (2, 5);

(3) (1, 5) (2, 4);   (4) (1, 2) (3, 5);   (4) (1, 3) (2, 5);   (4) (1, 5) (2, 3);

(5) (1, 2) (3, 4);   (5) (1, 3) (2, 4);   (5) (1, 4) (2, 3).

The number of all partitions of a set X with n elements is denoted by B_n and is known as the *Bell number*. Clearly $B_n = S(n, 1) + S(n, 2) + \dots + S(n, n)$, for every partition of a set X with n objects has one class or two classes ... or n classes.

PROPOSITION 4. *The Bell numbers satisfy the following recurrence relation:*

$$B_{n+1} = \sum_{k=0}^{n} \binom{n}{k} B_k, \qquad (4.7)$$

where $B_0 = 1$ by definition.

In order to prove this proposition, we notice that $B_{n+1} = \sum\limits_{k=1}^{n+1} S(n+1, k)$ and, taking into account (4.5), we get

$$B_{n+1} = 1 + \sum_{k=2}^{n+1} S(n+1, k) = 1 + \sum_{k=2}^{n+1} \sum_{i=k-1}^{n} \binom{n}{i} S(i, k-1)$$

$$= 1 + \sum_{i=1}^{n} \binom{n}{i} \sum_{k=2}^{i+1} S(i, k-1)$$

by changing the order of the summation. But

$$\sum_{k=2}^{i+1} S(i, k-1) = B_i, \text{ hence } B_{n+1} = 1 + \sum_{i=1}^{n} \binom{n}{i} B_i = \sum_{i=0}^{n} \binom{n}{i} B_i$$

if we define $B_0 = 1$.

This property can be established without any computation, by a proof similar to that of property (4.5) of the Stirling numbers. It is clear that B_n is also the number of equivalence relations which can be defined on a set with n elements.

An interesting property of the numbers B_n, discovered by E. T. Bell, states that these numbers are the coefficients of the Taylor series of the function $e^{e^t - 1}$, also called the *generating function of the numbers B_n*.

PROPOSITION 5.

$$\sum_{n=0}^{\infty} \frac{B_n}{n!} t^n = e^{e^t - 1}. \tag{4.8}$$

Given a sequence $(a_n)_{n \in N}$ of real numbers, let us denote by $f(x)$ the sum of the series $\sum\limits_{n=0}^{\infty} a_n [x]_n$ for those values of x for which the series is convergent. Let us define an operator L by the relation $L(f(x)) = \sum\limits_{n=0}^{\infty} a_n$, which has a sense only if the series $\sum\limits_{n=0}^{\infty} a_n$ is convergent [12]. We shall first prove that if $g(x) = \sum\limits_{n=0}^{\infty} a_n x^n$ then $L(g(x)) = \sum\limits_{n=0}^{\infty} a_n B_n$ (where $B_0 = 1$). (That the operator L can be applied to the function g, can be verified by expressing x^n as a function of the polynomials $[x]_k$ using (4.3)).

From (4.3), we obtain the partial sums

$$s_m = \sum_{n=0}^{m} a_n x^n = \sum_{n=0}^{m} a_n \sum_{k=0}^{n} [x]_k S(n, k) = \sum_{k=0}^{m} \left(\sum_{n=k}^{m} a_n S(n, k) \right) [x]_k,$$

where we have set $[x]_0 = 1$. Therefore

$$L(g(x)) = \lim_{m \to \infty} \sum_{k=0}^{m} \left(\sum_{n=k}^{m} a_n S(n, k) \right) = \lim_{m \to \infty} \sum_{n=0}^{m} a_n \left(\sum_{k=1}^{n} S(n, k) \right) =$$

$$= \lim_{m \to \infty} \sum_{n=0}^{m} a_n B_n = \sum_{n=0}^{\infty} a_n B_n$$

(since L can be applied to g, the series is convergent).

In order to prove Bell's exponential formula (4.8), consider Taylor's development

$$e^{tx} = 1 + \frac{tx}{1!} + \frac{t^2 x^2}{2!} + \dots + \frac{t^n x^n}{n!} + \dots$$

Using the substitution $e^t = u + 1$, we obtain

$$e^{tx} = (u + 1)^x = \sum_{n=0}^{\infty} \frac{[x]_n}{n!} u^n,$$

hence

$$L(e^{tx}) = \sum_{n=0}^{\infty} \frac{u^n}{n!} = e^u = e^{e^t - 1}.$$

According to the above remark, we get

$$L(e^{tx}) = \sum_{n=0}^{\infty} \frac{t^n B_n}{n!}$$

and this series is convergent. By comparing the two expressions of $L(e^{tx})$, we obtain precisely the relation (4.8).

In order to introduce the Fibonacci numbers, we shall solve the following problem:

PROPOSITION 6 (*Kaplansky*). *If* $f(n, k)$ *stands for the number of subsets of* $X=\{1, 2, ..., n\}$ *which have k elements and do not contain two consecutive integers, then*

$$f(n, k) = \binom{n - k + 1}{k} \tag{4.9}$$

where $k \leqslant \left[\dfrac{n + 1}{2}\right].$

In order to prove this result, we associate with every subset $S \subset X$, a binary word $\alpha_1 \alpha_2 \dots \alpha_n$ with $\alpha_i = 1$ if $i \in S$ and $\alpha_i = 0$ if $i \notin S$. We have seen in Chap. 1 that this correspondence is a bijection from the family of subsets of X onto the set of binary words of length $n = |X|$. The 0-1 word associated with the set S will not contain two consecutive 1's due to the restriction that S must not contain two consecutive integers. Now the mapping defined in this way, from the subsets of X which have k elements and do not contain two consecutive integers, onto the words $\alpha_1 \alpha_2 \dots \alpha_n$ with n binary positions, among which there are k ones and $n - k$ zeros, with no two consecutive 1's, is a bijection; therefore we shall count these words, because two sets which have a bijection between them, have the same number of elements. To this end, consider $n - k$ digits equal to 0, numbered from 1 to $n - k$, and add k digits equal to 1 such that no two 1's be placed one after another. Each digit 1 can be characterized by the ordinal number of the digit 0 which precedes it. Hence we must choose k integers from the set $\{0, 1, 2, ..., n - k\}$, which is possible in $f(n, k) = \binom{n - k + 1}{k}$ distinct ways.

The set $\{1, 2, ..., n - k\}$ of ordinal numbers of the digits equal to zero has been augmented with 0, which corresponds to the case when a digit 1 is situated in the first position of the word $\alpha_1 \alpha_2 \dots \alpha_n$.

Thus the number of subsets of X which do not contain two consecutive integers, taking into account the empty set which corresponds to the word $\alpha_1 \alpha_2 \dots \alpha_n$ with $\alpha_1 = \alpha_2 = \dots = \alpha_n = 0$, is equal to

$$F_{n+1} = \sum_{k=0}^{\left[\frac{n+1}{2}\right]} \binom{n - k + 1}{k}. \tag{4.10}$$

(The index k takes values up to $\left[\dfrac{n + 1}{2}\right]$, because the symbol $\binom{n - k + 1}{k}$ makes sense only if $n - k + 1 \geqslant k$, i.e. $k \leqslant \dfrac{n + 1}{2}$). From (4.10) we infer that $F_0 = 1$, $F_1 = 1$, $F_2 = 2$, $F_3 = 3$, $F_4 = 5$ and we shall prove that $F_{n+1} = F_n + F_{n-1}$ for every $n \geqslant 1$.

The numbers F_n are called *Fibonacci numbers*.

In order to prove the relation $F_{n+1} = F_n + F_{n-1}$ notice that every word of length n, consisting of 0's and 1's, and which does not contain two consecutive integers, either has a 0 in the last position, or has the digits 01 in the last two positions; the words obtained by deleting the termination 0 or 01 have lengths $n-1$ and $n-2$, respectively, and do not contain two consecutive 1's. Hence there exists a bijection from the set of 0-1 words of length n which do not contain two consecutive 1's to the union of the disjoint sets consisting of the words of length $n-1$ which do not contain two consecutive 1's, to which we add a 0 in the n-th position, and the words of length $n-2$ which do not contain two consecutive 1's, to which we add the digits 01 in this order in the last two positions. Therefore $F_{n+1} = F_n + F_{n-1}$.

The generating function of the Fibonacci numbers is $\dfrac{1}{1 - x - x^2}$ *which means* that the Maclaurin expansion of this function is:

$$\frac{1}{1 - x - x^2} = F_0 + F_1 x + F_2 x^2 + F_3 x^3 + \dots .$$

For if $|x(1 + x)| < 1$, then

$$\frac{1}{1 - x - x^2} = \frac{1}{1 - x(1 + x)} = 1 + x(1 + x) + x^2(1 + x)^2 + x^3(1 + x)^3 + \dots$$

and, taking into account Newton's binomial formula, we see that the coefficient of x^n in the above expression is $\displaystyle\sum_{k \geqslant 0} \binom{n - k}{k} = F_n$.

PROPOSITION 7 (*Kaplansky*). *Denote by* $f^*(n, k)$ *the number of subsets of* $X = \{1, 2, ..., n\}$ *which have k elements and contain neither two consecutive integers, nor the numbers 1 and n simultaneously. Then*

$$f^*(n, k) = \frac{n}{n - k} \binom{n - k}{k}, \tag{4.11}$$

where $k \leqslant \left[\dfrac{n}{2}\right]$.

In order to prove this result, notice that the subsets with the above formulated property and which contain the number n cannot contain either of the numbers $n-1$ and 1. Their number is $f(n - 3, k - 1)$, while the number of those which do not contain the number n is $f(n - 1, k)$, in both cases we assume they do not contain two consecutive integers. For example, the subsets which contain the number n can be obtained by adding n to the $f(n - 3, k - 1)$ subsets of $X \setminus \{1, n - 1, n\}$ which have $k - 1$ elements and do not contain two consecutive integers.

As every k-element subset of X which contains neither two consecutive integers nor the numbers 1 and n simultaneously, belongs to the union of the two disjoint families of k-element subsets of X which contains only subsets with the required properties, it follows that

$$f^*(n, k) = f(n - 3, \ k - 1) + f(n - 1, k) = \binom{n - k - 1}{k - 1} + \binom{n - k}{k} =$$

$$= \left(\frac{k}{n - k} + 1 \right) \binom{n - k}{k} = \frac{n}{n - k} \binom{n - k}{k}.$$

In this case we consider the numbers $1, 2, \ldots, n$ to be situated on a circle in this order, the number 1 becoming a neighbour of n.

The numbers $F_n^* = \sum\limits_{k=0}^{\left[\frac{n}{2}\right]} f^*(n, k)$ are known as the *Lucas numbers*. They also satisfy the recurrence relation $F_{n+1}^* = F_n^* + F_{n-1}^*$.

Example 3. The problem of married couples *(Lucas)*. We are now in a position to formulate and solve the problem of married couples, which consists of finding the number $T(n)$ of ways of seating n husbands (numbered $1, 2, \ldots, n$) and their wives (numbered $\bar{1}, \bar{2}, \ldots, \bar{n}$, respectively) at a circular table with men and women in alternate positions and such that no wife sits next to her husband.

This is a model for a class of combinatorial problems of the same type. We shall see later that the numbers $T(n)$ occur for example in a formula for counting Latin rectangles.

A bijection f from the set $\{1, 2, \ldots, n\}$ onto the set $\{\bar{1}, \bar{2}, \ldots, \bar{n}\}$ defines an arrangement in the following way:

We seat arbitrarily the man number 1, at his right the woman number $f(1)$, at her right the man number 2, at his right the woman number $f(2)$, etc. For every $i = 1, 2, \ldots, n$, let A_{2i-1} be the set of bijections f with $f(i) = \bar{i}$ and for $i \neq n$ let A_{2i} denote the set of bijections f with $f(i) = \overline{i + 1}$, while for $i = n$ let A_{2n} stand for the set of bijections f with $f(n) = \bar{1}$.

A bijection defines an arrangement which meets the requirements of the problem, if and only if it belongs to none of the sets $A_1, A_2, \ldots, A_{2n}$, hence $T(n) = n! - \left| \bigcup\limits_{i=1}^{2n} A_i \right|$. Now the principle of inclusion and exclusion yields

$$\left| \bigcup_{i=1}^{2n} A_i \right| = \sum_{i=1}^{2n} |A_i| - \sum_{1 \leqslant i < j \leqslant 2n} |A_i \cap A_j| + \ldots - \left| \bigcap_{i=1}^{2n} A_i \right| =$$

$$= \sum_{K \subset \{1, 2, \ldots, 2n\}} (-1)^{|K| - 1} \left| \bigcap_{i \in K} A_i \right|.$$

If the set $K \subset \{1, 2, \ldots, 2n\}$ contains k numbers, then $\left| \bigcap\limits_{i \in K} A_i \right| = (n - k)!$ if K does not contain two consecutive integers from the system $(1, 2, \ldots, 2n, 1)$ and $\left| \bigcap\limits_{i \in K} A_i \right| = 0$ otherwise.

In fact, if K contains two consecutive integers from the system $(1, 2, \ldots, 2n, 1)$, say $2j$ and $2j + 1$, then if there is a bijection $f \in \bigcap_{i \in K} A_i$, this implies $f \in A_{2j} \cap A_{2j+1}$ hence $f(j) = \bar{j} + 1$ and also $f(j + 1) = \bar{j} + 1$. Thus f is not an injection, hence it cannot be a bijection. If K contains the integers $2j - 1$ and $2j$, then from $f \in \bigcap_{i \in K} A_i$ we deduce $f(j) = \bar{j}$ and $f(j) = \overline{j + 1}$, which is absurd for the image by f of an element from the domain of the function f is uniquely defined. Therefore $| \bigcap_{i \in K} A_i | = 0$ if K contains two consecutive integers from the system $(1, 2, \ldots, 2n, 1)$, because a reasoning similar to the above one holds for the numbers $2n$ and 1. In the opposite case, that is, if K does not contain two consecutive integers from the system $(1, 2, \ldots, 2n, 1)$, then $f \in \bigcap_{i \in K} A_i$ implies that the images of $k = |K|$ elements from the set $\{1, 2, \ldots, n\}$ are well defined, while the images of the other $n - k$ elements can be chosen in $(n - k)!$ ways, such that f is a bijection of the set $\{1, 2, \ldots, n\}$ onto the set $\{\bar{1}, \bar{2}, \ldots, \bar{n}\}$.

But we have seen that the number of k-element sets K which do not contain two consecutive integers from the system $(1, 2, \ldots, 2n, 1)$ equals $f^*(2n, k) = \dfrac{2n}{2n - k} \dbinom{2n - k}{k}$. We thus obtain

$$T(n) = n! - \sum_{i=1}^{2n} |A_i| + \sum_{1 \leqslant i < j \leqslant 2n} |A_i \cap A_j| + \ldots + \left| \bigcap_{i=1}^{2n} A_i \right| =$$

$$= n! - \frac{2n}{2n - 1} \binom{2n - 1}{1} (n - 1)! + \frac{2n}{2n - 2} \binom{2n - 2}{2} (n - 2)! + \ldots + (-1)^n \frac{2n}{n} \binom{n}{n} 0!$$

or

$$T(n) = n! - \frac{2n}{2n - 1} \binom{2n - 1}{1} (n - 1)! + \frac{2n}{2n - 2} \binom{2n - 2}{2} (n - 2)! + \ldots + (- 1)^n \cdot 2, \quad (4.12)$$

which is the number of solutions of the problem of married couples.

The problem makes sense for $n \geqslant 2$, so we obtain

$$T(2) = 2! - \frac{4}{3} \binom{3}{1} 1! + 2 = 0,$$

$$T(3) = 3! - \frac{6}{5} \binom{5}{1} 2! + \frac{6}{4} \binom{4}{2} 1! - 2 = 1,$$

$$T(4) = 4! - \frac{8}{7} \binom{7}{1} 3! + \frac{8}{6} \binom{6}{2} 2! - \frac{8}{5} \binom{5}{3} 1! + 2 = 2,$$

$$T(5) = 5! - \frac{10}{9} \binom{9}{1} 4! + \frac{10}{8} \binom{8}{2} 3! - \frac{10}{7} \binom{7}{3} 2! + \frac{10}{6} \binom{6}{4} 1! - 2 = 13,$$

. . . .

The solutions for $n = 3$ and $n = 4$ are given in Fig. 4.1. We notice that two solutions differ one from another only in the order of the elements $\bar{1}, \bar{2}, \ldots, \bar{n}$, the elements $1, 2, \ldots, n$ being arranged in this order in the trigonometric sense.

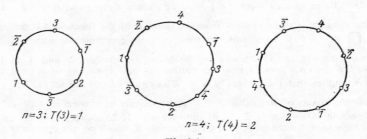

$$n=3; \ T(3)=1 \qquad n=4; \ T(4)=2$$

Fig.4.1

Problems

1.   A partial partition of X is a partition of a subset $Y \subset X$, $Y \neq \varnothing$.

Prove that the number of partial partitions of a set X with $|X| = n$ elements equals $B_{n+1} - 1$.

H i n t: Use Proposition 4.

2.   Let $S_i(n, k)$ be the number of partitions of the set X with $|X| = n$ elements in k blocks, each of cardinal greater than or equal to i ($n \geqslant ki$).

Prove that:

$$\text{i)} \ \ S_i(n, k) = kS_i(n-1, k) + \binom{n-1}{i-1} S_i(n-i, k-1);$$

$$\text{ii)} \ \ S_i(n, k) = \frac{1}{k!} \sum_{(j_1, \ldots, j_k)} \frac{n!}{j_1! \ldots j_k!} \ ;$$

where the sum is over all solutions $j_1, \ldots, j_k$ of the equation $j_1 + \ldots + j_k = n$ and $j_s \geqslant i$ for $s = 1, 2, \ldots, k$.

3.   Prove that $S(n, 2) = 2^{n-1} - 1$; $S(n, n-1) = \binom{n}{2}$;

$$S(n, n-2) = \binom{n}{3} + 3\binom{n}{4}; \ \ S(n, n-3) = \binom{n}{4} + 10\binom{n}{5} + 15\binom{n}{6}.$$

4.   Prove that the Stirling numbers of the second kind $S(n, k)$ with $n \geqslant k$ satisfy $k^{n-k} \leqslant S(n, k) \leqslant \binom{n-1}{k-1} k^{n-k}$.

5. Show that the generating function for the Stirling numbers of the second kind is $\dfrac{1}{k!}(e^t - 1)^k$, i.e.

$$\frac{1}{k!}(e^t - 1)^k = \sum_{n=k}^{\infty} \frac{S(n, k)}{n!} t^n.$$

6. Prove that for $n > 1$:

$$1 - S(n, 2) + 2!\, S(n, 3) - 3!\, S(n, 4) + \ldots + (-1)^{n-1}(n-1)! = 0.$$

H i n t: $k!\, S(n, k)$ is equal to the coefficient of x^n in the expansion of $(e^x - 1)^k$, multiplied by $n!$.

7. The number of ways of filling up a $2 \times n$ rectangle with dominoes (i.e. with 1×2 rectangles) is well known. On page 139 of his book, *Polyominoes*, S. W. Golomb asks for the corresponding result for $3 \times n$ rectangles. Let $U(n)$ be the number of ways of covering a $3 \times n$ rectangle with dominoes. Obviously $U(n) = 0$ if n is odd. Show that

$$U(2m) = \frac{1}{2\sqrt{3}}\,[(\sqrt{3} + 1)(2 + \sqrt{3})^m + (\sqrt{3} - 1)(2 - \sqrt{3})^m].$$

(Tomescu, 1973)

H i n t: Show that $U(n + 1) = 4U(n) - U(n - 1)$ with $U(1) = 3$, $U(2) = 11$.

8. For a recurrence relation

$$f(n + 2) = af(n + 1) + bf(n)$$

where a, b are real numbers, and $n = 1, 2, 3, \ldots$, let us write the quadratic equation

$$r^2 = ar + b$$

which is called the characteristic equation of the given relation.

Prove that if this characteristic equation has two distinct roots r_1 and r_2, then the general solution of the recurrence relation is of the form $f(n) = C_1 r_1^n + C_2 r_2^n$ where C_1 and C_2 are determined by the initial conditions $f(1)$ and $f(2)$ in the form:

$$C_1 + C_2 = f(1)$$
$$C_1 r_1 + C_2 r_2 = f(2).$$

9. Show that the Fibonacci numbers are given by

$$F_n = \frac{1}{\sqrt{5}}\left[\left(\frac{1 + \sqrt{5}}{2}\right)^{n+1} - \left(\frac{1 - \sqrt{5}}{2}\right)^{n+1}\right].$$

10. A bi-covering with k blocks of an n-set X is a family of distinct subsets of X: $A_1, A_2, ..., A_k$ such that:

i) $A_i \neq \varnothing$ for $i = 1, 2, ..., k$;

ii) $|\{A_i | x \in A_i\}| = 2$ for any $x \in X$.

Let $c(n, k)$ denote the number of bi-coverings with k blocks of an n-set X. Prove that

$$c(n, 3) = \frac{1}{2}(3^{n-1} - 1); \quad c(n, 4) = \frac{1}{2}(3^{n-1} - 1)(2^{n-2} - 1).$$

(Comtet, 1968)

11. Prove the following generalization of Proposition 6:

Let X be the set $\{1, 2, ..., n\}$ and let $f_a(n, k)$ denote the number of k-subsets $Y = \{i_1, i_2, ..., i_k\} \subset X$, such that $|i - j| \geqslant a + 1$ for any $i, j \in Y$.

Show that

$$f_a(n, k) = \binom{n - ak + a}{k}.$$

12. Prove the following generalization of Proposition 7:

Let X be the set $\{1, 2, ..., n\}$ and let $f_a^*(n, k)$ denote the number of k-subset $Y = \{i_1, i_2, ..., i_k\} \subset X$ such that for any $i, j \in Y$

$$\min(|i - j|, n - |i - j|) \geqslant a + 1.$$

Show that $f_a^*(n, k) = \dfrac{n}{n - ak} \dbinom{n - ak}{k}$.

(Yamamoto, 1956)

H i n t: Prove that $f_a^*(n, k) = f_a(n - a, k) + a f_a(n - 1 - 2a, k - 1)$.

For further generalizations see [10].

13. Let us consider the graphs which are constructed by connecting n hexagons so that no more than two hexagons are permitted to meet at an edge.

The graphs of this class for which no three hexagons are allowed to have a point in common correspond to "tree-like polyhexes".

Prove that the generating function of the number of tree-like polyhexes rooted at an edge is equal to

$$U(x) = \frac{1}{2x}(1 - 3x - \sqrt{(1 - x)(1 - 5x)}) = x + 3x^2 + 10x^3 + 36x^4 + ...$$

(Harary, Read, 1970)

14. Let $a = (a_0, a_1, a_2, ...)$ be an infinite sequence of real numbers. The generalized Stirling numbers are defined as follows:

i) The Stirling numbers of the first kind $s_a(n, k)$ by

$$(x|a)_n = (x - a_0)(x - a_1)...(x - a_{n-1}) = \sum_{k=0}^{n} s_a(n, k)x^k; \quad (x | a)_0 = 1;$$

ii) The Stirling numbers of the second kind $S_a(n, k)$ by

$$x^n = \sum_{k=0}^{n} S_a(n, k) (x - a_0) (x - a_1) \dots (x - a_{k-1}) = \sum_{k=0}^{n} S_a(n, k) (x|a)_k.$$

Prove that:

1) $$s_a(n, k) = s_a(n - 1, k - 1) - a_{n-1} s_a(n - 1, k);$$

2) $$S_a(n, k) = S_a(n - 1, k - 1) + a_k S_a(n - 1, k);$$

3) $$s_a(n, k) = \sum_{l=k}^{n} s_a(n + 1, l + 1) a_n^{l-k};$$

4) $$S_a(n, k) = \sum_{l=k}^{n} S_a(l - 1, k - 1) a_k^{n-l};$$

5) $$s_a(n, k) = \sum_{l=k}^{n} (-1)^{n-l} s_a(l - 1, k - 1) \prod_{j=l}^{n-1} a_j;$$

6) $$\sum_{k=0}^{n} s_a(n, k) S_a(k, m) = \sum_{k=0}^{n} S_a(n, k) s_a(k, m) = \delta_{n,m} \text{ (Kronecker's } \delta).$$

(Comtet, 1972)

5. For the generalized Stirling number of the second kind $S_a(n, k)$ associated with the sequence $(a_0, a_1, a_2, \dots)$ prove that

$$\sum_{n=k}^{\infty} S_a(n, k) t^n = \frac{t^k}{(1 - a_0 t) (1 - a_1 t) \dots (1 - a_k t)}.$$

(Comtet, 1972)

Hint: If we put $f_k = \sum_{n=k}^{\infty} S_a(n, k) t^n$, we have $f_k = \sum_{n \geq k} (S_a(n - 1, k - 1) +$

$- a_k S_a(n - 1, k)) t^n = t f_{k-1} + a_k t f_k$, hence $f_k = \dfrac{f_{k-1} t}{1 - a_k t}$; $f_0 = \sum_{n=0}^{\infty} a_0^n t^n = \dfrac{1}{1 - a_0 t}$.

6. Prove the following recurrence relations:

i) $$\binom{i + j}{i} s(n, i + j) = \sum_{k=0}^{n} \binom{n}{k} s(k, i) s(n - k, j);$$

ii) $$\binom{i + j}{i} S(n, i + j) = \sum_{k=0}^{n} \binom{n}{k} S(k, i) S(n - k, j).$$

H i n t: Use Vandermonde's formula (2.5) and the identity (4.3), by counting in two different ways $[x + y]_n$ for i) and $(x + y)^n$ for ii).

17.   Prove that $F_{n+1}F_{n-1} - F_n^2 = (-1)^{n+1}$.

H i n t: Use the matrix identity $\begin{pmatrix} 1 & 1 \\ 1 & 0 \end{pmatrix}^{n+1} = \begin{pmatrix} F_{n+1} & F_n \\ F_n & F_{n-1} \end{pmatrix}$ and take the determinant of both sides of this equality.

18.   Show that $\sum_{k=0}^{n} \binom{n}{k} F_{m+k} = F_{m+2n}$.

H i n t: Use (4.10).

19.   Let $X = A_1 \cup A_2 \cup \ldots \cup A_r$ be a partition of an n-set X, having the property that all the numbers: $|A_1|, |A_2|, \ldots, |A_r|$ are pairwise different. Prove that
$$r \leqslant \left[\frac{-1 + \sqrt{8n + 1}}{2} \right].$$

20.   Show that $[x]^n = \sum_{k=0}^{n} |s(n, k)| x^k$.

BIBLIOGRAPHY

1. Bell, E. T., *Exponential polynomials*, Ann. of Math. Ser. II, **35**, 1934, 258—277.
2. Comtet, L., *Birecouvrements et birevêtements d'un ensemble fini*, Studia Sci. Math. Hungar. **3**, 1968, 137—152.
3. Comtet, L., *Nombres de Stirling généraux et fonctions symétriques*, C. R. Acad. Sci. Paris Ser. A, **275**, 1972, 747—750.
4. Frucht, R., *A combinatorial approach to the Bell polynomials and their generalisations*, Recent Progress in Combinatorics, Proc. third Waterloo conference on combinatorics, 1968, 69—7 Academic Press, New York, 1969.
5. Golomb, S. W., *Polyominoes*, Allen, 1966.
6. Harary, F., Read, R. C., *The enumeration of tree-like polyhexes*, Proc. Edinburgh Math Soc., **17**, 1970, 1—13.
7. Kaplansky, I., *Solution of the "Problème des ménages"*, Bull. Amer. Math. Soc., **49**, 194 784—785.
8. Kreweras, G., *Une famille d identités mettant en jeu toutes les partitions d'un ensemble fini a variables en un nombre donné de classes*, C. R. Acad. Sci. Paris, Sér. A, **270**, 1970, 1140—114.
9. Lucas, E., *Théorie des nombres*, Paris, 1891.
10. Moser, W. O. J., Abramson, M., *Enumerations of combinations with restricted difference and cospan*, J. Combinatorial Theory, **7**, 1969, 162—170.
11. Rennie, B. C., Dobson, A. J., *On Stirling numbers of the second kind*, J. Combinatorial Theory, **7**, 1969, 116—121.
12. Rota, G.-C., *The number of partitions of a set*, Amer. Math. Monthly, **71**, 1964, 498—50
13. Tomescu, I., Problem N° E2417, Amer. Math. Monthly, **80**, 1973, 559—560.
14. Touchard, J., *Sur un problème de permutations*, C. R. Acad. Sci. Paris, **198**, 1934, 631—63
15. Vilenkin, N., *Combinatorial mathematics (for recreation)*, Mir Publishers, Moscow, 197.
16. Webb, W. A., Parberry, E. A., *Divisibility properties of Fibonacci polynomials*, Fibonac Quart., **7**, 1969, 457—463.
17. Yamamoto, K., *Structure polynomial of Latin rectangles and its application to a combinatori problem*, Mem. Fac. Sci. Kyushu Univ., Ser. A, **10**, 1956, 1—13.

CHAPTER 5

Partitions

Problems concerning the representation of a natural number as a sum of natural numbers or as a sum of powers of natural numbers, belong to number theory and go back to Euler. Thus we recall the famous conjecture of Waring, proved by Hilbert in 1909 [9], which states that for each $k=1, 2, 3, ...$, every natural number n can be written as a sum of at most $g(k)$ k-th powers of natural numbers, the number $g(k)$ depending only on k, not on the represented number n [5].

In the sequel we shall present certain properties and computational algorithms for the number of representations of a positive integer as a sum of a given number of integers subject to certain supplementary conditions.

A *partition* of the natural number n into m parts is, by definition, a representation of n in the form

$$n = a_1 + a_2 + ... + a_m, \tag{5.1}$$

where the natural numbers a_i satisfy $a_1 \geqslant a_2 \geqslant ... \geqslant a_m \geqslant 1$.

In view of the latter condition, we notice that two representations of n as a sum of natural numbers differ only in respect of the numbers $a_1, a_2, ..., a_m$, not in their order. The numbers $a_1, a_2, ..., a_m$ will be called *the parts of n* in the partition (5.1) and we shall denote by $P(n, m)$ the number of partitions of n into m parts.

If we consider a set X with n elements, a system $(a_1, a_2, ..., a_m)$ with $a_1 + a_2 + ... + a_m = n$ and $a_i \geqslant 1$ for $i = 1, ..., m$, defines the type of a partition of X into m sets. As the order of classes in a partition is immaterial, it follows that the order of the integers $a_1, a_2, ..., a_m$ is immaterial in the representation of n, hence the number $P(n, m)$ also represents the number of types of partitions of a set with n elements into m classes. Table 5.1 enumerates all possible partitions of n into m parts $(1 \leqslant m \leqslant n)$ for $n = 1, 2, 3, 4, 5$.

Taking into account the property from example 2, Chap. 2, we deduce that

$$P(n, m) \geqslant \frac{1}{m!} \binom{n-1}{m-1}.$$

PROPOSITION 1. *The numbers $P(n, m)$ satisfy the recurrence relation:*

$$P(n, 1) + P(n, 2) + ... + P(n, k) = P(n + k, k)$$

and

$$P(n, 1) = P(n, n) = 1. \tag{5.2}$$

The relations $P(n, 1) = P(n, n) = 1$ are obvious. It remains to prove the first equality of (5.2). To this end consider the set of partitions of the number n into at most k parts; their number is $P(n, 1) + P(n, 2) + ... + P(n, k)$. Each such

TABLE 5.1

$n = 1$	1	$P(1, 1) = 1$
2	2	$P(2, 1) = 1$
	$1 + 1$	$P(2, 2) = 1$
3	3	$P(3, 1) = 1$
	$2 + 1$	$P(3, 2) = 1$
	$1 + 1 + 1$	$P(3, 3) = 1$
4	4	$P(4, 1) = 1$
	$3 + 1$ $2 + 2$	$P(4, 2) = 2$
	$2 + 1 + 1$	$P(4, 3) = 1$
	$1 + 1 + 1 + 1$	$P(4, 4) = 1$
5	5	$P(5, 1) = 1$
	$4 + 1$ $3 + 2$	$P(5, 2) = 2$
	$3 + 1 + 1$ $2 + 2 + 1$	$P(5, 3) = 2$
	$2 + 1 + 1 + 1$	$P(5, 4) = 1$
	$1 + 1 + 1 + 1 + 1$	$P(5, 5) = 1$

partition may be written $n = a_1 + a_2 + ... + a_m + 0 + ... + 0$, where the sum contains k terms and $a_1 \geqslant a_2 \geqslant ... \geqslant a_m \geqslant 1$ $(1 \leqslant m \leqslant k)$. From this representation of n we obtain a partition of $n + k$ into k parts as follows:

$$n + k = (a_1 + 1) + (a_2 + 1) + ... + (a_m + 1) + 1 + 1 + ... + 1,$$

where the sum contains k terms and $a_1 + 1 \geqslant a_2 + 1 \geqslant ... \geqslant a_m + 1 \geqslant 1$.

The mapping thus defined is an injection, because to different partitions of n into at most k parts correspond different partitions of $n + k$ into k parts. This mapping is also a surjection, and hence a bijection, because every partition of $n+k$ into k parts is the image of that partition of n into $m \leqslant k$ parts which is obtained by subtracting a unit from each term of the partition of $n+k$ and by retaining the terms different from zero obtained in this way. Therefore, there is a bijection between the set of partitions of n into at most k parts and the set of partitions of $n + k$ into k parts.

Formulae (5.2), where by definition we have set $P(n, m) = 0$ for $m > n$, enable a recursive computation of the numbers $P(n, k)$, row by row (Table 5.2). We have

TABLE 5.2

$P(n, m)$	$m = 1$	2	3	4	5	6	...	$P(n)$
$n = 1$	1	0	0	0	0	0		1
2	1	1	0	0	0	0		2
3	1	1	1	0	0	0		3
4	1	2	1	1	0	0		5
5	1	2	2	1	1	0		7
6	1	3	3	2	1	1		11

denoted the number of all the partitions of n by $P(n)$, so $P(n) = P(n, 1) + P(n, 2) + \ldots + P(n, n)$. An asymptotic evaluation of the number $\log P(n)$, has been obtained by Hardy and Ramanujan [8], namely $\log P(n) \sim \pi \sqrt{\dfrac{2n}{3}} - \log(4n\sqrt{3})$.

PROPOSITION 2. *The number of partitions of n for which the greatest part is equal to k, equals the number of partitions of n into k parts.*

For instance, in the case $n = 7$ and $k = 4$ we have the following partitions of 7 for which the greatest part is 4:

$$4 + 1 + 1 + 1, \; 4 + 2 + 1, \; 4 + 3$$

and the following partitions of 7 into 4 parts:

$$4 + 1 + 1 + 1, \; 3 + 2 + 1 + 1, \; 2 + 2 + 2 + 1.$$

In order to prove this property, we associate with each partition of n a Ferrers diagram which will make the reasoning more suggestive. The Ferrers diagram associated with the partition (5.1) of n contains a_1 cells in the first row, a_2 cells in the second row, ..., a_m cells in the m-th row. The Ferrers diagram associated with the partition $7 = 3 + 2 + 1 + 1$ is drawn in Fig. 5.1. The total number of cells of the Ferrers diagram corresponding to a partition of n will be n. If we count the cells column by column, we shall of course obtain the same total number n of cells and row k will contain as many cells as the number of parts greater than or equal to k in the original partition of n. We thus obtain a new partition of n of the form $n = a_1^* + a_2^* + ...$, where a_k^* represents the number of parts in the previous partition of n which are greater than or equal to k. This partition of n is said to be the partition *conjugate* to the partition (5.1) and its number of parts will be a_1, i.e. the greatest part in the partition (5.1) of n. Thus the partition conjugate to the partition $3 + 2 + 1 + 1$ is $4 + 2 + 1$ and the Ferrers diagram of this partition is obtained from the diagram in Fig. 5.1 by a

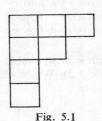

Fig. 5.1

reflection in the diagonal of the upper left cell. Notice that the operation of taking the conjugate is involutive, i.e. the conjugate of the conjugate coincides with the original partition. So, conjugation may be viewed as a mapping defined on the set of partitions of n into k parts, with values in the set of partitions of n for which the greatest part is k. This function is injective, because two distinct partitions of n into k parts generate distinct partitions of n with the greatest part equal to k. Moreover, since this mapping is involutive, it is a surjection, hence a bijection.

PROPOSITION 3. *The number of partitions of n into pairwise distinct odd parts is the same as the number of partitions of n which coincide with their conjugates.*

In order to prove this property we shall define a bijection from the set of partitions of n into pairwise distinct odd parts onto the set of partitions of n which coincide with their conjugates, by means of the language of Ferrers diagrams. We associate with the partition $n = (2k_1 + 1) + (2k_2 + 1) + ...$, where $k_1 \geqslant k_2 \geqslant ...$, a symmetric Ferrers diagram associated with a partition of n which coincides with its conjugate, defined as follows: on the first row as well as on the first column we arrange $k_1 + 1$ cells, then the $2k_2 + 1$ cells are arranged with $k_2 + 1$ on the second row and second column etc. This construction is possible, because all the parts of the partition of n are pairwise distinct and the resulting diagram will be symmetric, because all the parts are odd. Thus, e.g. to the partition $7 + 5 + 3 + 1$ into pairwise distinct odd parts corresponds the self-conjugate partition $4 + 4 + 4 + 4$ by the above described procedure. The correspondence thus defined is obviously injective. To prove its surjectivity, consider a self-conjugate partition P of n. Denote by a_1 the number of cells in the first row and first column of the Ferrers diagram associated with P. Let a_2 be the number of cells in the first row and first column of the diagram obtained by deleting the first row and first column of the diagram of P; etc. We thus obtain a partition Q of n into pairwise distinct odd parts, which is precisely the image of P by the given function.

PROPOSITION 4. *The number of partitions of n into pairwise distinct parts is equal to the number of partitions of P into odd parts.*

To prove this property, we shall establish a bijection from the set of partitions of n into odd parts onto the set of partitions of n into pairwise distinct parts.

If in a partition of n into odd parts the number $2k + 1$ occurs p times, we write the number p as a sum of powers of 2:

$$p = 2^{i_1} + 2^{i_2} + \ldots,$$

the representation being unique. The $(2k + 1)p$ cells of the Ferrers diagram associated with the given partition of n will be re-arranged as follows: the numbers of cells in the various rows will be $(2k + 1)2^{i_1}$, $(2k + 1)2^{i_2}$, ... etc., in decreasing order of the numbers of cells from top to bottom. In this way the partition

$$7 + 5 + 5 + 3 + 3 + 3 + 1 + 1 + 1 + 1$$

will correspond to a new partition, whose Ferrers diagram contains 7 cells, 10 cells, 6 cells, 3 cells and 4 cells on its rows, hence it corresponds to the partition $10 + 7 + 6 + 4 + 3$ into pairwise distinct parts.

The partition of n defined in this way contains only pairwise distinct parts, for every integer is uniquely written as the product of an odd number and a power of 2.

The correspondence thus defined is a bijection. Indeed, the uniqueness of the representation of a natural number as a sum of powers of 2 shows the injectivity of this mapping. To prove that this correspondence is also a surjection, consider a partition P of n into pairwise distinct parts. Each part is a natural number, hence it can be written as the product of an odd number and a power of 2. By a convenient rearrangement of similar terms, we obtain only terms of the form $(2k+1)p$ with $k=0, 1, \ldots$, which will generate p terms equal to $2k + 1$. By writing these terms in decreasing order of their growths, we obtain a partition Q of n into odd parts. Now if we apply to Q the above defined transformation from the partitions of n into odd parts onto the partitions of n into pairwise distinct parts, we obtain precisely P.

PROPOSITION 5 *(Euler). Let $Q_e(n)$ be the number of partitions of n into an even number of distinct summands and $Q_o(n)$ be the number of partitions into an odd number of distinct summands.*

If n cannot be represented in the form $n = \dfrac{3k^2 \pm k}{2}$, then $Q_e(n) = Q_o(n)$; otherwise $Q_e(n) = Q_o(n) + (-1)^k$.

In order to prove Euler's theorem, let us illustrate a transformation of an array with an even number of rows into an array with the same number of cells having an odd number of rows, and conversely. Since we are considering only partitions into distinct parts, the arrays of such partitions consist of several trapezoids on top of each other.

Denote the number of cells of the lower row of the array by m, the number of rows of the upper trapezoid by k.

If $m \leqslant k$ we discard the last row (at the South of array) and extend the first m rows of the upper trapezoid by one cell (on the line of $45°$ at the East of the diagram). This transformation does not alter the total number of cells; all rows are then of different length, but the parity of the number of rows will change.

Figures 5.2 and 5.3 illustrate such a transformation.

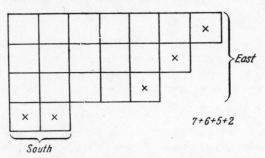

Fig. 5.2

Now let the array contain at least two trapezoids and $m > k$. Then take one cell from each row of the upper trapezoid and use these cells to make the South row of a new array.

This can be done because $m > k$ and therefore the generated row is shorter than the first row of the original array. Besides, since we took all the rows of the upper trapezoid, all the rows in the newly obtained array will have unequal lengths. Finally, the new array contains as many cells as the original one, but the parity of the number of rows has changed: the new array contains one more row. The same operation can be performed on arrays consisting of one trapezoid (i.e. the East line contains k cells and k equals the number of parts of n) if $k \geqslant m - 2$.

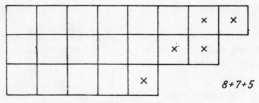

Fig. 5.3

Note that the above-described transformation is involutive: performing this transformation twice we get the original diagram, hence it is a bijection (see exercise 6, Chap. 1). Thus, arrays of partitions of n which permit this transformation split up into the same number of arrays with an even and an odd number of rows. It now remains to find out which arrays do not permit such a transformation. Clearly, they consist of one trapezoid, and for them we either have $m = k$ or $m = k + 1$ (Fig. 5.4). In the former instance, the array contains $\dfrac{3k^2 - k}{2}$ cells, in the latter, $\dfrac{3k^2 + k}{2}$ cells.

The foregoing reasoning shows that if n is not a number of the form $\dfrac{3k^2 \pm k}{2}$, then it has an equal number of partitions into an even and into an odd number

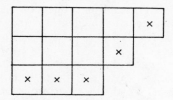 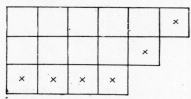

Fig. 5.4

of distinct summands. If $n = \dfrac{3k^2 \pm k}{2}$ and k is even, then there remains one array that does not permit the transformation and has an even number of rows. And so there will be one more partition into an even number of summands than into an odd number. But if $n = \dfrac{3k^2 \pm k}{2}$ and k is odd, then there will be one more partition into an odd number of summands. The proof of the proposition is complete.

COROLLARY *(Euler's identity). If the infinite product*

$$(1 - x)(1 - x^2)(1 - x^3) \dots (1 - x^n) \dots$$

is converted into a series, we obtain

$$1 + \sum_{n=1}^{\infty} \psi(n)x^n = 1 - x - x^2 + x^5 + x^7 - x^{12} - x^{15} + x^{22} + \dots,$$

where $\psi(n) = (-1)^k$ *if* $n = \dfrac{3k^2 \pm k}{2}$ *and* $\psi(n) = 0$ *if* n *cannot be represented in the form* $n = \dfrac{3k^2 \pm k}{2}$ *with* k *an integer.*

Indeed, $\psi(n) = Q_e(n) - Q_o(n)$ and this result is implied by proposition 5.

Let us denote by $P(n, m|h)$ the number of partitions of n into m parts for which the least part is equal to h, and by $P(n|h)$ the number of partitions of n for which the least part is equal to h. It follows from this definition that:

$$P(n, m) = P(n, m|1) + P(n, m|2) + \dots + P\left(n, m \left| \left[\frac{n}{m} \right] \right. \right)$$

and

$$P(n|h) = P(n, 1|h) + P(n, 2|h) + \dots + P(n, n|h).$$

PROPOSITION 6. *The relations*

$$P(n, m|1) = P(n - 1, m - 1)$$

and

$$P(n, m|h) = P(n - m, m|h - 1) \tag{5.3}$$

hold for $h > 1$.

In order to prove the relations (5.3), we shall establish a bijection from the set of partitions of n into m parts with least part equal to h, onto the set of partitions of $n-m$ into m parts with least part equal to $h-1$, for $h > 1$, and onto the set of partitions of $n - 1$ into $m - 1$ parts, for $h = 1$. For $h > 1$, we shall associate with a partition $n = a_1 + a_2 + ... + a_m$ with $a_1 \geqslant a_2 \geqslant ... \geqslant a_m = h$, the partition $n - m = (a_1 - 1) + (a_2 - 1) + ... + (a_{m-1} - 1) + (h - 1)$, which has m parts among which the least one is $h - 1$. To distinct partitions of n will correspond distinct partitions of $n - m$ obtained by subtracting one unit from each part. Given a partition Q of $n - m$ into m parts with the least part equal to $h - 1$, this is the image by the above defined transformation of that partition of n into m parts with the least part equal to h, which is obtained from Q by adding one unit to each part. So our transformation is a bijection. In the case $h = 1$, the bijection is defined as follows: with every partition Q of n into m parts among which the least one is equal to 1, we associate the partition of $n - 1$ into $m - 1$ parts obtained by suppressing the part of Q equal to 1.

PROPOSITION 7. *The numbers $P(n|h)$ satisfy the recurrence relations:*

$$P(n|1) = P(n - 1)$$

and

$$P(n|h) = P(n - 1|h - 1) - P(n - h|h - 1) \tag{5.4}$$

for $h > 1$.

The first relation (5.4) is proved by associating with each partition Q of n with least part equal to 1, a partition of $n - 1$ obtained from Q by deleting the part equal to 1; this mapping is a bijection.

To prove the second relation (5.4), write $n = h + (n - h)$, hence $P(n|h) = \sum_{k \geqslant h} P(n - h|k)$. If $h > 1$, we also obtain $P(n - 1|h - 1) = \sum_{k \geqslant h-1} P(n - h|k)$. By comparing these two relations, we obtain the required result.

Using the recurrence relations (5.4), we can compute the numbers $P(n|h)$ and obtain an array analogous to Table 5.2. Another method for obtaining the number of partitions of an integer into several parts with certain properties consists of the utilization of generating functions: these are the functions for which the coefficients of the Maclaurin expansion represent precisely the numbers of partitions of the integers into a given number of parts with certain properties. This method goes back to Euler and Laplace.

Let us show that the generating function of the number $P(n)$ of all partitions of the positive integer n is

$$F_1(x) = (1 - x)^{-1}(1 - x^2)^{-1}(1 - x^3)^{-1} \dots$$

which means that the relation

$$\frac{1}{(1 - x)(1 - x^2)(1 - x^3)\dots} = \sum_{n=0}^{\infty} P(n)x^n,$$

where by definition $P(0) = 1$, holds in a certain neighbourhood of the origin.

In a neighbourhood of the origin we can write

$$\frac{1}{(1 - a_1 x)(1 - a_2 x^2)\dots(1 - a_k x^k)\dots}$$

$$= (1 + a_1 x + a_1^2 x^2 + \dots)(1 + a_2 x^2 + a_2^2 x^4 + \dots)\dots(1 + a_k x^k + a_k^2 x^{2k} + \dots)\dots$$

$$= 1 + a_1 x + (a_1^2 + a_2)x^2 + \dots + (a_1^{\lambda_1} a_2^{\lambda_2} \dots a_k^{\lambda_k} + \dots)x^n + \dots$$

We notice that the term $a_1^{\lambda_1} a_2^{\lambda_2} \dots a_k^{\lambda_k}$ which occurs in the coefficient of x^n, has the property that $\lambda_1 + 2\lambda_2 + \dots + k\lambda_k = n$, therefore it defines a partition of n, namely

$$n = \underbrace{(1 + 1 + \dots + 1)}_{\lambda_1} + \underbrace{(2 + 2 + \dots + 2)}_{\lambda_2} + \dots + \underbrace{(k + k + \dots + k)}_{\lambda_k}.$$

Taking into account the distributive law, we note that the exponents of the letters which occur in the coefficient of x^n generate repetition-free all the partitions of n, hence if we take $a_1 = a_2 = \dots = 1$, the coefficient of x^n will be precisely the number of partitions of n, i.e. $P(n)$. It can be shown that the generating function of the numbers $P(n, m)$ is $F_2(x) = x^m(1 - x)^{-1}(1 - x^2)^{-1} \dots (1 - x^m)^{-1}$, the generating function of the number of partitions of n into odd parts is $F_3(x) = \prod_{k=1}^{\infty} (1 - x^{2k-1})^{-1}$, that of the number of partitions of n into pairwise distinct parts is $F_4(x) = \prod_{k=1}^{\infty} (1 + x^k)$, the generating function of the number of partitions of n

into an odd number of parts is $F_5(x) = \prod_{k=1}^{\infty} (1 - x^{2k+1})^{-1}$, while that of the partitions of n into pairwise distinct odd parts is $F_6(x) = \prod_{k=1}^{\infty} (1 + x^{2k+1})$. It follows from proposition 4 that $F_3(x) = F_4(x)$. This result was obtained by Euler:

$$F_3(x) = \frac{1}{1-x} \cdot \frac{1}{1-x^3} \cdot \frac{1}{1-x^5} \cdots = \frac{1-x^2}{1-x} \cdot \frac{1-x^4}{1-x^2} \cdot \frac{1-x^6}{1-x^3} \cdots =$$

$$= (1 + x)(1 + x^2)(1 + x^3) \ldots = F_4(x).$$

For a presentation of the various generating functions which occur in counting problems see Riordan [13]. For a generalization to multipartitions see [3].

Problems

1. Prove that $\sum_{k \geq 0} (-1)^{\binom{k+1}{2}} P\left(2n + 1 - \binom{k+1}{2}\right) = 0$ and

$\sum_{k \geq 0} (-1)^{\binom{k+1}{2}} P\left(2n - \binom{k+1}{2}\right) = Q(n)$, the number of partitions of n into unequal parts.

(Karpe, 1969; Ewell, 1973)

Hint: Use the following identity

$$\frac{\prod_{n=1}^{\infty} (1 - x^{2n})}{\prod_{n=1}^{\infty} (1 + x^{2n-1})} = \sum_{n=0}^{\infty} (-x)^{\binom{n+1}{2}}$$

obtained from an identity of Gauss by replacing x by $-x$. Since $1 - x^{2n} = (1 - x^n)(1 + x^n)$, this equation is equivalent to $\prod_{n=1}^{\infty} (1 - x^n) \prod_{n=1}^{\infty} (1 + x^{2n}) = \sum_{n=0}^{\infty} (-x)^{\binom{n+1}{2}}$.

2. In how many ways can three numbers be chosen from $3n$ consecutive integers so that their sum is divisible by 3?

Hint: The answer is $\dfrac{n}{2}(3n^2 - 3n + 2)$.

3. Prove that the number of partitions of the integer n is equal to the number of partitions of the integer $2n$ into n parts.

Hint: Apply Proposition 1.

4.   Verify the following equalities:

$$P(n, m|2) = P(n - 2, m - 1) - P(n - 2, m - 1|1)$$

$$= P(n - 2, m - 1) - P(n - 3, m - 2).$$

$$P(n, m|3) = P(n - 3, m - 1) - P(n - 3, m - 1|1) - P(n - 3, m - 1|2)$$

$$= P(n - 3, m - 1) - P(n-4, m-2) - P(n-5, m-2) + P(n-6, m-3).$$

(Targhetta, 1967)

5.   If $h = \left[\dfrac{n}{m} \right]$, prove that

$$P(n, m|h) = P(n - 1, m - 1) - P(n - h, m - 1|1)$$

$$- P(n - h, m - 1|2) - \ldots - P(n - h, m - 1|h - 1).$$

(Targhetta, 1967)

6.   If we denote by $Q(n, m)$ the number of partitions of n into m unequal (positive, integer) parts, prove that

$$Q(n, m) = Q(n - m, m) + Q(n - m, m - 1).$$

(Platner, 1888)

7.   Prove that $Q(n, 3) = \left[\dfrac{n^2 - 6n + 12}{12} \right]$.

8.   Prove the following property of the generalized binomial coefficients

$$\left[\begin{matrix} n \\ k \end{matrix} \right]_q = \frac{(q^n - 1) \ldots (q^{n-k+1} - 1)}{(q^k - 1) \ldots (q - 1)},$$

where q is a power of a prime:

$$\left[\begin{matrix} n \\ k \end{matrix} \right]_q = a_0 + a_1 q + a_2 q^2 + \ldots + a_p q^p + \ldots$$

where a_p enumerates the partitions of p into at most k parts not exceeding $n - k$.

(Sylvester, 1882)

H i n t: See Knuth [11].

9. Show that the number of partitions of n satisfies the following recurrence relation:

$$P(n) = \sum_{\substack{0 \leqslant i \leqslant m \\ m < j \leqslant n}} e(j-i)\, P(i)\, P(n-j)$$

for any $0 < m < n$, where $e(j-i) = (-1)^{k+1}$ if $j - i = \dfrac{3k^2 \pm k}{2}$ and k is a nonnegative integer and $e(j-i) = 0$ otherwise.

(Arkin, Pollack, 1970)

H i n t: Use Euler's theorem in the form:

$$\sum_{n=0}^{\infty} P(n)x^n = \frac{1}{(1-x)(1-x^2)(1-x^3)\ldots} = \frac{(1-x)(1-x^2)\ldots}{(1-x)(1-x^2)\ldots} \cdot \frac{1}{(1-x)(1-x^2)\ldots}$$

$$= \sum_{k \geqslant 0} (-1)^k x^{\frac{3k^2 \pm k}{2}} \sum_{i=0}^{\infty} P(i)x^i \sum_{j=0}^{\infty} P(j)x^j.$$

10. Show that the number of ways of partitioning n into at most m parts is equal to $P(n + m, m)$.
H i n t: Use Ferrers diagrams.

11. Prove that the number of ways of partitioning n into at most m parts is equal to the number of partitions of $n + \dbinom{m+1}{2}$ into m unequal parts.

H i n t: To each array of n cells containing no more than m rows adjoin an isosceles right triangle of m rows, and then reduce the diagram to its normal form.

12. Let $Q_k(n, m)$ denote the number of partitions of n into m unequal parts, each part being chosen from the set $\{k, k+1, \ldots, n\}$, $k \geqslant 1$.
Prove that:

i) $Q_k(n, m) = Q_1(n - (k-1)m, m)$;

ii) $Q_1(n, m) = \sum_q Q_1(n - qm, m - 1)$,

where the sum is over all q such that q is the first term in the partition of n, i.e $n = q + x_2 + \ldots$ and $q < x_2 < \ldots$;

iii) $Q_1(n, m) = Q_1(n - m, m) + Q_1(n - m, m - 1)$.

(Platner, 1888)

13. Prove that $\displaystyle\sum_{k_1 + 2k_2 + \ldots + nk_n = n} 1 = P(n)$.

14. Let $p(n)$ and $q(n)$ be the number of unrestricted partitions of n, and the number of partitions of n into odd parts, respectively.

Prove that:

i) $q(n) = \sum_{i \geqslant 0} (-1)^i q(i) q(2n - i)$, where $q(0) = 1$;

ii) $p(n) = \sum_{i \geqslant 0} p(i) q(n - 2i)$, where $p(0) = q(0) = 1$.

H i n t: Use generating functions.

(Karpe, 1969)

15. Prove Euler's pentagonal theorem:

$$P(n) = P(n - 1) + P(n - 2) - P(n - 5) - P(n - 7) + \ldots$$

$$= \sum_{k \geqslant 1} (-1)^{k-1} \left(P\left(n - \frac{3k^2 - k}{2}\right) + P\left(n - \frac{3k^2 + k}{2}\right) \right).$$

16. Show that $P(n, m) = P(n - m)$ for $m \geqslant \dfrac{n}{2}$.

BIBLIOGRAPHY

1. Arkin, J., Pollack, R., *Recurrence formulas*, Fibonacci Quart., **8**, 1970, 4—5.
2. Berge, C., *Principes de combinatoire*, Dunod, Paris, 1968.
   English edition: *Principles of combinatorics*, Academic Press, New York, 1971.
3. Cheema, M. S., Motzkin, T. S., *Multipartitions and multipermutations*, Combinatorics, Proc. Sympos. Pure Math., **19**, 1971, 39—70.
4. Dickson, L. E., *History of the theory of numbers*, Vol. II, 1920, 145.
5. Ellison, W. J., *Waring's problem*, Amer. Math. Monthly, **78**, 1971, 10—36.
6. Ewell, J. A., *Partition recurrences*, J. Combinatorial Theory (A), **14**, 1973, 125—127.
7. Gupta, H., *Partitions — a survey*, J. Res. Nat. Bur. Standards Sect. B, **74**, 1970, 1—29.
8. Hardy, G., Ramanujan, S., *Asymptotic formulae in combinatorial analysis*, Proc. London Math. Soc., **17**, 1918, 75—115.
9. Hilbert, D., *Beweis für die Darstellbarkeit der ganzen Zahlen durch eine feste Anzahl n-ter Potenzen (Waringsches Problem)*, Math. Ann., **67**, 1909, 281—300.
10. Karpe, R., *Some relations between combinatorial numbers p(n)* (Czech), Časopis Pěst Mat., **94**, 1969, 108—114.
11. Knuth, D. E., *Subspaces, subsets and partitions*, J. Combinatorial Theory (A), **10**, 1971, 178—180.
12. Narayana, T. V., Mathsen, R. M., Sarangi, J., *An algorithm for generating partitions and its applications*, J. Combinatorial Theory (A), **11**, 1971, 54—61.
13. Riordan, J., *Combinatorial identities*, Wiley, New York, 1968.
14. Sylvester, J. J., with insertions by F. Franklin, *A constructive theory of partitions, arranged in three acts, an interact and an exodion*, Amer. J. Math., **5**, 1882, 251—330 and **6**, 1884, 334—336.
15. Targhetta, M. L., *Construzione e computo delle partizioni di un numero naturale*, Rend. Sem. Fac. Sci. Univ. Cagliari, **37**, 1967, 1—7.
16. Vilenkin, N. Ya., *Combinatorial mathematics (for recreation)*, Mir Publishers, Moscow, 1972.

CHAPTER 6

Counting Trees

Consider an undirected graph $G = (X, U)$, where X is the set of vertices and U is the set of edges. This graph can be drawn in the plane, by representing its vertices as points in the plane and its edges as line segments joining certain pairs of points. A *chain* is a sequence of vertices such that every two consecutive members of the chain are joined by an edge. A *cycle* is a chain for which the initial vertex and the terminal vertex coincide.

A graph is *connected* if every pair of vertices is connected by a chain. A *subgraph* G_1 of the graph G has a set $X_1 \subset X$ as its set of vertices and all the edges of U joining pairs of vertices from X_1 as its set of edges. *The degree of a vertex* $x \in X$, denoted by $d(x)$, is the number of edges having x as an extremity.

The maximum number of independent cycles of the graph G, that is, the *cyclomatic number* of G, denoted by $k(G)$, is equal to $m - n + p$, where m is the number of edges, n is the number of vertices and p is the number of connected components of the graph G [5].

A *tree* is defined as a cycle-free connected graph. Fig. 6.1 depicts two trees. Such trees are encountered in cycle-free connected communication networks, in the theory of classification, in genetics, etc. Proposition 1 furnishes various characterizations of trees [5] and we shall see later the connection between trees and permutation groups.

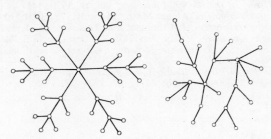

Fig. 6.1

PROPOSITION 1. *Let G be a graph with n vertices $(n \geqslant 2)$. The following properties are equivalent and characterize a tree:*

(1) *G is connected and cycle-free.*

(2) *G is cycle-free and has $n - 1$ edges.*

(3) *G is connected and has n — 1 edges.*

(4) *G is cycle-free and a unique cycle is obtained by adding one supplementary edge between any two vertices not joined by an edge.*

(5) *G is connected and any graph obtained from G by deleting an arbitrary edge, is not connected.*

(6) *For every two vertices of G there exists one chain and only one which connects them.*

We shall prove the following implications:

$$(1) \Rightarrow (2) \Rightarrow (3) \Rightarrow (4) \Rightarrow (5) \Rightarrow (6) \Rightarrow (1).$$

If (1) holds then G being connected and cycle-free, the number of connected components is $p = 1$ and $k(G) = 0$, therefore $m - n + 1 = 0$ and $m = n - 1$. If (2) holds, then $k(G) = 0$, and $m = n - 1$, so $n - 1 - n + p = 0$. Thus $p = 1$, that is, G has a single connected component, hence it is connected, thus proving (3). If (3) holds, then $p = 1$, $m = n - 1$, hence $k(G) = m - n + p = 0$, so that G is cycle-free. If we add one edge, we obtain a new graph G' with $k(G') = 1$, therefore G' contains a unique cycle, thus proving (4). If (4) holds, then G is necessarily connected, for otherwise two vertices x and y would exist which are not connected by any chain, so that no cycle can be created by adding an edge between x and y. Let us now prove that under the hypothesis (4), if an arbitrary edge of G is deleted, then the resulting graph is no longer connected. Assume, on the contrary, that by deleting the edge joining two vertices x and y, a new connected graph G' is obtained. Then the graph G' contains a chain from x to y which, together with the deleted edge between x and y, would generate a cycle in the graph G, thus contradicting (4). So (4) $\Rightarrow$ (5). We shall also prove (5) $\Rightarrow$ (6) by reductio ad absurdum. If two vertices x and y are connected by two distinct chains in G, then one of the chains contains an edge which does not belong to the other; by deleting this edge we get a graph which is connected, thus contradicting (5). Finally (6) $\Rightarrow$ (1), for the hypothesis (6) implies that G is connected and the existence of a cycle would imply that two arbitrary vertices of the cycle are connected by two distinct chains, thus contradicting (6).

We now prove the following corollary of the above proposition: *a graph $G = (X, U)$ has a partial graph $H = (X, V)$, where $V \subset U$, which is a tree, if and only if it is connected.*

If G is not connected, none of its partial graphs is connected; a fortiori none of them is a tree. If G is connected, let us look for an edge the deletion of which does not disconnect the graph G. If such an edge does not exist, then it follows from (5) that G is a tree. If such an edge does exist, then we delete it and obtain a connected partial graph of G, for which we repeat the above process. After a finite number of steps we obtain a partial graph H from which no edge can be deleted; H is then a partial tree of G.

In the following we shall present some results on counting trees with labelled vertices — or, briefly, labelled trees.

Fig. 6.2 illustrates three trees with the vertices denoted by x_1, x_2, x_3; these trees are pairwise distinct, since for example the vertices x_1, x_2, x_3 have different degrees in the three trees. By erasing the labels associated with the vertices we obtain the three identical trees in Fig. 6.3.

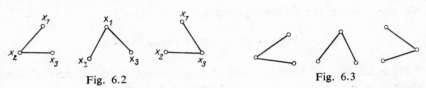

Fig. 6.2   Fig. 6.3

The problem of counting non labelled graphs will be treated later on in this book, by means of the Pólya — de Bruijn method. Using a generalization of the Lagrange expansion for multivariate analytic functions, I. J. Good [14] obtained a method for counting certain classes of trees.

PROPOSITION 2 *(Moon). If* $T(n; d_1, d_2, ..., d_n)$ *stands for the number of labelled trees whose vertices have the degrees* $d(x_1) = d_1$, $d(x_2) = d_2, ..., d(x_n) = d_n$, *respectively, then*

$$T(n; d_1, d_2, ..., d_n) = \binom{n-2}{d_1 - 1, d_2 - 1, ..., d_n - 1}. \qquad (6.1)$$

We shall show that if $n \geqslant 3$ there is a bijection between the trees T_n with n labelled nodes and the n^{n-2} sequences $(a_1, a_2, ..., a_{n-2})$ that can be formed from the numbers $1, 2, ..., n$ (Prüfer, 1918).

From any tree T_n remove the endnode (and its incident edge) with the smallest label to form a smaller tree T_{n-1} and let a_1 denote the label of the node that was joined to the removed node; repeat this process on T_{n-1} to determine a_2 and continue until only two nodes, joined by an edge, are left. This mapping is one-to-one. In order to show that each such sequence $(a_1, a_2, ..., a_{n-2})$ where $1 \leqslant a_i \leqslant n$ corresponds to some tree T_n, let b_1 denote the smallest positive integer that does not occur in the sequence, let $(c_2, ..., c_{n-2})$ denote the sequence obtained from $(a_2, ..., a_{n-2})$ by decreasing all terms larger than b_1 by one. Then $(c_2, ..., c_{n-2})$ is a sequence of length $n - 3$ formed from the numbers $1, 2, ..., n - 1$ and we may assume there exists a tree T_{n-1} with nodes $1, 2, ..., n - 1$ that corresponds to this sequence.

Relabel the nodes of T_{n-1} by adding one to each label that is not less than b_1; if we introduce an n-th node labelled b_1 and join it to the node labelled a_1 in T_{n-1}, we obtain a tree T_n that corresponds to the original sequence $(a_1, a_2, ..., a_{n-2})$.

Now the result follows from the property that every vertex a_i appears $d_i - 1$ times in the sequence $(a_1, a_2, ..., a_{n-2})$, where d_i is the degree of a_i.

PROPOSITION 3 *(Menon)*. *The integers* $d_1, d_2, ..., d_n \geqslant 1$ *are the degrees of the vertices of a tree if and only if*

$$\sum_{i=1}^{n} d_i = 2(n-1).$$

From (6.1) this condition is equivalent to $T(n; d_1, d_2, ..., d_n) \neq 0$.

PROPOSITION 4 *(Cayley's formula)*. *The number of trees with vertices* $x_1, x_2, ..., x_n$ *is equal to* n^{n-2}.

To determine the number of trees with n labelled vertices, we evaluate the sum

$$\sum_{\substack{d_1, ..., d_n \geqslant 1 \\ d_1 + ... + d_n = 2(n-1)}} \binom{n-2}{d_1-1, ..., d_n-1} = (1+1+...+1)^{n-2} = n^{n-2},$$

by using the multinomial formula. For $n = 4$ these trees are drawn in Fig. 6.4; we have taken into account that $d_1 + d_2 + d_3 + d_4 = 6$ and that the number of partitions of 6 into 4 parts is $P(6, 4) = 2$, the two partitions being $2 + 2 + 1 + 1$ and $3 + 1 + 1 + 1$.

There are several proofs of Cayley's formula, which can be found in [20].

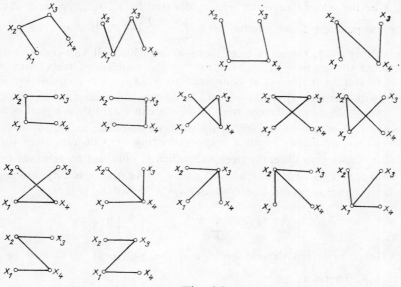

Fig. 6.4

PROPOSITION 5 *(Clarke)*. *The number of trees with vertices* $x_1, x_2, ..., x_n$ *and* $d(x_1) = k$ *is*

$$\binom{n-2}{k-1}(n-1)^{n-k-1}. \tag{6.2}$$

The number of trees we are looking for is equal to

$$\sum_{d_2, d_3, \ldots, d_n \geqslant 1} \binom{n-2}{k-1,\, d_2-1,\, \ldots,\, d_n-1}$$

$$= \frac{(n-2)!}{(k-1)!\,(n-k-1)!} \sum_{d_2, d_3, \ldots, d_n \geqslant 1} \binom{n-k-1}{d_2-1,\, d_3-1,\, \ldots,\, d_n-1}$$

$$= \binom{n-2}{k-1}(n-1)^{n-k-1},$$

where in both sums the indices $d_2, d_3, \ldots, d_n$ are such that

$$d_2 + d_3 + \ldots + d_n = 2n - k - 2.$$

PROPOSITION 6 *(Cayley)*. *The number of graphs with vertices* $x_1, x_2, \ldots, x_n$ *consisting of p trees with no common vertices, such that* $x_1, x_2, \ldots, x_p$ *belong to p distinct trees, is equal to*

$$T(n, p) = pn^{n-p-1}. \tag{6.3}$$

Let A be the set of trees with vertices denoted by $x_0, x_1, \ldots, x_n$ and $d(x_0) = p$. By using proposition 5 we obtain $|A| = \binom{n-1}{p-1} n^{n-p}$. If $P \subset \{1, 2, \ldots, n\}$ and $|P| = p$, denote by A_P the set of trees from A for which all vertices x_i with $i \in P$ are connected to x_0. But $|A_P| = T(n, p)$, for the number of trees with vertices $x_0, x_1, \ldots, x_p$ and such that x_0 is connected to $x_{i_1}, x_{i_2}, \ldots, x_{i_p}$, equals the number of graphs with vertices $x_1, x_2, \ldots, x_n$ and consisting of p disjoint trees such that $x_{i_1}, x_{i_2}, \ldots, x_{i_p}$ belong to different trees. For, we can establish a bijection between these two sets of graphs in the following way: we associate a tree for which x_0 is connected to $x_{i_1}, x_{i_2}, \ldots, x_{i_p}$, with a graph consisting of p disjoint trees for which $x_{i_1}, x_{i_2}, \ldots, x_{i_p}$ belong to different trees and which is obtained by removal of vertex x_0 and of the p edges that join it to $x_{i_1}, \ldots, x_{i_p}$. As the sets A_P are pairwise disjoint for the various choices of the index set P, it follows that

$$|A| = \sum_P |A_P| = \sum_P T(n, p) = \binom{n}{p} T(n, p),$$

because $T(n, p)$ does not depend on P and the p-element set P can be chosen in $\binom{n}{p}$ ways. Therefore

$$T(n, p) = \frac{|A|}{\binom{n}{p}} = \frac{\binom{n-1}{p-1} n^{n-p}}{\binom{n}{p}} = pn^{n-p-1}.$$

For $p = 2$ and $n = 4$ the resulting graphs are drawn in Fig. 6.5.

PROPOSITION 7 *(Moon)*. *Let* $G_1 = (X_1, U_1)$, $G_2 = (X_2, U_2)$, ..., $G_p = (X_p, U_p)$ *be trees whose sets of vertices are pairwise disjoint and contain* $n_1, n_2, ..., n_p$ *vertices,*

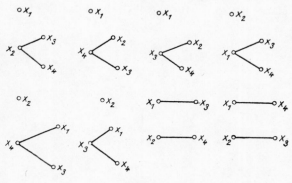

Fig. 6.5

respectively. The number of trees constructed on the set of vertices $\bigcup\limits_{i=1}^{p} X_i$ *and which have the* G_i $(i = 1, ..., p)$ *as subtrees is equal to*

$$T(G_1, G_2, ..., G_p) = n_1 n_2 ... n_p (n_1 + n_2 + ... + n_p)^{p-2}. \qquad (6.4)$$

Assume that each vertex set X_i reduces to one vertex y_i. The number of trees with p vertices $y_1, y_2, ..., y_p$ of degrees $d(y_i) = d_i$ is equal to $\begin{pmatrix} p-2 \\ d_1-1, ..., d_p-1 \end{pmatrix}$, in view of proposition 2. But the d_1 edges having y_1 as an extremity can be joined with the n_1 vertices from X_1 in $\underbrace{n_1 \cdot n_1 \cdot ... \cdot n_1}_{d_1} = n_1^{d_1}$ distinct ways, for each of the d_1 edges can be joined with any of the n_1 vertices. Similarly there are $n_2^{d_2}$ possibilities for the tree G_2, ..., $n_p^{d_p}$ possibilities for the tree G_p. Therefore, by fixing the numbers $d_1, d_2, ..., d_p \geqslant 1$ with $d_1 + d_2 + ... + d_p = 2(p-1)$, there exist

$$\begin{pmatrix} p-2 \\ d_1-1, ..., d_p-1 \end{pmatrix} n_1^{d_1} ... n_p^{d_p}$$

possibilities for constructing trees on the vertex set $\bigcup\limits_{i=1}^{p} X_i$ so that the trees G_i are subgraphs. Notice also that if we have constructed a tree on the vertex set $\{y_1, y_2, ..., y_p\}$ by the above procedure of redistributing the edges with extremity y_i to the vertices of the tree G_i, we obtain connected cycle-free graphs, i.e. trees, which contain $G_1, ..., G_p$ as subgraphs. In this way we obtain irredundantly all the trees with the property indicated in the formulation of the proposition. Indeed, if G is a tree with the vertex set $\bigcup\limits_{i=1}^{p} X_i$ and which contains the trees G_i as subgraphs, denote by d_i the number of edges which join the vertices from the set X_i with the vertices of G which are not in X_i, for $i = 1, ..., p$. Then this tree will be one of the

$$\binom{p-2}{d_1 - 1, ..., d_p - 1} n_1^{d_1} ... n_p^{d_p}$$

trees constructed above. But

$$T(G_1, G_2, ..., G_p) = \sum_{\substack{d_1, ..., d_p \geqslant 1 \\ d_1 + ... + d_p = 2(p-1)}} \binom{p-2}{d_1 - 1, ..., d_p - 1} n_1^{d_1} n_2^{d_2} ... n_p^{d_p}$$

$$= n_1 n_2 ... n_p (n_1 + n_2 + ... + n_p)^{p-2}.$$

In the sequel we obtain formulae for counting trees by using the principle of inclusion and exclusion (3.3).

Let $X = \{x_1, x_2, ..., x_n\}$ be a set of vertices and $U = \{u_1, ..., u_q\}$ a set of $q \leqslant \binom{n}{2}$ edges which join pairs of vertices from X. Denote by $T(X, U)$ the number of trees which can be formed with vertices from X and edges which do not belong to U.

Given a graph $G = (X, V)$ with n vertices and p connected components which contain $n_1, n_2, ..., n_p$ vertices, respectively, set $m(V) = n_1 n_2 ... n_p$ if each of the p connected components of the graph G is a tree, and $m(V) = 0$ in the opposite case, i.e. if the graph G contains at least one cycle.

PROPOSITION 8 *(Temperley). The number of trees constructed on the n-element set X as vertex set, the edges of which do not belong to the set U, is equal to*

$$T(X, U) = n^{n-2} \sum_{V \subset U} m(V) \left(\frac{-1}{n} \right)^{|V|}. \tag{6.5}$$

If $v \in U$, denote by A_v the set of trees which contain the edge v and have X as vertex set. If the graph (X, V) is cycle-free and has p connected components (which are trees), proposition 7 yields the number of trees which use all the edges of V, namely

$$\left| \bigcap_{v \in V} A_v \right| = n_1 n_2 ... n_p (n_1 + n_2 + ... + n_p)^{p-2} = m(V) n^{p-2}.$$

If the graph (X, V) contains a cycle, the above formula is valid, for both sides of the equality vanish. In the case of a cycle-free graph the cyclomatic number is null, hence $p = n - |V|$ and

$$\left| \bigcap_{v \in V} A_v \right| = m(V) n^{n-2-|V|}.$$

By applying Sylvester's formula, we obtain the number of trees which belong to none of the sets A_v with $v \in U$:

$$T(X, U) = n^{n-2} + \sum_{\substack{V \subset U \\ V \neq \varnothing}} (-1)^{|V|} \left| \bigcap_{v \in V} A_v \right|$$

$$= n^{n-2} + \sum_{\substack{V \subset U \\ V \neq \varnothing}} (-1)^{|V|} m(V) n^{n-2-|V|} = n^{n-2} \sum_{V \subset U} m(V) \left(\frac{-1}{n} \right)^{|V|},$$

where we have set $m(\varnothing) = 1$ by definition.

The subsequent propositions are consequences of this result.

PROPOSITION 9 *(Weinberg). If U is a set of q edges, no two of which have a common extremity, then*

$$T(X, U) = n^{n-2} \left(1 - \frac{2}{n} \right)^q . \tag{6.6}$$

In this case for $V \subset U$ we get $m(V) = 2^{|V|}$ since the graph is cycle-free, the number of its connected components is equal to the number of edges, i.e. to $|V|$, and every connected component has two vertices. Then formula (6.5) yields

$$T(X, U) = n^{n-2} \sum_{k=0}^{q} \binom{q}{k} \left(-\frac{2}{n} \right)^k = n^{n-2} \left(1 - \frac{2}{n} \right)^q,$$

because the set $V \subset U$ such that $|V| = k$ can be chosen in $\binom{q}{k}$ distinct ways.

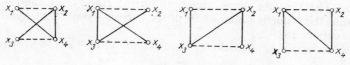

Fig. 6.6

Thus, for $n = 4$, $q = 2$ and the set U consisting of the edges $[x_1, x_2]$ and $[x_3, x_4]$, there are four trees with vertices x_1, x_2, x_3, x_4 and with edges not in U; they are drawn in Fig. 6.6 (continuous lines).

Proposition 10 *(O'Neil). If U is a set of q edges with a common extremity, then*

$$T(X, U) = n^{n-2} \left(1 - \frac{1}{n}\right)^{q-1} \left(1 - \frac{q+1}{n}\right). \tag{6.7}$$

In this case, for every $V \subset U$ with $|V| = k$, the graph (X, V) is a tree with $k + 1$ vertices, hence $m(V) = k + 1$. Since the edge set $V \subset U$ can be chosen in $\binom{q}{k}$ distinct ways, formula (6.5) yields

$$\frac{T(X, U)}{n^{n-2}} = \sum_{k=0}^{q} \binom{q}{k} (k+1) \left(-\frac{1}{n}\right)^{k}$$

$$= \sum_{k=0}^{q} \binom{q}{k} \left(-\frac{1}{n}\right)^{k} + \sum_{k-1=0}^{q-1} \left(-\frac{q}{n}\right) \binom{q-1}{k-1} \left(-\frac{1}{n}\right)^{k-1}$$

$$= \left(1 - \frac{1}{n}\right)^{q} - \frac{q}{n} \left(1 - \frac{1}{n}\right)^{q-1} = \left(1 - \frac{1}{n}\right)^{q-1} \left(1 - \frac{q+1}{n}\right),$$

thus implying (6.7).

Proposition 11 *(Austin). If U is a set of edges which join in all possible ways the vertices from the set $S \subset X$ with $|S| = s$, i.e. if (S, U) is a complete graph, then*

$$T(X, U) = n^{n-2} \left(1 - \frac{s}{n}\right)^{s-1}. \tag{6.8}$$

Denote by V_p the family of those sets of edges $V \subset U$ for which the graph (S, V) is cycle-free and has p connected components. In this case $|V| = s - p$ for $V \in V_p$ hence, summing first with respect to $p = 1, ..., s$, we obtain

$$\sum_{V \subset U} m(V) \left(\frac{-1}{n}\right)^{|V|} = \sum_{p=1}^{s} \left(\frac{-1}{n}\right)^{s-p} \sum_{V \in V_p} m(V).$$

The index $m(V)$ has the same value both for the graph (S, V) and for the graph (X, V), because the latter is obtained from the former by adding several one-vertex connected components.

We now make use of a trick for calculating $\sum_{V \in V_p} m(V)$. The graph (S, V) with $V \in V_p$ consists of p trees with no common edges and having $s_1, s_2, ..., s_p$ vertices, with $s_1 + s_2 + ... + s_p = s$. Notice that

$$m(V) = s_1 s_2 ... s_p = \underbrace{(1 + 1 + ... + 1)}_{s_1} ... \underbrace{(1 + 1 + ... + 1)}_{s_p}.$$

Each term of the sum resulting from the expansion of the above product can be associated with a graph consisting of p disjoint trees, with a marked vertex in each tree; namely, if the i-th unit from the expansion of s_k occurs in the considered term, then we mark vertex i from the component having s_k vertices. Let P be the set of marked vertices. As (S, U) is a complete graph, $\sum_{V \in V_p} m(V)$ counts irredundantly all cycle-free graphs with p connected components and s vertices, with a vertex of P marked in each connected component, for all choices of $P \subset S$ with $|P| = p$. It follows that

$$\sum_{V \in V_p} m(V) = \sum_{\substack{P \subset S \\ |P| = p}} ps^{s-p-1} = \binom{s}{p} ps^{s-p-1}.$$

We have taken into account proposition 6 and the fact that the set $P \subset S$ can be chosen in $\binom{s}{p}$ distinct ways. Therefore in this case

$$T(X, U) = n^{n-2} \sum_{p=1}^{s} \left(\frac{-1}{n} \right)^{s-p} \binom{s}{p} ps^{s-p-1}$$

$$= n^{n-2} \sum_{p-1=0}^{s-1} \left(\frac{-s}{n} \right)^{s-p} \binom{s-1}{p-1} = n^{n-2} \left(1 - \frac{s}{n} \right)^{s-1}.$$

PROPOSITION 12 *(Moon). If U consists of $m - 1$ edges that form an m-vertex elementary chain (not passing twice through the same vertex), then the number of trees with vertices from the n-vertex set X and which do not contain any edge in U, is equal to*

$$T(X, U) = n^{n-2} \sum_{p=1}^{m} \binom{m+p-1}{m-p} \left(\frac{-1}{n} \right)^{m-p}. \tag{6.9}$$

We apply formula (6.5), the index $m(V)$ being determined as before but in this case the graph is (Y, V), where Y is the set of m vertices of the cycle-free chain. As (Y, U) is cycle-free, its partial graph (Y, V) is also cycle-free, hence its cyclomatic number is zero. Therefore, if (Y, V) has p connected components, the number of its edges is $|V| = |Y| - p = m - p$. If these connected components contain $m_1, m_2, \ldots, m_p$ vertices, respectively, then $m_1 + m_2 + \ldots + m_p = m$. We now establish a bijection between the decompositions of the elementary chain in the statement of the proposition (with Y as vertex-set) into p connected components obtained by removal of certain edges and the representations of m in the form $n = m_1 + m_2 + \ldots + m_p$ with the integers $m_1, m_2, \ldots, m_p \geq 1$, two representations being distinct if they differ either in the nature of the terms, or in their order.

We associate a decomposition $m = m_1 + \ldots + m_p$ with the decomposition of the chain $[x_1, x_2, \ldots, x_m]$ into p connected components that is obtained by removing the edges $[x_{m_1}, x_{m_1+1}]$, $[x_{m_1+m_2}, x_{m_1+m_2+1}]$, $\ldots$, $[x_{m_1+\ldots+m_{p-1}}, x_{m_1+\ldots+m_{p-1}+1}]$, so that (Y, V) has p connected components, having $m_1, m_2, \ldots, m_p$ vertices, respectively.

If the chain constructed on the vertex set Y is a cycle or contains cycles, then the correspondence described above is no longer a bijection. Therefore

$$\sum_{V \subset U} m(V) \left(\frac{-1}{n} \right)^{|V|} = \sum_{p=1}^{m} \left(\frac{-1}{n} \right)^{m-p} \sum_{\substack{V \subset U \\ |V| = m-p}} m(V)$$

$$= \sum_{p=1}^{m} \left(\frac{-1}{n} \right)^{m-p} \sum_{\substack{m_1, \ldots, m_p \geqslant 1 \\ m_1 + \ldots + m_p = m}} m_1 m_2 \ldots m_p.$$

To compute the latter sum, notice that it is equal to the coefficient of x^m from the expansion

$$(x + 2x^2 + 3x^3 + \ldots)^p = x^p (1 - x)^{-2p}$$

(which is valid in a neighbourhood of the origin), in view of the expansion

$$1 + 2x + 3x^2 + \ldots = \frac{d}{dx} \left(\frac{1}{1-x} \right) = \frac{1}{(1-x)^2}.$$

By using the Taylor series of $(1 - x)^{-2p}$ given by the generalized binomial formula, we see that this coefficient is

$$\frac{(-1)^{m-p} (-2p)(-2p-1) \ldots (-2p - (m-p-1))}{(m-p)!}$$

$$= \frac{(m+p-1)(m+p-2) \ldots (2p+1) 2p}{(m-p)!} = \binom{m+p-1}{m-p}.$$

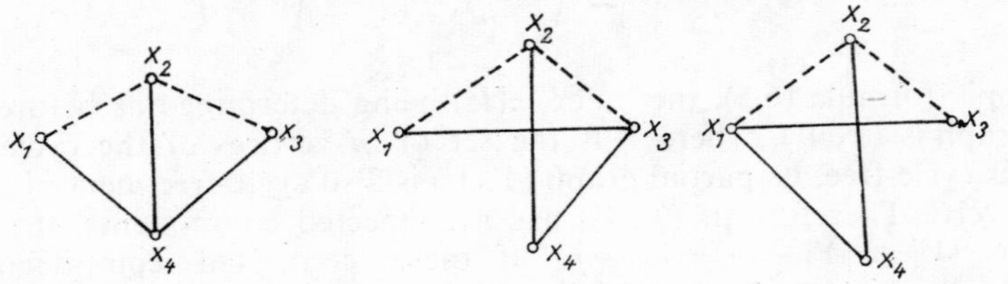

Fig. 6.7

For example, if $m = 3$, $n = 4$ and U consists of the edges $[x_1, x_2]$ and $[x_2, x_3]$ then $T(X, U) = 3$ by formula (6.9), i.e. there are 3 trees which do not use the edges of U; they are represented by dotted lines in Fig. 6.7.

PROPOSITION 13 (*Scoins, Glicksman*). *If the graph* (X, U) *is the join of two disjoint complete graphs* (S, V) *and* (T, W), *that is*, $S \cup T = X$, $S \cap T = \varnothing$ *and* $V \cup W = U$, *with* $|S| = s$ *and* $|T| = t$, *then the number of trees constructed on the vertex set* X *and which do not contain any edge from* U, *is equal to*

$$T(X, U) = s^{t-1} t^{s-1}. \tag{6.10}$$

By using proposition 8, we shall prove that

$$\frac{T(X, V \cup W)}{n^{n-2}} = \frac{T(X, V)}{n^{n-2}} \cdot \frac{T(X, W)}{n^{n-2}} \tag{6.11}$$

which is the same as proving that

$$\sum_{Z \subset V \cup W} m(Z) \left(\frac{-1}{n} \right)^{|Z|} = \sum_{Z_1 \subset V} m(Z_1) \left(\frac{-1}{n} \right)^{|Z_1|} \sum_{Z_2 \subset W} m(Z_2) \left(\frac{-1}{n} \right)^{|Z_2|}.$$

Indeed, if $Z \subset V \cup W$ and $V \cap W = \varnothing$, then $Z = Z_1 \cup Z_2$ with $Z_1 \subset V$ and $Z_2 \subset W$; in this case

$$m(Z_1) \left(\frac{-1}{n} \right)^{|Z_1|} m(Z_2) \left(\frac{-1}{n} \right)^{|Z_2|} = m(Z) \left(\frac{-1}{n} \right)^{|Z|}. \tag{6.12}$$

If $Z_1 = \varnothing$ or $Z_2 = \varnothing$, then $m(Z_1) = 1$ or $m(Z_2) = 1$, respectively, so that (6.12) holds. If (X, Z_1) or (X, Z_2) contains cycles, then so does (X, Z) and conversely; hence formula (6.12) is satisfied, for both its sides vanish.

If (X, Z_1) has p connected components which are trees with $n_1, n_2, \ldots, n_p$ vertices and (X, Z_2) has q connected components which are trees with $m_1, m_2, \ldots, m_q$ vertices, Z_1 and Z_2 being disjoint, it follows that (X, Z) has $p + q$ connected components which are trees and contain $n_1, \ldots, n_p, m_1, \ldots, m_q$ vertices, respectively. In this case $m(Z) = m(Z_1) m(Z_2) = n_1 \ldots n_p m_1 \ldots m_q$ and $|Z| = |Z_1| + |Z_2|$, which implies (6.12) and a fortiori (6.11). By using proposition 11, we obtain

$$T(X, U) = \frac{1}{n^{n-2}} T(X, V) T(X, W) = n^{n-2} \left(1 - \frac{s}{n} \right)^{s-1} \left(1 - \frac{t}{n} \right)^{t-1} = s^{t-1} t^{s-1},$$

which is precisely (6.10). For $s = t = 2$ we obtain four trees drawn in Fig. 6.6, where the forbidden edges $[x_1, x_2]$ and $[x_3, x_4]$ have been drawn in dotted lines.

Austin [1] generalized this result as follows: if the graph (X, U) is the join of p disjoint complete graphs which contain $l_1, l_2, \ldots, l_p$ vertices, respectively, where $l_1 + l_2 + \ldots + l_p = n$, then the number of trees having the vertex set X and containing no edge from U is

$$T(X, U) = n^{p-2} \prod_{i=1}^{p} (n - l_i)^{l_i - 1}.$$

Good [14] used a multivariate generating function to give another derivation of this result and Oláh [24] gave a proof based on Prüfer's method. For the determination of the partial trees of a connected graph a formula based on the calculation of certain determinants can also be used [4].

Let us now determine the number of connected graphs with n labelled vertices and n edges. These graphs have cyclomatic number equal to 1, hence they contain a single cycle and are obtained from a tree with n vertices by adding a single edge between two vertices which are not already joined by an edge.

PROPOSITION 14 *(Rényi). The number of connected graphs with n labelled vertices and which contain a single cycle is equal to*

$$C(n, n) = \frac{1}{2} \sum_{p=3}^{n} [n - 1]_{p-1} n^{n-p}.$$ (6.13)

Let us count the connected graphs with n labelled vertices which contain a single cycle of length p. For $3 \leqslant p \leqslant n$, Clarke's formula (6.2) implies that the number of trees with $n - p + 1$ vertices labelled $x_0, x_{p+1}, \ldots, x_n$, with the degree $d(x_0) = r$, is equal to

$$\binom{n - p - 1}{r - 1} (n - p)^{n-p-r} \text{ for } r = 1, \ldots, n - p.$$

If vertex x_0 is replaced by a cycle having the p vertices $x_1, x_2, \ldots, x_p$, then r edges can be assigned to the labelled vertices of the cycle in p^r different ways and the graph thus obtained is connected, has n labelled vertices and contains a single cycle of length p.

Therefore, for every choice of the cycle with p elements, the number of connected graphs with n labelled vertices which contain only this cycle is equal to

$$\sum_{r=1}^{n-p} p^r \binom{n - p - 1}{r - 1} (n - p)^{n-p-r} = pn^{n-p-1}.$$

The vertices of the cycle of length p can be chosen in $\binom{n}{p}$ different ways and for every choice for the vertices of the cycle there exist $(p - 1)!/2$ possibilities of choosing the edges of the cycle, because the number of Hamiltonian cycles of a complete graph with p vertices is $(p-1)!/2$ (§7.5). Therefore the number of connected graphs with n labelled vertices, which contain a single cycle with p vertices, is equal to

$$\frac{1}{2} \binom{n}{p} (p - 1)! pn^{n-p-1} = \frac{1}{2} [n - 1]_{p-1} n^{n-p},$$

because we have irredundantly counted these graphs. The sum of these numbers for $p = 3, \ldots, n$ yields formula (6.13) which gives the number $C(n, n)$ of connected graphs with n labelled vertices and n edges.

Problems

1. A spanning tree of a connected nonoriented graph $G = (X, U)$ is a tree $T = (X, V)$ where $V \subset U$.

Prove that the number of spanning trees of the graph illustrated in Fig. 6.8 is equal to

$$\frac{1}{2\sqrt{3}} \left((2 + \sqrt{3})^m - (2 - \sqrt{3})^m \right).$$

(Sedláček, 1969)

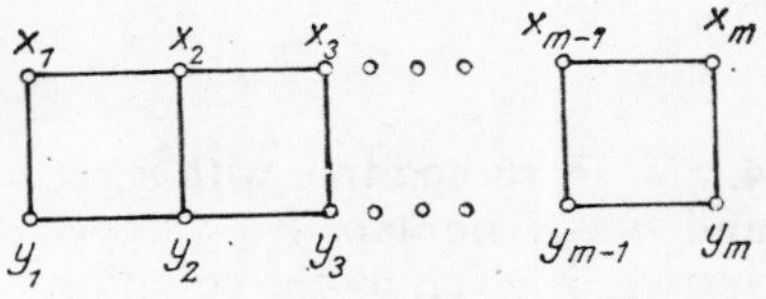

Fig. 6.8

H i n t: Prove that this number, denoted by S_m, satisfies the following recurrence relation: $S_{m+1} = 4S_m - S_{m-1}$ with initial values $S_1 = 1$, $S_2 = 4$.

2. A 2-dimensional tree or a 2-tree is defined inductively as follows:

Two vertices joined by an edge form a 2-tree; a 2-tree T_{n+1} with $n + 1$ vertices is obtained from a 2-tree T_n with n vertices by adding a new vertex x_{n+1} and the triangle which contains that vertex and two already adjacent vertices.

Hence a 2-tree with n vertices has $2n - 3$ edges and $n - 2$ triangles.

Prove that the number of labelled 2-trees with n vertices is equal to

$$\binom{n}{2} (2n - 3)^{n-4}.$$

(Beineke, Moon, Palmer, 1969)

H i n t: Let $T_2(n)$ be the number of labelled 2-trees with n vertices and let $R_2(n)$ be the number of such trees rooted by an edge $[u, v]$. We have $\binom{n}{2} R_2(n) = (2n-3) T_2(n)$.

If we denote by $C(n, k)$ the number of labelled 2-trees with n vertices in which the rooted edge $[u, v]$ belongs to k triangles, we get $2k (n - 2) C(n, k + 1) = (n-k-2) C(n, k)$. We deduce by induction that $C(n, k) = \binom{n - 3}{k - 1} (2n - 4)^{n-k-2}$, hence $R_2(n) = \sum_{k=1}^{n-2} C(n, k) = (2n - 3)^{n-3}$.

The number of k-dimensional trees with n labelled vertices equals $\binom{n}{k} (kn - k^2 + 1)^{n-k-2}$ [3], [22] and the number of plane, labelled 2-trees is given by $n(n - 1)^2 \dfrac{(5n - 10)!}{(4n - 6)!}$ [27].

3. Prove that the number $C(n, k)$ of connected graphs with k vertex-disjoint cycles and n labelled vertices is equal to

$$C(n, k) = \frac{n! \, n^{k-2}}{2^k} \sum_{i=0}^{n-3k} \frac{n^i}{i!} \sum_{(\lambda_3, \ldots, \lambda_p)} \frac{1}{\lambda_3! \ldots \lambda_p!}$$

where the last sum extends over all solutions in nonnegative integers of the equations

$$3\lambda_3 + \dots + p\lambda_p = n - i$$
$$\lambda_3 + \dots + \lambda_p = k.$$

(Tomescu, 1972)

4.   A rooted tree with n vertices is a tree with one distinguished vertex (the root) and $n - 1$ nonlabelled vertices. Prove that the number r_n of rooted trees with n vertices is given by the equation

$$r_n = \sum_{m=1}^{n-1} \sum_{(k_1, \dots, k_{n-1})} \binom{r_1 + k_1 - 1}{k_1} \binom{r_2 + k_2 - 1}{k_2} \dots \binom{r_{n-1} + k_{n-1} - 1}{k_{n-1}},$$

where the second sum is over all nonnegative integers $k_1, \dots, k_{n-1}$ satisfying $k_1 + k_2 + \dots + k_{n-1} = m$ and $k_1 + 2k_2 + \dots + (n-1)k_{n-1} = n - 1$.

(Cayley, 1897)

H i n t: A tree T_n with n vertices has $n - 1$ edges. If the root is adjacent to m edges, we may join k_p of these edges to the roots of trees T_p in as many ways as there are combinations with repetitions of r_p objects taken k_p at a time, i.e. in $\binom{r_p + k_p - 1}{k_p}$ ways, and the formula follows.

5.   By a plane tree is meant a realization of a tree by points and edges in the plane.

By an isomorphism between two plane trees is meant an isomorphism in the usual sense for such trees which preserves the clockwise cyclic order of the edges about each node.

A planted tree is a rooted tree with the root at a node of degree one.

Prove that the number of nonisomorphic planted plane trees with $n \geqslant 1$ edges equals the Catalan number $\dfrac{1}{n} \binom{2n - 2}{n - 1}$.

(Harary, Prins, Tutte, 1963)

H i n t: Prove that this number, denoted by P_n, satisfies the recurrence relation:

$$P_n = P_1 P_{n-1} + P_2 P_{n-2} + P_3 P_{n-3} + \dots + P_{n-1} P_1,$$

where $P_1 = 1$.

6.   Show that the number of nonisomorphic rooted plane trees with n edges is equal to

$$\frac{1}{2n} \sum_{s|n} \varphi\left(\frac{n}{s}\right) \binom{2s}{s}.$$

(Walkup, 1972)

H i n t: Apply Pólya's theorem (Chap. 8).

7. Let $r(x) = r_1 x + r_2 x^2 + \dots$ denote the generating function for the number of rooted trees. Prove that the generating function $t(x)$ for the number of non-labelled trees is equal to

$$t(x) = r(x) - \frac{1}{2} r^2(x) + \frac{1}{2} r(x^2).$$

(Otter, 1948)

8. An endnode of a graph is a node of degree one. Let $R(n, k)$ denote the number of trees T_n with n labelled nodes and k endnodes. If $2 \leqslant k \leqslant n$ prove that $R(n, k) = \frac{n!}{k!} S(n - 2, n - k)$. Derive from this the number of labelled trees T_n.

(Rényi, 1959)

H i n t : Prove that $kR(n, k) = n(n - k) R(n - 1, k - 1) + nR(n - 1, k)$, for $k = 2, 3, \dots, n$ and $n \geqslant 3$. The result now follows by induction, using a recurrence relation for the Stirling numbers of the second kind.

9. Show that there are n^{n-3} different trees with n unlabelled vertices and $n - 1$ edges labelled $1, 2, \dots, n - 1$ for $n \geqslant 3$.

(Palmer, 1969)

10. Prove that the number $T(n)$ of trees with n labelled vertices satisfies the following recurrence relation

$$2(n - 1) T(n) = \sum_{i=1}^{n-1} \binom{n}{i} T(i) T(n - i) i(n - i).$$

(Dziobek, 1917; Bol, 1938)

11. Let G denote a graph with $n \geqslant 2$ labelled vertices $1, 2, \dots, n$ and m edges. Suppose we label the edges of G with the numbers $1, 2, \dots, m$ and orient each edge arbitrarily. The vertex-edge or vertex-branch incidence matrix of G is the n by m matrix $A = \{a_{ij}\}_{\substack{i=1,\dots,n \\ j=1,\dots,m}}$ in which a_{ij} equals $+1$ or -1 if the edge j is oriented away from or towards the vertex i and zero otherwise.

Prove that if the graph G has n vertices and is connected, then the rank of its incidence matrix A is $n - 1$.

(Kirchhoff, 1847)

12. The reduced incidence matrix A_n of a connected graph G is the matrix obtained from the incidence matrix A by deleting some row, say the n-th.

Prove that if B is any non-singular square submatrix of A (or A_n), then its determinant is ± 1.

(Poincaré, 1901)

H i n t : If B is non-singular, then there is a column of B which contains just one non-zero entry. The required result follows by induction if we expand the determinant of B along this column.

13. Prove that a submatrix B of order $n - 1$ of A_n is non-singular if and only if the edges corresponding to the columns of B determine a spanning subtree of G.

(Chuard, 1922)

H i n t: If H denotes the spanning subgraph of G whose $n - 1$ edges correspond to the columns of B, then B is the reduced incidence matrix of H. Show that H is connected if and only if rank $(B) = n - 1$.

14.   Prove the Matrix Tree Theorem: If A_n is a reduced incidence matrix of the graph G, then the number of spanning trees of G equals the determinant of $A_n A_n^T$, where T denotes transpose.

(Brooks, Smith, Stone, Tutte, 1940)

H i n t: Use the Binet-Cauchy theorem.

Assuming that $m \geqslant n - 1$ we obtain $\det(A_n A_n^T) = \sum \det B \det B^T = \sum (\det B)^2 = \sum 1 =$ the number of spanning trees of G.

15.   For a nonoriented finite graph $G = (X, U)$ prove that the matrix $A_n A_n^T$ is the matrix obtained from the matrix C by deleting some row, say the n-th and the column with the same index, where $C = \{c_{ij}\}_{i,j=1,\dots,n}$ is defined as follows: c_{ii} is equal to the number of vertices adjacent to i in G, $c_{ij} = -1$ if $i \neq j$ and $[i, j] \in U$; $c_{ij} = 0$ if $i \neq j$ and $[i, j] \notin U$.

16.   Obtain the number $T(n)$ of trees with n labelled nodes as the number of spanning trees of the complete graph K_n.

17.   Let G denote a directed graph with $n \geqslant 2$ labelled nodes and m arcs. Let $D = \{d_{ij}\}$ denote the n by n matrix in which $d_{ij} = -1$ if there is an arc directed from vertex i to vertex j, $d_{ij} = 0$ if there is no such arc (for $i \neq j$), and d_{ii} equals the number of arcs directed from the other vertices to i. Let D_i denote the cofactor of d_{ii} in D.

If the directed graph G has $m = n - 1$ arcs, prove that it is an arborescence rooted at node n if and only if $D_n = 1$; otherwise $D_n = 0$.

H i n t: If G is an arborescence, we can label its vertices with the numbers $1, 2, \dots, n$ so that D_n is a determinant of an upper triangular matrix having its entries on the main diagonal equal to one.

If $D_n \neq 0$, prove that G is an arborescence rooted at node n because G is connected and a unique arc goes to any given vertex i ($i = 1, \dots, n - 1$).

18.   Prove the Matrix Tree Theorem for Directed Graphs: For an oriented graph G the number of all spanning arborescences of G that are rooted at the n-th vertex is equal to D_n.

(Tutte, 1948)

H i n t: Let $a_1, \dots, a_n$ denote the rows of the adjacency matrix A of G and let $D_n(a_1, \dots, a_{n-1}) = D_n$. If the vector $e_k = (\delta_{ik})_{i=1,\dots,n}$ where $\delta_{ik} = 1$ if $i = k$ and $\delta_{ik} = 0$ otherwise, denote $K_i = \{k | (k, i)$ is an arc of $G\}$. Then

$$D_n(a_1, \dots, a_{n-1}) = D_n\left(\sum_{k_1 \in K_1} e_{k_1}, \dots, \sum_{k_{n-1} \in K_{n-1}} e_{k_{n-1}}\right) = \sum_{\substack{k_1 \in K_1 \\ \dots\dots \\ k_{n-1} \in K_{n-1}}} D_n(e_{k_1}, \dots, e_{k_{n-1}}).$$

By using the previous result, $D_n(e_{k_1}, \dots, e_{k_{n-1}})$ is equal to 1 if the $n - 1$ arcs $(k_1, 1), (k_2, 2), \dots, (k_{n-1}, n - 1)$ span an arborescence rooted at the vertex n and it is equal to 0 otherwise.

19. An r-th generation symmetric n-ary tree $T_{n,r}$ will mean a rooted tree, where for $0 \leqslant j < r$ all vertices at distance j from the root are adjacent to exactly n (unordered) vertices at distance $j + 1$ from the root, and all vertices at distance r are terminal.

Prove that for any $n \geqslant 1$, the number of nonisomorphic rooted subtrees $\tau(n, r)$ of the r-th generation symmetric n-ary tree $T_{n,r}$ is given by the recursive equations:

$$\tau(n, 0) = 1 \qquad \text{for all } n \geqslant 1,$$

$$\tau(n, r) = \binom{\tau(n, r-1) + n}{n} \quad \text{for all } r \geqslant 1, \, n \geqslant 1.$$

In particular $\tau(n, 1) = n + 1$ and $\tau(n, 2) = \binom{2n+1}{n}$ for all $n \geqslant 1$.

(Matula, 1970)

20. Prove that the number of arborescences having n labelled vertices is equal to n^{n-1}.

21. Let $A = (X, \Gamma)$ be an arborescence rooted at $x_1 \in X$. We associate each endnode with a variable x; further if $\Gamma x_k = \{x_{k_1}, ..., x_{k_r}\}$ and the vertices $x_{k_1}, ..., x_{k_r}$ are associated with the polynomials $P_{k_1}(x), ..., P_{k_r}(x)$, then the vertex x_k is associated by definition with the polynomial $P_k(x) = \prod_{i=1}^{r} P_{k_i}(x) + 1$.

Denoting by $P_A(x) = P_1(x)$ the polynomial associated in this way with the root x_1, prove that the number of subarborescences of A which have the same root x_1 is equal to $P_A(2) - 1$.

(Tomescu, 1971)

BIBLIOGRAPHY

1. Austin, T. L., *The enumeration of point labelled chromatic graphs and trees*, Canad. J. Math., **12**, 1960, 535−545.
2. Beineke, L. W., Moon, J. W., *Several proofs of the number of labelled 2-dimensional trees*, Proof Techniques in Graph Theory (ed: F. Harary), Academic Press, New York, 1969.
3. Beineke, L. W., Pippert, R. E., *The number of labelled k-dimensional trees*, J. Combinatorial Theory, **6**, 1969, 200−205.
4. Berge, C., *Principes de combinatoire*, Dunod, Paris, 1968.
   English edition: *Principles of combinatorics*, Academic Press, New York, 1971.
5. Berge, C., *Graphes et hypergraphes*, Dunod, Paris, 1970.
   English edition: *Graphs and hypergraphs*, North-Holland, Amsterdam, 1973.
6. Bol, G., *Über eine kombinatorische Frage*, Abh. Math. Sem. Univ. Hamburg, **12**, 1938, 242−245.
7. Brooks, R. L., Smith, C. A. B., Stone, A. H., Tutte, W. T., *Dissections of a rectangle into squares*, Duke. Math. J., **7**, 1940, 312−340.
8. Cayley, A., *A theorem on trees*, Quart. J. Math., **23**, 1889, 376−378.
9. Cayley, A., *Collected mathematical papers*, Cambridge, 1889−1897.
10. Chuard, J., *Questions d'analysis situs*, Rend. Circ. Mat. Palermo, **46**, 1922, 185−224.
11. Clarke, L. E., *On Cayley's formula for counting trees*, J. London Math. Soc., **33**, 1958, 471−475.

12. Dziobek, O., *Eine Formel der Substitutionstheorie*, Sitzungsberichte der Berliner Mathematischen Gesellschaft, **17**, 1917, 64—67.

13. Glicksman, S., *On the representation and enumeration of trees*, Proc. Cambridge Philos. Soc., **59**, 1963, 509—517.

14. Good, I. J., *The generalization of Lagrange's expansion and the enumeration of trees*, Proc. Cambridge Philos. Soc., **61**, 1965, 499—517.

15. Harary, F., Prins, G., Tutte, W. T., *The number of plane trees*, Nederl. Akad. Wetensch. Proc. Ser. A, **67** (Indag. Math., 26), 1963, 319—329.

16. Kirchhoff, G., *Über die Auflösung der Gleichungen, auf welche man bei der Untersuchung der linearen Verteilung galvanischer Ströme geführt wird*, Ann. Phys. Chem., **72**, 1847, 497—508. English translation: J. B. O'Toole, IRE Trans. Circuit Theory, CT—5, 1958, 4—7.

17. Klarner, D. A., *A correspondence between two sets of trees*, Nederl. Akad. Wetensch. Proc. Ser. A, **72** (Indag. Math., 31), 1969, 292—296.

18. Matula, D. W., *On the number of subtrees of a symmetric n-ary tree*, SIAM J. Appl. Math., **18**, 1970, 688—703.

19. Menon, V. V., *On the existence of trees with given degrees*, Sankhya, Ser. A, **26**, 1964, 63—68.

20. Moon, J. W., *Various proofs of Cayley's formula for counting trees*, A Seminar on Graph Theory (ed: F. Harary), Holt, Rinehart and Winston, New York, 1967, 70—78.

21. Moon, J. W., *Counting labelled trees*, Canadian Mathematical Monographs, Nº 1, Canadian Mathematical Congress, 1970.

22. Moon, J. W., *The number of k-trees (labelled)*, J. Combinatorial Theory, **6**, 1969, 196—199.

23. O'Neil, P. V., *The number of trees in a certain network*, Notices Amer. Math. Soc., **10**, 1963, 569.

24. Oláh, G., *A problem on the enumeration of certain trees* (Russian), Studia Sci. Math. Hungar., **3**, 1968, 71—80.

25. Otter, R., *The number of trees*, Ann. of Math., **49**, 1948, 583—599.

26. Palmer, E. M., *On the number of labelled 2-trees*, J. Combinatorial Theory, **6**, 1969, 206—207.

27. Palmer, E. M., Read, R. C., *On the number of plane 2-tress*, J. London Math. Soc., (2), **6**, 1973, 583—592.

28. Paul, A. J., *Generation of directed trees and 2-trees without duplication*, IEEE Trans. Circuit Theory, CT—14, 1967, 354—356.

29. Poincaré, H., *Second complément à l'analysis situs*, Proc. London Math. Soc., **32**, 1901, 277—308.

30. Prüfer, H., *Neuer Beweis eines Satzes über Permutationen*, Arch. Math. Phys. **27**, 1918, 742—744.

31. Rényi, A., *On connected graphs*, Magyar Tud. Akad. Mat. Kutató Int. Közl., **4**, 1959, 385—388.

32. Rényi, A., *Some remarks on the theory of trees*, Magyar Tud. Akad. Mat. Kutató Int. Közl., **4**, 1959, 73—85.

33. Riordan, J., *An introduction to combinatorial analysis*, John Wiley, New York, 1958.

34. Riordan, J., Sloane, N. J. A., *The enumeration of rooted trees by total height*, J. Austral. Math. Soc., **10**, 1969, 278—282.

35. Sedláček, J., *On the number of spanning trees of finite graphs*, Časopis Pěst. Mat., **94**, 1969, 217—221.

36. Scoins, H. I., *The number of trees with nodes of alternate parity*, Proc. Cambridge Philos. Soc., **58**, 1962, 12—16.

37. Temperley, H. N. V., *On the mutual cancellation of cluster integrals in Mayer's fugacity series*, Proc. Phys. Soc., **83**, 1964, 3—16.

38. Tomescu, I., *Le nombre des graphes connexes k-cycliques aux sommets étiquetés*, Calcolo (Pisa), **9**, 1972, 71—74.

39. Tomescu, I., *Le nombre des sous-arborescences d'une arborescence donnée*, An. Univ. Bucureşti, Mat.-Mec., **20**, 1971, 141—145.

40. Tutte, W. T., *The dissection of equilateral triangles into equilateral triangles*, Proc. Cambridge Philos. Soc., **44**, 1948, 463—482.

41. Walkup, D. W., *The number of plane trees*, Mathematika, **19**, 1972, 200—204.

42. Weinberg, L., *Number of trees in a graph*, Proc. IRE, **46**, 1958, 1954—1955.

CHAPTER 7

Permutation Groups and Burnside's Theorem

7.1. PERMUTATION GROUPS

We recall that a permutation of degree n is a bijection of the set $X = \{1, 2, ..., n\}$ onto itself. If f is a permutation of the set X, we write

$$f = \begin{pmatrix} 1 & 2 & & n \\ f(1) & f(2) & \cdots & f(n) \end{pmatrix},$$

where the numbers $f(1), f(2), ..., f(n)$ are the numbers $1, 2, ..., n$, possibly written in another order.

We have seen in Chap. 2 that the number of permutations of degree n is $n!$. We now show that the permutations of degree n form a group with respect to composition of functions.

PROPOSITION 1. *The permutations of degree n form a group, denoted by S_n and called the symmetric group of n variables.*

Given two permutations f and g of the set $X = \{1, 2, ..., n\}$, the product fg is, by definition, the composition of the two mappings, i.e. the operation which consists of performing first the permutation g, then the permutation f, that is, $fg(i) = f(g(i))$ for every $i \in X$.

For example, if $f = \begin{pmatrix} 1 & 2 & 3 & 4 & 5 \\ i_1 & i_2 & i_3 & i_4 & i_5 \end{pmatrix}$ and $g = \begin{pmatrix} 1 & 2 & 3 & 4 & 5 \\ 3 & 1 & 5 & 4 & 2 \end{pmatrix}$, then

$$fg = \begin{pmatrix} 1 & 2 & 3 & 4 & 5 \\ i_3 & i_1 & i_5 & i_4 & i_2 \end{pmatrix}.$$

The definition is permissible because the composition of two bijections from X to X is also a bijection from X to X.

The product of permutations of n elements is not commutative for $n \geqslant 3$, i. e. in general $fg \neq gf$. For example

$$\begin{pmatrix} 1 & 2 & 3 & 4 & 5 \\ 2 & 3 & 1 & 5 & 4 \end{pmatrix} \begin{pmatrix} 1 & 2 & 3 & 4 & 5 \\ 1 & 3 & 4 & 2 & 5 \end{pmatrix} = \begin{pmatrix} 1 & 2 & 3 & 4 & 5 \\ 2 & 1 & 5 & 3 & 4 \end{pmatrix}$$

and

$$\begin{pmatrix} 1 & 2 & 3 & 4 & 5 \\ 1 & 3 & 4 & 2 & 5 \end{pmatrix} \begin{pmatrix} 1 & 2 & 3 & 4 & 5 \\ 2 & 3 & 1 & 5 & 4 \end{pmatrix} = \begin{pmatrix} 1 & 2 & 3 & 4 & 5 \\ 3 & 4 & 1 & 5 & 2 \end{pmatrix}.$$

Let us show that the set S_n of permutations of degree n is a group with respect to composition, i.e. the following axioms are satisfied:

1) associativity: $f(gh) = (fg)h$,

2) existence of the unit element e, such that $fe = ef = f$ for every $f \in S_n$,

3) existence of inverse elements: for every $f \in S_n$ there is an element in S_n denoted by f^{-1}, such that $ff^{-1} = f^{-1}f = e$.

Axiom 1) follows from the fact that composition of functions is an associative operation, i.e.

$$(f(gh))\,(i) = f(g(h(i))) = (fg)\,(h(i)) = ((fg)h)(i) \quad \text{for every} \quad i \in X.$$

The unit element e is the identity permutation defined by $e(i) = i$ for every $i \in X$ or $e = \begin{pmatrix} 1\,2\,...\,n \\ 1\,2\,...\,n \end{pmatrix}$. Now

$$fe(i) = f(i) \quad \text{and} \quad ef(i) = f(i) \quad \text{for every} \quad i \in X,$$

hence

$$fe = ef = f \quad \text{for every} \quad f \in S_n.$$

The inverse of the permutation $f = \begin{pmatrix} 1\,2\,...\,n \\ i_1\,i_2\,...\,i_n \end{pmatrix}$ is the permutation $f^{-1} = \begin{pmatrix} i_1\,i_2\,...\,i_n \\ 1\,2\,...\,n \end{pmatrix}$, for which we rearrange the columns so that the numbers of the first row are in increasing order. In other words, f^{-1} is the inverse of the bijection f regarded as a function. For $ff^{-1}(i_k) = f(k) = i_k$ and $f^{-1}f(k) = f^{-1}\,(i_k) = k$, hence $ff^{-1} = f^{-1}f = e$.

From the axioms 1), 2) and 3) one deduces immediately the uniqueness of the unit element e and for every $f \in S_n$, the uniqueness of its inverse element f^{-1}.

If n is a positive integer, we denote by f^n the permutation $\underbrace{ff \dots f}_{n\,\text{factors}}$ and by f^{-n} the permutation $(f^n)^{-1}$.

A subgroup G of S_n is a subset of permutations of S_n with the property that G satisfies the axioms 1), 2) and 3) with respect to the above defined product of permutations. It can be seen easily that the unit of the subgroup G coincides with the unit e of the group S_n.

The following proposition gives a characterization of subgroups in the case of finite groups.

PROPOSITION 2. *A subset $H \subset G$, where G is a finite group, is a subgroup of G if and only if $h_1 \in H$ and $h_2 \in H$ imply $h_1 h_2 \in H$ for every $h_1, h_2 \in H$.*

We must show that the set H is a group with respect to the restriction to H of the composition operation in G. This restriction, which we denote in the same way as the operation in G, is defined as follows: for every $h_1, h_2 \in H$, $h_1 h_2 = h_3$ if the result of the composition operation in G applied to h_1 and h_2 in this order is h_3 and $h_3 \in H$. Moreover, the restriction to H of the composition operation from G must satisfy axioms 1), 2) and 3).

Necessity results from the fact that if H is a subgroup of G, then the restriction to H of the composition operation of G being defined for every pair of elements h_1 and h_2 from H, it follows that $h_1h_2 \in H$.

To prove sufficiency, associate with every $g \in H$ the mapping $\varphi_g : H \to H$ defined by

$$\varphi_g(h) = gh \text{ for } h \in H;$$

the function φ_g is well defined since the product of two elements from H is in H.

The mapping φ_g is injective, because $\varphi_g(h_1) = \varphi_g(h_2)$ means $gh_1 = gh_2$, therefore $h_1 = g^{-1}(gh_2) = eh_2 = h_2$. Since φ_g is an injection from H to H and H is a finite set (for so is G), it follows that φ_g is also surjective, hence it is a bijection. Since φ_g is surjective, there exists an element h_0 in H such that $\varphi_g(h_0) = g$, therefore $g = gh_0$ and $h_0 = g^{-1}g = e$, the unit element in the group G.

Similarly there exists an element h_1 in H such that $\varphi_g(h_1) = e$, hence $e = gh_1$ and $h_1 = g^{-1} \in H$. As the element g has been taken arbitrarily in H, it follows that its inverse $g^{-1} \in H$. Therefore H is a subgroup of G, because the associativity axiom being satisfied in G, is satisfied in particular for the elements of H, then H contains the unit element e and the inverse h^{-1} of every element $h \in H$.

Given a group G, a subgroup $H \subset G$ is said to be a *normal subgroup* if $Hg = gH$ for every $g \in G$, where $Hg = \{x = hg | h \in H\}$ and similarly for gH. This condition is equivalent to $H = gHg^{-1}$ for every $g \in G$.

The normal subgroups commute with all the subgroups of G, that is, if A is a subgroup and H is a normal subgroup of G then $AH = \bigcup_{a \in A} aH = \bigcup_{a \in A} Ha = HA$, where we have set $AH = \{ah \mid a \in A, h \in H\}$.

PROPOSITION 3. *If H is a subgroup of G, the sets of the form $Ha = \{ha | h \in H\}$ with $a \in G$ form a partition of the set G. The set of these classes is known as the quotient set, denoted by G/H. The number of classes is* $|G/H| = \dfrac{|G|}{|H|}$. *If, moreover, H is a normal subgroup, then the quotient set is a group with respect to the multiplication defined by:*

$$C_1C_2 = \{x_1x_2 \mid x_1 \in C_1 \text{ and } x_2 \in C_2\}$$

and the mapping $f(a) = Ha$ of G onto the quotient set G/H is a homomorphism, i.e. $f(ab) = f(a)f(b)$.

We define the binary relation $\sim$ by $x \sim y$ if $x \in Hy$. This is actually a reflexive, symmetric and transitive relation:

$x \sim x$, for $x = ex$ hence $x \in Hx$;

$x \sim y$ implies $y \sim x$, for $x \sim y$ means $x \in Hy$ or, equivalently, there exists an element $h \in H$ such that $x = hy$, hence $y = h^{-1}x$, therefore $y \in Hx$ (because h being an element of the subgroup H, it follows that $h^{-1} \in H$); but this means $y \sim x$.

$x \sim y$ and $y \sim z$ imply $x \in Hy$ and $y \in Hz$ or, equivalently, there exist $h_1, h_2 \in H$, such that $x = h_1 y$ and $y = h_2 z$, therefore $x = (h_1 h_2)z$ or $x \in Hz$, which means $x \sim z$ (because $h_1 h_2 \in H$ as H is a subgroup).

The classes of this equivalence are the sets Ha with $a \in G$, for, if $x \in Ha$ and $y \in Ha$ then $x \sim a$ and $y \sim a$, hence $x \sim y$, and if $x \in Ha$ and $y \sim x$, it follows that $x \sim a$ and $y \sim x$, hence $y \sim a$ and $y \in Ha$.

Since the classes of an equivalence relation defined on a set form a partition of that set, it follows that the sets Ha determine a partition of G. But each class Ha has exactly $|H|$ elements, because there exists a bijection $g \colon H \to Ha$ defined by $g(h) = ha$. The mapping g is injective, for $h_1 a = h_2 a$ implies $h_1 = h_2 aa^{-1} = h_2 e = h_2$, and surjective, because for every element $y \in Ha$ there exists $h \in H$ such that $y = ha$, hence $y = g(h)$. The classes of this equivalence being disjoint and having the same number $|H|$ of elements, it follows that $|H| \cdot |G/H| = |G|$, hence the number of classes is $|G/H| = \dfrac{|G|}{|H|}$.

If the subgroup H is normal in G, then the set of equivalence classes is a group if we define the product of classes as in the statement of the proposition. The subgroup H being normal, we have $Ha = aH$, therefore

$$(Ha)\,(Hb) = H^2(ab) = H(ab),$$

$$(Ha)\,(He) = H^2 a = Ha,$$

$$(Ha)\,(Ha^{-1}) = H(aa^{-1})H = H^2 = H = He.$$

The first of the above three equations shows that the product of two equivalence classes is an equivalence class as well, and the product thus defined is associative, for

$$((Ha)\,(Hb))\,(Hc) = H(ab)c = Ha(bc) = Ha((Hb)\,(Hc)).$$

The latter equality, together with the second and the third equality of the above three, show that the set G/H is a group. The mapping $f \colon G \to G/H$, defined by $f(a) = Ha$ is a group homomorphism, because

$$f(ab) = H(ab) = (Ha)\,(Hb) = f(a)f(b).$$

The first part of this proposition is in fact Lagrange's theorem which states that if H is a subgroup of the finite group G, then the order of H is a divisor of the order of G, the order of a finite group being by definition the number of its elements. For example, the group S_3 of permutations of the set $\{1, 2, 3\}$ contains the permutations

$$e = \begin{pmatrix} 1 & 2 & 3 \\ 1 & 2 & 3 \end{pmatrix}, \quad a = \begin{pmatrix} 1 & 2 & 3 \\ 2 & 1 & 3 \end{pmatrix}, \quad b = \begin{pmatrix} 1 & 2 & 3 \\ 3 & 2 & 1 \end{pmatrix}, \quad c = \begin{pmatrix} 1 & 2 & 3 \\ 1 & 3 & 2 \end{pmatrix},$$

$$d = \begin{pmatrix} 1 & 2 & 3 \\ 2 & 3 & 1 \end{pmatrix} \text{ and } f = \begin{pmatrix} 1 & 2 & 3 \\ 3 & 1 & 2 \end{pmatrix}.$$

According to Lagrange's theorem, the subgroups of S_3 may contain 1, 2, 3 or 6 elements. The one-element subgroup is $\{e\}$, the six-element subgroup is S_3 itself. The two-element subgroups are $\{e, a\}$, $\{e, b\}$ and $\{e, c\}$. Every subgroup must contain the unit e and the other subsets $\{e, d\}$ and $\{e, f\}$ are not subgroups, because $d^2 = f$ and $f^2 = d$. The single subgroup with three elements is $\{e, d, f\}$, for $df = fd = e$ whereas the other nine pairs of non-identical permutations together with the unit e do not form subgroups, because $ab = f$, $ac = d$, $ad = c$, $af = b$, $bc = f$, $bd = a$, $bf = c$, $cd = b$, $cf = a$ so that the product of two elements from the subset does not belong to the subset. Among the six subgroups of S_3, only $\{e\}$, S_3 and $\{e, d, f\}$ are normal subgroups.

7.2. The Cycles of a Permutation

A notion which will be essential in the Pólya-de Bruijn enumeration theorem, is that of a cycle of a permutation.

Consider a permutation f of degree n and a number k with $1 \leqslant k \leqslant n$. The elements of the sequence $k = f^0(k), f(k), f^2(k), f^3(k), \ldots$ cannot be pairwise distinct, because the set X has n elements. The first element which occurs twice in the above sequence is k, for if the first such element is $f^p(k)$ where $p \geqslant 1$, then $f^p(k) = f^q(k)$ for some $q > p \geqslant 1$, hence $f(f^{p-1}(k)) = f(f^{q-1}(k))$ where $f^{p-1}(k) \neq f^{q-1}(k)$, thus contradicting the injectivity of the function f.

Let r be the least positive integer for which $f^r(k) = k$. The permutation $\begin{pmatrix} k & f(k) & \ldots & f^{r-1}(k) \\ f(k) & f^2(k) & \ldots & k \end{pmatrix}$ is said to be a *cycle* and is denoted by $[k, f(k), \ldots, f^{r-1}(k)]$. It can be shown that every permutation may be uniquely written (up to the order of factors) as a product of cycles with no common elements. The product of cycles with no common elements is commutative, hence the order of the cycles from the decomposition of the permutation is immaterial. This decomposition can be obtained in the following way. Consider the sequence $1, f(1), f^2(1), \ldots$ and take, as above, the cycle which contains 1. If $f(1) = 1$, this cycle consists of the single element 1 and is denoted by [1]. Let i_1 denote the least element in the set obtained from $X = \{1, 2, \ldots, n\}$ by removal of the elements of the cycle already constructed. Consider the sequence $i_1, f(i_1), f^2(i_1), \ldots$, and take the cycle which contains i_1 etc. The set X being finite, the above procedure terminates after a finite number of steps and we thus obtain the decomposition of the permutation f into a product of cycles with no common elements. A cycle $[i_1, i_2, \ldots, i_k]$ is regarded as a permutation of degree n defined as follows: $f(i_1) = i_2, f(i_2) = i_3, \ldots, f(i_k) = i_1$ and $f(j) = j$ for every j different from $i_1, i_2, \ldots, i_k$. For example:

$$f = \begin{pmatrix} 1 & 2 & 3 & 4 & 5 & 6 & 7 & 8 & 9 & 10 \\ 3 & 4 & 7 & 9 & 5 & 1 & 6 & 2 & 10 & 8 \end{pmatrix} = [1, 3, 7, 6] [2, 4, 9, 10, 8] [5].$$

This permutation can be represented by a directed graph with the vertex set X, where an arrow from i to j means that $j = f(i)$. Each connected component of this graph is a circuit which corresponds to a cycle from the decomposition of the permutation f, as can be seen in Fig. 7.1.

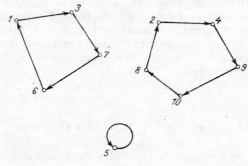

Fig. 7.1

By a *cyclic permutation* we mean a permutation with exactly one cycle of length greater than 1. For example:

$$f = \begin{pmatrix} 1 & 2 & 3 & 4 & 5 & 6 \\ 3 & 1 & 2 & 4 & 5 & 6 \end{pmatrix} = [1, 3, 2]\,[4]\,[5]\,[6]$$

is a cyclic permutation of the elements 1, 3, 2.

A cyclic permutation which contains a single cycle of length 2, say $[i, j]$, the other cycles being of length 1, is said to be a *transposition*. Notice that the transposition $[i, j]$ interchanges the elements i, j and leaves the other elements of X unchanged.

If G is a subgroup of S_n, the permutations s and t are said to be *conjugate* with respect to G, if there exists an element $g \in G$ such that $s = gtg^{-1}$. The relation of being conjugate is an equivalence because it is:

— reflexive: $s = ese^{-1}$ with $e =$ the identity permutation;

— symmetric: $s = gtg^{-1}$ implies $t = g^{-1}sg = hsh^{-1}$ where we have set $h = g^{-1} \in G$;

— transitive: $s = gtg^{-1}$ and $t = huh^{-1}$ imply $s = ghuh^{-1}g^{-1} = (gh)u(gh)^{-1}$ with $gh \in G$ since G is a subgroup of S_n.

PROPOSITION 4. *Two permutations s and t are conjugate with respect to S_n if and only if the number of their cycles is the same, say m, and their cycles have respectively the same length n_i for $i = 1, 2, ..., m$.*

To prove necessity, assume that s and t are conjugate, i.e. there exists $g \in S_n$ such that $s = gtg^{-1}$. Let

$$t = [t_{11}t_{12} \ldots t_{1i}]\,[t_{21}\,t_{22} \ldots t_{2j}] \ldots [t_{m1}\,t_{m2} \ldots t_{mk}]$$

be the cycle decomposition of the permutation t and set $g(t_{pq}) = s_{pq}$. Then $s = gtg^{-1}$ implies

$$s(s_{11}) = gtg^{-1}(s_{11}) = gt(t_{11}) = g(t_{12}) = s_{12}, \ldots$$

so the cycle decomposition of the permutation s is

$$s = [s_{11} \, s_{12} \ldots s_{1i}] \, [s_{21} \, s_{22} \ldots s_{2j}] \ldots [s_{m1} \, s_{m2} \ldots s_{mk}];$$

therefore two conjugate permutations are decomposed into cycles of equal lengths.

To prove sufficiency, define the permutation g starting from the cycle decompositions of the permutations s and t as follows: $g(t_{pq}) = s_{pq}$. Since both the decomposition of s into cycles with no common elements and the decomposition of t contain each of the numbers $1, 2, \ldots, n$ once and only once, it follows that $g \in S_n$ and a calculation similar to the above shows that $s = gtg^{-1}$.

Let us now determine the number of permutations from an equivalence class with respect to conjugation viewed as an equivalence relation in S_n; in other words, let us find the number of permutations having λ_1 cycles of length 1, λ_2 cycles of length 2, ..., λ_k cycles of length k. As the cycles have no common elements, $\lambda_1 + 2\lambda_2 + \ldots + k\lambda_k = n$. A permutation which contains λ_1 cycles of length 1, ..., λ_k cycles of length k will be said to be of type $1^{\lambda_1} \, 2^{\lambda_2} \ldots k^{\lambda_k}$, by analogy with the partitions which contain λ_1 classes with one element, ..., λ_k classes with k elements.

PROPOSITION 5. *The number $h(\lambda_1, \lambda_2, \ldots, \lambda_k)$ of permutations of type $1^{\lambda_1} 2^{\lambda_2} \ldots k^{\lambda_k}$ is*

$$h(\lambda_1, \lambda_2, \ldots, \lambda_k) = \frac{n!}{\lambda_1! \, \lambda_2! \ldots \lambda_k! \, 1^{\lambda_1} \, 2^{\lambda_2} \ldots k^{\lambda_k}} ; \qquad (7.1)$$

this result is known as Cauchy's formula.

In order to prove this proposition, let us write a permutation f of type $1^{\lambda_1} 2^{\lambda_2} \ldots k^{\lambda_k}$ with the cycles in increasing order of their lengths:

$$f = \underbrace{[*] \ldots [*]}_{\lambda_1} \, \underbrace{[**] \ldots [**]}_{\lambda_2} \ldots \underbrace{\overbrace{\ldots [** \ldots *]}^{k}}_{\lambda_k},$$

where the stars stand for the numbers $1, 2, \ldots, n$. By omitting the parentheses, we obtain a word of length n made up of distinct letters from the alphabet $\{1, 2, \ldots, n\}$. The number of these words is $n!$. But the same permutation generates $\lambda_1! \, \lambda_2! \ldots \lambda_k! \, 1^{\lambda_1} \, 2^{\lambda_2} \ldots k^{\lambda_k}$ different words. For we can permute the λ_i cycles of length i $(i = 1, 2, \ldots, k)$, and this is possible in $\lambda_1! \, \lambda_2! \ldots \lambda_k!$ distinct ways. On the other hand, a cycle of length i can be written in i distinct ways, taking as initial letter each of the i elements, thus yielding $1^{\lambda_1} 2^{\lambda_2} \ldots k^{\lambda_k}$ possibilities. In the same way as for the proof of formula (4.6) which counts the partitions of a given type, one can show that the above procedure furnishes irredundantly the $n!$ words of length n, made up out of the numbers $1, 2, \ldots, n$, hence

$$n! = \lambda_1! \, \lambda_2! \ldots \lambda_k! \, 1^{\lambda_1} 2^{\lambda_2} \ldots k^{\lambda_k} \, h(\lambda_1, \lambda_2, \ldots, \lambda_k),$$

thus implying (7.1).

For example, for $n = 4$ there are 5 classes of conjugate permutations, corresponding to the 5 partitions of 4 as a sum of positive integers:

$$4 = 1 + 1 + 1 + 1 = 2 + 1 + 1 = 2 + 2 = 3 + 1.$$

The number of permutations of type $1^2 2^1$ is equal to:

$$h(2, 1) = \frac{4!}{1^2 2! 2^1 1!} = 6$$

namely

$$[1]\,[2]\,[3, 4], \quad [1]\,[3]\,[2, 4], \quad [1]\,[4]\,[2, 3], \quad [2]\,[3]\,[1, 4], \quad [2]\,[4]\,[1, 3]$$

and $[3]\,[4]\,[1, 2]$.

7.3. THE PARITY OF A PERMUTATION

An important notion in the theory of permutation groups is that of *parity* of a permutation.

Consider an arbitrary permutation, say $f = \begin{pmatrix} 1 & 2 & 3 & 4 & 5 & 6 \\ 3 & 1 & 5 & 4 & 6 & 2 \end{pmatrix}$. We say that the symbol 1 presents an *inversion*, because in the second row the number 3 which is greater than 1 is situated before 1; the number 2 presents four inversions, because the numbers 3, 5, 4 and 6 are greater than 2 and situated before 2, etc. The total number of inversions of f is $I(f) = 6$.

The *signature* of a permutation f is by definition the number $p(f) = (-1)^{I(f)}$. If $p(f) = 1$, we say that the permutation is *even*, while if $p(f) = -1$ the permutation is said to be *odd*. We shall see that the parity of a permutation can be deduced easily from the decomposition of that permutation into a product of cycles with no common elements.

We recall that a transposition is a permutation whose cycle decomposition contains a single cycle $[i, j]$ of length 2, the other cycles being of length 1.

A transposition is an odd permutation. For let $[i, j]$ be a transposition with $i < j$. The second row of this permutation is the sequence $1, 2, \ldots, i-1, j, i+1, \ldots, j-1, i, j+1, \ldots, n$. The numbers $1, 2, \ldots, i-1, j, j+1, \ldots, n$ do not present any inversions, while the numbers $i+1, i+2, \ldots, j-1$ present a single inversion in j and the number i presents inversions in $j, i+1, \ldots, j-1$, therefore the total number of inversions is equal to $j - i - 1 + j - i = 2(j - i) - 1$, i.e. an odd number.

If t is a transposition, then $t^2 = e$, hence $t = t^{-1}$. By multiplying a permutation f by a transposition $[i, j]$, the terms located in places i and j in the second row interchange, the other terms remaining unchanged. For example, if

$f = \begin{pmatrix} 1 & 2 & 3 & 4 & 5 & 6 \\ 3 & 1 & 5 & 4 & 6 & 2 \end{pmatrix}$ and $t = [3, 4]$, then $ft = \begin{pmatrix} 1 & 2 & 3 & 4 & 5 & 6 \\ 3 & 1 & 4 & 5 & 6 & 2 \end{pmatrix}$ and the number of inversions has been reduced by one, for $I(ft) = 5$.

PROPOSITION 6. *The minimum number of transpositions of the form $[i, i + 1]$, necessary for transforming a permutation f of degree n into the identity permutation e is equal to the number $I(f)$ of inversions of the permutation f. If the permutation f is transformed into the identity permutation after q transpositions of the form $[i, i+1]$, then q has the same parity as $I(f)$.*

We first prove that the permutation f can be transformed into the identity permutation, which presents no inversions, by $I(f)$ transpositions. We can bring the element 1 into the first place of the second row of the permutation f by interchanging certain pairs of consecutive terms as many times as the number of inversions presented by 1; but this transformation is in fact a repeated multiplication of f with certain transpositions of the form $[i, i + 1]$. Similarly we can bring the element 2 into the second place of the second row by interchanging certain pairs of consecutive terms as many times as the number of inversions of 2, etc. Therefore, by multiplying the permutation f by $I(f)$ transpositions of the form $[i, i + 1]$, we transform f into the identity permutation e. But the multiplication of the permutation f with the transposition $[i, i + 1]$ results in a new permutation f' which differs from f in the places of the neighbouring elements $f(i)$ and $f(i + 1)$, therefore $I(f') = I(f) + 1$ if $f(i) < f(i + 1)$ and $I(f') = I(f) - 1$ if $f(i) > f(i + 1)$. This implies that $I(f)$ is the minimum number of transpositions which transform f into e, for at each multiplication with a transposition of the form $[i, i + 1]$ the number of inversions is reduced by at most one.

If q transpositions of this form are needed, this means that the parity of the number $I(f)$ of inversions has been changed q times in order to obtain $I(f) = 0$, hence $I(f)$ and q have the same parity.

If follows from the above proposition that the symmetric group S_n is generated by the $n-1$ transpositions $[1, 2], [2, 3], ..., [n-1, n]$, i.e. every permutation $f \in S_n$ can be written as a product of such permutations. Indeed, we have seen that by multiplying the permutation f with the transpositions $t_1, t_2, ..., t_q$ of the form $[i, i + 1]$, by $q = I(f)$, we obtain the identity permutation, i.e. $ft_1t_2 ... t_q = e$. By right multiplication of the latter equation by $t_q^{-1} ... t_2^{-1}t_1^{-1}$ we obtain $f = et_q^{-1} ... ...t_2^{-1}t_1^{-1} = t_q ... t_2t_1$. If the permutation f can be written as a product of q transpositions of the form $[i, i + 1]$, i.e. $f = t_1t_2 ... t_q$, then we infer that $ft_qt_{q-1} ... t_1 = e$, but we have seen in the previous proposition that q and $I(f)$ have the same parity.

PROPOSITION 7. *The signature of a permutation f, i.e. the function $p : S_n \to \{-1, +1\}$ defined for every $f \in S_n$ by the relation $p(f) = (-1)^{I(f)}$ is a homomorphism of the group S_n into the multiplicative group $\{1, -1\}$, i.e. $p(f_1f_2) = p(f_1)p(f_2)$.*

The set $\{1, -1\}$ is a group with respect to ordinary multiplication of integers. Let us show that p is a homomorphism from S_n to $\{1, -1\}$. We have seen that every permutation f from S_n can be decomposed into a product of $I(f)$ transpositions of the form $[i, i + 1]$. Then $f_1f_2 = (t_1t_2 ... t_{I(f_1)})(t_1't_2' ... t_{I(f_2)}')$. This product consists of $I(f_1) + I(f_2)$ transpositions of the form $[i, i + 1]$, therefore the number $I(f_1) + I(f_2)$ has the same parity as $I(f_1 f_2)$ in view of the previous remark. Hence

$$(-1)^{I(f_1 f_2)} = (-1)^{I(f_1) + I(f_2)} = (-1)^{I(f_1)}(-1)^{I(f_2)}$$

thus proving that p is a group homomorphism.

If follows from the previous discussion that if the permutation f can be expressed as a product of q transpositions, then q has the parity of $I(f)$, in other words, f is even or odd as q is even or odd. Indeed, let $f = t_1 t_2 \ldots t_q$, where $t_1, t_2, \ldots, t_q$ are transpositions, so they are of odd parity. Then proposition 7 implies that $(-1)^{I(f)} = p(f) = p(t_1)p(t_2) \ldots p(t_q) = (-1)^q$, therefore $I(f)$ and q have the same parity.

PROPOSITION 8. *If f is a permutation of type $1^{\lambda_1} 2^{\lambda_2} \ldots k^{\lambda_k}$, its parity coincides with that of the sum $\lambda_2 + \lambda_4 + \lambda_6 + \ldots$, i.e. the parity of a permutation is equal to the number of its even cycles.*

We first prove that a cycle consisting of m elements can be decomposed into a product of $m - 1$ transpositions. Now, $f = [k_1, k_2, \ldots, k_m] = [k_1, k_2][k_2, k_3] \ldots$ $\ldots [k_{m-1}, k_m]$, which is a product of $m - 1$ transpositions. Therefore, if the permutation f is of type $1^{\lambda_1} 2^{\lambda_2} \ldots k^{\lambda_k}$, it can be written as a product of $\lambda_2 + 2\lambda_3 + \ldots + (k - 1)\lambda_k$ transpositions, hence f has the parity of the sum $\lambda_2 + \lambda_4 + \lambda_6 + \ldots$

PROPOSITION 9. *In a permutation group $G \subset S_n$, either all permutations are even, or the number of odd permutations equals the number of even permutations.*

If in the subgroup G there exists an odd permutation h, consider the bijection φ_h of the group G onto itself, defined by $\varphi_h(g) = hg$ for every $g \in G$. Since $p(h) = = -1$, we get

$$\sum_{g \in G} p(g) = \sum_{g \in G} p(hg) = \sum_{g \in G} p(h)p(g) = - \sum_{g \in G} p(g)$$

hence $\sum_{g \in G} p(g) = 0$, which means that there are as many even permutations as odd permutations, because $p(g) = 1$ for an even permutation and $p(g) = -1$ for an odd permutation.

PROPOSITION 10. *The set*

$$A_n = \{g | g \in S_n, p(g) = 1\}$$

of even permutations is a normal subgroup of S_n, known as the alternating group. It contains $\dfrac{1}{2} n!$ permutations.

We shall prove a more general result, namely if $p : G \to G_1$ is a homomorphism of the group G into the group G_1, then the set Ker $(p) = \{g | g \in G, p(g) = e_1\}$, where e_1 is the unit of the group G_1, is a normal subgroup (known as the *kernel of the homomorphism p*), of the group G. Ker(p) is a subgroup, because if $g, g' \in$ Ker(p), then $p(g) = p(g') = e_1$, hence $p(gg') = p(g)p(g') = e_1 e_1 = e_1$, therefore $gg' \in$ Ker(p). If the group G is finite, then Ker(p) is a subgroup of G. In the opposite case, let us still prove that the unit e of G and the inverse of every element of Ker(p) belong to Ker(p). As p is a group homomorphism, $p(ge) = p(g)p(e) = p(g)$ since $ge = g$. By left multiplying with $[p(g)]^{-1}$, we get $p(e) = e_1$ hence $e \in$ Ker(p). As $gg^{-1} = e$, it follows that $p(g)p(g^{-1}) = p(e) = e_1$, therefore $[p(g)]^{-1} = p(g^{-1})$. If $g \in$ Ker(p), then $p(g^{-1}) = [p(g)]^{-1} = e_1^{-1} = e_1$, hence $g^{-1} \in$ Ker(p). So the kernel of the homomorphism p is a subgroup of G.

Let us prove that the subgroup Ker(p) is normal, that is, Ker(p)$g = g$ Ker(p) for every $g \in G$. If $x \in$ Ker(p)g, then there exists $k \in$ Ker(p) such that $x = kg$. If we set $h = g^{-1}x$, we get $p(h) = p(g^{-1})p(x) = [p(g)]^{-1}p(k)p(g) = [p(g)]^{-1}p(g) = e_1$, since $p(k) = e_1$. But $h = g^{-1}x$ implies $x = gh$ and $p(h) = e_1$, hence $h \in$ Ker(p) or $x \in g$ Ker(p). We have thus obtained Ker(p)$g \subset g$ Ker(p); the converse inclusion is proved similarly. So the kernel of p is a normal subgroup of G.

To prove proposition 10, take as G the symmetric group S_n, as G_1 the multiplicative group $\{1, -1\}$, and as homomorphism p the signature of the permutations of degree n. Then proposition 7 implies that the set A_n of even permutations is a normal subgroup of S_n, because it is the kernel of the homomorphism $p(g) = (-1)^{I(g)}$. Finally $|A_n| = \dfrac{1}{2}|S_n| = \dfrac{1}{2}n!$ because the even permutations are as numerous as the odd ones.

7.4. THE ORBITS OF A PERMUTATION GROUP

If $G \subset S_n$ is a permutation group and $x, y \in X = \{1, 2, ..., n\}$, we shall write $x \sim y(G)$ if there exists a permutation $f \in G$ such that $g = f(x)$. In this case we say that x is equivalent to y modulo G. The relation defined in this way is an equivalence relation, because it is:

— reflexive: $x \sim x(G)$ since $x = e(x)$, where e is the identity permutation, which belongs to G;

— symmetric: $x \sim y(G)$ means there exists $g \in G$ such that $y = g(x)$, hence $x = g^{-1}(y)$ and so $y \sim x(G)$ because $g^{-1} \in G$;

— transitive: $x \sim y(G)$ and $y \sim z(G)$ means there exist $g_1, g_2 \in G$ such that $y = g_1(x)$ and $z = g_2(y)$, hence $g_2 g_1(x) = z$ and $x \sim z(G)$ for $g_2 g_1 \in G$.

The classes of this equivalence relation are called the *orbits* of the group G. The orbits of a permutation are a generalization of the notion of cycle of a permutation, because if we take the subgroup $\{e, f, f^2, f^3, ...\}$ generated by a permutation $f \in S_n$ in the role of the group G, the orbits of G coincide with the cycles of the permutation f.

The problem we want to solve is that of determining the number of orbits of a permutation group G. We denote by $G_k = \{g | g \in G, g(k) = k\}$ the set of permutations which leave the element k unchanged.

For every $k = 1, 2, ..., n$, the set G_k is a subgroup of G, because if $f, g \in G_k$, then $fg(k) = f(k) = k$, so $fg \in G_k$. The sets G_k are non-empty since the identity permutation satisfies $e(k) = k$ for every $k = 1, ..., n$.

PROPOSITION 11. *If O_k is the orbit of the group G which contains the element k and G_k is the subgroup of G which leaves k unchanged, then*

$$|G_k| \cdot |O_k| = |G|. \tag{7.2}$$

Since G_k is a subgroup of G, we can construct the quotient set G/G_k. Then $|G/G_k| = \dfrac{|G|}{|G_k|}$ by proposition 3. We shall prove that this number is equal to $|O_k|$ by constructing a bijection from O_k onto G/G_k. Notice that $g(i) = h(i) = k$ with $g, h \in G$ implies $hg^{-1}(k) = h(i) = k$, that is, $hg^{-1} \in G_k$ or $h \in G_k g$. For every element $i \in O_k$ there exists a permutation $g_i \in G$ such that $g_i(i) = k$, for $i \sim k(G)$.

We define a bijection $\varphi : O_k \to G/G_k$ as follows: $\varphi(i) = G_k g_i \in G/G_k$, where the permutation g_i satisfies the relation $g_i(i) = k$. The mapping φ is well defined for if there exist two $h, g \in G$ such that $g(i) = h(i) = k$, then we have seen that $h \in G_k g$, which is equivalent to the fact that the class of h coincides with the class of g, or $G_k h = G_k g$. The mapping φ is injective, because $i, j \in O_k$ and $i \neq j$ imply the existence of two permutations $h, g \in G$ such that $g(i) = h(j) = k$ hence $hg^{-1}(k) = = h(i) \neq k = h(j)$, for the permutation h is injective and $i \neq j$, thus implying $h(i) \neq h(j)$. But $hg^{-1}(k) \neq k$ implies $hg^{-1} \notin G_k$ or $h \notin G_k g$, therefore $G_k g \neq G_k h$, i.e. $\varphi(j) \neq \varphi(i)$. The mapping φ is also surjective, since the class $G_k g$ is the image of the element $g^{-1}(k) = l \in O_k$, due to the fact that $g(l) = k$ and $l \sim k(G)$. The mapping φ being a bijection, it follows that $|O_k| = |G/G_k| = \dfrac{|G|}{|G_k|}$.

PROPOSITION 12 *(Burnside). If $\lambda_1(g)$ is the number of cycles of length 1 of the permutation g or the number of elements of X which are left invariant by the permutation g, the number of orbits of a group $G \subset S_n$ is*

$$|C_G| = \frac{1}{|G|} \sum_{g \in G} \lambda_1(g). \tag{7.3}$$

The idea of the proof consists of counting in two different ways the elements of X invariant with respect to the permutations $g \in G$:

$$\sum_{g \in G} \lambda_1(g) = \sum_{k \in X} |G_k| = \sum_{i=1}^{q} \sum_{k \in O_i} |G_k|, \tag{7.4}$$

where we have denoted by $O_1, O_2, ..., O_q$ the orbits of the group G. But if two elements j and k belong to the same orbit O_i, proposition 11 implies $|G_j| = |G_k| = = \dfrac{|G|}{|O_i|}$ and the equality (7.4) becomes

$$\sum_{g \in G} \lambda_1(g) = \sum_{i=1}^{q} |O_i| \frac{|G|}{|O_i|} = q|G|,$$

thus implying (7.3), where we have denoted by $|C_G|$ the number of equivalence classes with respect to the group G or, equivalently, the number of orbits of the group G. This result will be used later in the proof of Pólya's theorem.

For instance, let $n = 6$ and the subgroup $G \subset S_6$ be generated by the permutation $f = [1] [2, 3] [4, 5, 6]$. The elements of G are the following:

$$f = [1] [2, 3] [4, 5, 6], \qquad f^2 = [1] [2] [3] [4, 6, 5],$$
$$f^3 = [1] [2, 3] [4] [5] [6], \qquad f^4 = [1] [2] [3] [4, 5, 6],$$
$$f^5 = [1] [2, 3] [4, 6, 5], \qquad f^6 = [1] [2] [3] [4] [5] [6] = e.$$

Notice that a cycle raised to the power equal to the number of elements in the cycle, is the identity permutation; moreover, this is the least exponent with this property. Therefore the least exponent r for which a permutation f satisfies $f^r = e$ is the least common multiple of the lengths of the cycles of the permutation f, the product of two cycles with no common elements being commutative. In this case the orbits coincide with the cycles of the permutation f, namely $O_1 = \{1\}$, $O_2 = \{2, 3\}$ and $O_3 = \{4, 5, 6\}$. For example, $O_2 = \{2, 3\}$ and $G_2 = \{f^2, f^4, f^6\}$. The relation $|G_2| \cdot |O_2| = 2 \cdot 3 = 6 = |G|$ is satisfied. Also $\lambda_1(f) = 1$, $\lambda_1(f^2) = 3$, $\lambda_1(f^3) = 4$, $\lambda_1(f^4) = 3$, $\lambda_1(f^5) = 1$, $\lambda_1(f^6) = 6$ and the number of orbits is

$$|C_G| = \frac{1}{6} (1 + 3 + 4 + 3 + 1 + 6) = 3.$$

In the case of the group S_3 and subgroup $G = \{e, d, f\}$, where $e = [1] [2] [3]$, $d = [1, 2, 3]$, $f = [1, 3, 2]$, we obtain $\lambda_1(e) = 3$, $\lambda_1(d) = 0$ and $\lambda_1(f) = 0$, while the number of orbits is $|C_G| = \frac{1}{3} (3 + 0 + 0) = 1$ and this orbit is $\{1, 2, 3\}$.

A (round-robin) tournament T_n consists of n vertices $p_1, p_2, ..., p_n$ such that each pair of distinct vertices p_i and p_j is joined by one and only one of the oriented arcs (p_i, p_j) or (p_j, p_i). If the arc (p_i, p_j) is in T_n, then we say that p_i dominates p_j. Two tournaments T_n and T'_n are isomorphic if there exists a one-to-one dominance-preserving correspondence between their vertices.

Davis [9] used proposition 12 to determine $T(n)$, the number of nonisomorphic tournaments T_n. We now derive his formula for $T(n)$.

Any permutation f that belongs to the symmetric group S_n can be expressed as a product of disjoint cycles. If the disjoint cycle representation of f contains d_k cycles of length k, for $k = 1, 2, ..., n$, then f is said to be of (cycle) type $(d) = 1^{d_1} 2^{d_2} ... n^{d_n}$ where $d_1 + 2d_2 + ... + nd_n = n$. In the present context, we think of the permutations f as acting on the vertices of tournaments T_n; these permutations f then, in effect, define permutations among the tournaments themselves.

We determine first $d(f)$, the number of tournaments T_n such that $f(T_n) = T_n$, i.e. the number of fixed points of the permutation f.

PROPOSITION 13. *If the permutation f is of type $1^{d_1}2^{d_2} \dots n^{d_n}$, then $d(f) = 0$ if f has any cycles of even length; otherwise $d(f) = 2^D$, where*

$$D = \frac{1}{2}\left[\sum_{k,l=1}^{n} (k,l)\, d_k d_l - \sum_{k=1}^{n} d_k \right] \tag{7.5}$$

and (k,l) denotes the greatest common divisor of k and l.

If f has a cycle of even length $[i_1, i_2, \dots, i_{2k}]$, suppose that the tournament T_n with the property $f(T_n) = T_n$ has an arc $(p_{i_1}, p_{i_{k+1}})$. Because T_n is invariant under f, T_n also has the oriented arcs: $(p_{i_2}, p_{i_{k+2}})$, $(p_{i_3}, p_{i_{k+3}})$, ..., $(p_{i_{k+1}}, p_{i_1})$. But the existence of both arcs $(p_{i_1}, p_{i_{k+1}})$ and $(p_{i_{k+1}}, p_{i_1})$ contradicts the definition of a tournament, hence in this case we obtain $d(f) = 0$.

Otherwise let T_n be any tournament such that $f(T_n) = T_n$. Then T_n can be partitioned into subtournaments $T^{(1)}$, $T^{(2)}$, ... in such a way that two vertices p_i and p_j belong to the same subtournament if and only if i and j belong to the same cycle of f. Among the subtournaments $T^{(1)}$, $T^{(2)}$, ..., there are d_1 that contain a single vertex, d_3 that contain three vertices, and so forth. Let $p_1, p_2, \dots, p_k$ and p_{k+1}, $p_{k+2}, \dots, p_{k+l}$ denote the vertices of two of these subtournaments, $T^{(1)}$ and $T^{(2)}$ say. We may suppose the permutation f contains the cycles $[1, 2, \dots, k]$ and $[k+1, k+2, \dots, k+l]$. If the $\dfrac{k-1}{2}$ arcs joining p_1 to p_i $\left(i = 2, 3, \dots, \dfrac{k+1}{2}\right)$ are oriented arbitrarily, then the orientations of the remaining arcs of $T^{(1)}$ are determined uniquely by the condition that $f(T_n) = T_n$. Furthermore, if the (k, l) arcs joining p_1 to p_{k+i} $(i = 1, 2, \dots, (k,l))$ are oriented arbitrarily, then the orientations of the remaining arcs joining $T^{(1)}$ and $T^{(2)}$ are also determined uniquely by the condition that $f(T_n) = T_n$. This argument can be repeated as often as necessary. Therefore, in constructing tournaments T_n such that $f(T_n) = T_n$, we are free to orient arbitrarily only

$$\sum_{k=1}^{n} \frac{1}{2} d_k(k-1) + \sum_{k=1}^{n} (k,k)\binom{d_k}{2} + \sum_{k<l} (k,l) d_k d_l =$$

$$= \frac{1}{2}\left[\sum_{k,l=1}^{n} (k,l)\, d_k d_l - \sum_{k=1}^{n} d_k \right]$$

arcs, the orientations of the remaining arcs being determined by the orientations of these. The proposition now follows.

PROPOSITION 14. *If $T(n)$ denotes the number of nonisomorphic tournaments T_n, then*

$$T(n) = \sum_{(d)} \frac{2^D}{N_d},$$

where D is as defined in (7.5), $N_d = 1^{d_1}d_1! \ 2^{d_2}d_2! \ ... \ n^{d_n}d_n!$ and where the sum is over all solutions (d) in nonnegative integers of the equation

$$d_1 + 3d_3 + 5d_5 + ... = n.$$

This result is implied by Burnside's theorem and proposition 13, because the number of permutations f of type $1^{d_1}2^{d_2} ... n^{d_n}$ is equal to $\dfrac{n!}{N_d}$. The values of $T(n)$ for $n = 1, 2, ..., 10$ are given in Table 7.1 [19]; the first eight of these values were given by Davis [10].

TABLE 7.1

n	$T(n)$
1	1
2	1
3	2
4	4
5	12
6	56
7	456
8	6,880
9	191,536
10	9,733,056

7.5. CYCLIC PERMUTATIONS AND TREES

Let $T = \{t_1, t_2, ..., t_k\}$ be a set of transpositions of the set $X = \{1, 2, ..., n\}$. The set T will be associated with an unoriented graph (X, T) with vertices $1, 2, ..., n$, the edges $[i, j]$ of which are precisely the transpositions of the set T. For example let $X = \{1, 2, 3\}$ and T be the set consisting of the transpositions $[1, 2]$ and $[1, 3]$. The corresponding graph (X, T) is drawn in Fig. 7.2. By performing products of transpositions from the set T in all possible ways, we obtain all the permutations of the symmetric group S_3. Using the notations introduced above for the elements of S_3, we obtain

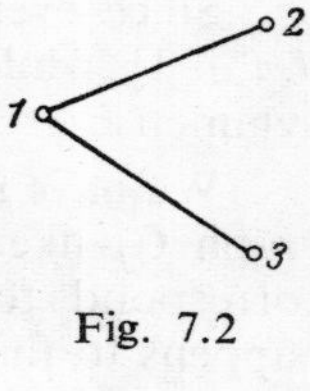

$a = [1, 2], \ b = [1, 3], \ e = [1, 2] [1, 2], \ c = [2, 3] = [1, 3] [1, 2] [1, 3],$

$$d = [1, 3] [1, 2] \text{ and } f = [1, 2] [1, 3].$$

Fig. 7.2

This result can be generalized as follows:

PROPOSITION 15 (Pólya). A set T consisting of $n - 1$ transpositions generates the symmetric group S_n if and only if the graph (X, T) is a tree.

To prove necessity, we proceed by reductio ad absurdum, i.e. we assume that (X, T) is not a tree. Since the graph (X, T) contains $n - 1$ edges and n vertices, it follows that its cyclomatic number is $p - 1$, where p is the number of its connected components. If $p = 1$ then the graph is connected and cycle-free, i.e. a tree, contrary to the assumed hypothesis. Therefore $p \geqslant 2$ and this graph contains at least two vertices a and b that belong to different connected components, say X_1 and X_2. Then the transposition $[a, b]$ cannot be written as a product of transpositions from the set T. Indeed, every product of transpositions from T can be written $f_1 f_2 \ldots f_p$, where f_1 is the product of transpositions which act only on the vertices of the connected component X_1, ..., f_p is the product of transpositions which act only on the vertices of the connected component X_p of the graph (X, T), since the product of transpositions with no common elements is commutative. Therefore f_i is a permutation which permutes only elements from the connected component X_i and leaves unchanged the vertices of the other connected components, whence we deduce that the permutation $f_1 f_2 \ldots f_p$, which is a product of transpositions in T, transforms every vertex of X into a vertex of the same connected component. This implies that the transposition $[a, b]$ cannot be written as a product of transpositions from T, thus contradicting the hypothesis. So necessity is established.

To prove sufficiency, assume that (X, T) is a tree, therefore a connected graph. In this case there exists a chain from vertex i to vertex $i + 1$ for every $i = 1, 2, \ldots, n - 1$. Let $[i, i_1], [i_1, i_2], \ldots, [i_k, i + 1]$ be the consecutive edges in T of this chain, where $i, i_1, i_2, \ldots, i_k, i + 1$ are pairwise distinct. But the transposition $[i, i+1]$ can be written

$$[i, i + 1] = [i + 1, i_k] [i_k, i_{k-1}] \ldots [i_3, i_2] [i_2, i_1] [i_1, i] [i_1, i_2] \ldots [i_{k-1}, i_k] [i_k, i+1],$$

i.e. it is a product of transpositions from T. Indeed, notice that $i + 1$ is transformed into i_k, then into $i_{k-1}, \ldots$, then into i_1 and i_1 into i, therefore $i + 1$ is transformed into i. Similarly i is transformed into i_1, then into $i_2, \ldots$, then into i_k and i_k into $i + 1$, therefore the image of i is $i + 1$. The other elements $i_1, i_2, \ldots, i_k$ remain unchanged, because the only non identical transformations are: i_1 goes to i_2 and i_2 goes to i_1, therefore i_1 goes to $i_1, \ldots, i_k$ goes to $i + 1$ and $i + 1$ goes to i_k, therefore i_k goes to i_k.

Since every permutation of S_n is generated by transpositions of the form $[i, i + 1]$, while these are generated by transpositions of T, it follows that the symmetric group S_n is generated by the transpositions of T.

We have seen that every permutation f of degree n is associated with a directed graph G_f like that in Fig. 7.1; the connected components of G_f are circuits that correspond to the cycles from the cycle decomposition of f. Let us see what happens to this graph when multiplying the permutation f by a transposition $[a, b]$, i.e. $g = f [a, b]$. Since the transposition $[a, b]$ transforms a into b and b into a, it follows that the graph G_g is obtained from the graph G_f by replacing the arcs $(a, f(a))$ and $(b, f(b))$ by the arcs $(a, f(b))$ and $(b, f(a))$, respectively.

If a and b are situated on two different cycles of the graph G_f, then these two cycles generate a single cycle in the graph G_g, while if a and b belong to the same cycle of G_f, then this cycle is decomposed into two cycles with no common vertices in the graph G_g, as is seen in Fig. 7.3.

PROPOSITION 16 *(Dénes). Let $T = \{t_1, t_2, ..., t_{n-1}\}$ be a set of $n - 1$ transpositions. The product $f = t_1 t_2 ... t_{n-1}$ is a cyclic permutation of degree n, that is, the graph G_f reduces to a single n-vertex circuit, if and only if the graph (X, T) is a tree.*

Necessity is proved as for proposition 15.

To prove sufficiency, assume that the graph (X, T) having the transpositions $t_1, t_2, ..., t_{n-1}$ as edges, is a tree; now consider the sequence of permutations $g_1 = t_1$, $g_2 = g_1 t_2$, $g_3 = g_2 t_3$, ..., $g_{n-1} = f$. We first show that the two elements i and

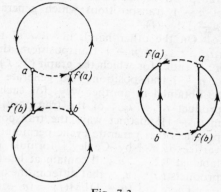

Fig. 7.3

j in the transposition $t_q = [i, j]$ are not located on the same circuit in the graph of the permutation $t_1 t_2 ... t_{q-1}$, for $q = 2, ..., n - 1$.

If i and j belong to the same circuit in the graph of the permutation $t_1 t_2 ... t_{q-1}$, this means that in this sequence of transpositions there exist transpositions of the form $[i, i_1]$, $[i_1, i_2]$, ..., $[i_k, j]$, with $k \geqslant 1$, which corresponds to a path from j to i. But in this case the unoriented graph (X, T) contains the edges $[i, i_1]$, ..., $[i_k, j]$ and $[i, j] = t_q$, hence it contains a cycle, thus contradicting the hypothesis that it is a tree.

Since i and j do not belong to the same circuit in the graph of the permutation $g_{q-1} = t_1 t_2 ... t_{q-1}$, it follows that the graph of the permutation $g_q = g_{q-1} t_q$ is obtained from the graph of the permutation g_{q-1} by concatenation of two circuits that contain the vertices i and j, respectively, hence the number of its connected components is $p(g_q) = p(g_{q-1}) - 1$. Now $p(G_{g_1}) = n - 1$, because G_{g_1} consists of one circuit with two vertices and $n - 2$ circuits containing one vertex each (loops), therefore $p(G_{g_2}) = n - 2$, $p(G_{g_3}) = n - 3$, ..., $p(G_f) = p(G_{g_{n-1}}) = 1$. Since the graph G_f is connected, having a unique connected component, it consists of a single circuit and f is a cyclic permutation of degree n.

PROPOSITION 17. *The number of possibilities of writing a cyclic permutation of degree n as a product of $n - 1$ transpositions is equal to n^{n-2}.*

In order to prove this result, notice that the number of cyclic permutations of degree n is $(n - 1)!$. The n symbols that form a cyclic permutation of degree n can be arranged in $n!$ distinct ways, but the same cyclic permutation f can be written in n distinct ways, choosing as first symbol of the cycle any of the n

symbols. Therefore the number of cyclic permutations of degree n is $\dfrac{n!}{n} = (n-1)!$; the number of possibilities of writing each cyclic permutation f as a product of transpositions will be denoted by $A(f)$. We thus obtain that the number of products of $n-1$ transpositions which generate all cyclic permutations of degree n is $(n-1)!\; A(f)$.

On the other hand, in view of the previous proposition, the only products $t_1 t_2 \ldots t_{n-1}$ of $n-1$ transpositions that generate a cyclic permutation of degree n are those for which the graph (X, T) is a tree. Therefore the number of products of $n-1$ transpositions that generate all cyclic permutations of degree n can also be obtained in another way: for each labelled tree (X, T) of n vertices, form the product $t_1 t_2 \ldots t_{n-1}$ of the transpositions associated with the edges and permute in $(n-1)!$ distinct ways the transpositions $t_1, t_2, \ldots, t_{n-1}$, thus obtaining distinct products that generate cyclic permutations. But the number of trees with n labelled vertices is n^{n-2} by Cayley's formula (Chap. 6). Two such trees, with vertices labelled $1, 2, \ldots, n$ will contain at least two distinct edges, hence they will generate products $t_1 t_2 \ldots t_{n-1}$ that differ in the nature of their factors.

Therefore $(n-1)!\; A(f) = (n-1)!\; n^{n-2}$, hence $A(f) = n^{n-2}$, that is, the number of ways of writing a cyclic permutation of degree n as a product of $n-1$ transpositions is equal to the number of trees with n labelled vertices.

Notice also, with respect to cyclic permutations, that the number of Hamiltonian cycles of a complete graph with n vertices, i.e. the number of cycles which pass exactly once through each of the n vertices, is equal to $\dfrac{1}{2}(n-1)!$. Indeed, each Hamiltonian cycle of a complete graph with n vertices generates two cyclic permutations of degree n, corresponding to the two possible senses of traversing the cycle, and this generates irredundantly all $(n-1)!$ cyclic permutations of degree n.

Problems

1.   Show that the number of nonisomorphic oriented loop-free graphs with n nodes is given by the formula

$$\sum_{(d)} \frac{2^{H_d}}{N_d};$$

where the sum is over all solutions (d) in nonnegative integers of the equation

$$d_1 + 2d_2 + \ldots + nd_n = n,$$

and

$$H_d = \sum_{k,\, l=1}^{n} d_k d_l (k, l) - \sum_{k=1}^{n} d_k;$$

$$N_d = 1^{d_1} d_1! \; 2^{d_2}\, d_2! \ldots n^{d_n} d_n!.$$

2. An oriented graph is complete if for any two different vertices x and y, there exists an arc (x, y) or (y, x) or both. Prove that the number of nonisomorphic oriented complete loop-free graphs with n nodes is given by the formula

$$\sum_{(d)} \frac{3^{F_d}}{N_d};$$

where the sum is over all solutions (d) in nonnegative integers of the equation

$$d_1 + 2d_2 + \ldots + nd_n = n,$$

and

$$F_d = \frac{1}{2}\left\{\sum_{k,l=1}^{n} d_k d_l(k, l) - \sum_{k \text{ odd}} d_k - 2\sum_{k \text{ even}} d_k\right\}.$$

(Harary, 1957)

3. Show that the number of nonisomorphic nonoriented loop-free graphs with n nodes is given by the formula

$$\sum_{(d)} \frac{2^{G_d}}{N_d};$$

where the sum is over all solutions (d) in nonnegative integers of the equation

$$d_1 + 2d_2 + \ldots + nd_n = n,$$

and

$$G_d = \frac{1}{2}\left\{\sum_{k,l=1}^{n} d_k d_l(k, l) - \sum_{k \text{ odd}} d_k\right\}.$$

4. Prove that the minimum number of transpositions necessary for expressing a permutation $f \in S_n$, $f \neq e$, as a product of transpositions, equals $n - c(f)$, where $c(f)$ is the number of cycles of f (including the cycles of length 1).
H i n t: Establish this property for cyclic permutations.

5. An up-down permutation of the set $\{1, 2, \ldots, n\}$ is a permutation $\begin{pmatrix} 1 & 2 & \ldots & n \\ a_1 & a_2 & \ldots & a_n \end{pmatrix}$ where $a_1 < a_2, a_2 > a_3, a_3 < a_4, a_4 > a_5, \ldots$

Prove that the number of up-down permutations of the set $\{1, 2, \ldots, n\}$, denoted by A_n, satisfies the equation:

$$\sum_{n=0}^{\infty} \frac{A_n x^n}{n!} = \sec x + \tan x$$

with $A_0 = A_1 = 1$.

(André, 1879)

H i n t: Show that:

a) $2A_n = \binom{n-1}{0} A_0 A_{n-1} + \binom{n-1}{1} A_1 A_{n-2} + \binom{n-1}{2} A_2 A_{n-3} + \ldots + \binom{n-1}{n-1} A_{n-1} A_0$,

where $A_0 = A_1 = A_2 = 1$.

b) If we set $a_k = \dfrac{A_k}{k!}$, then $2na_n = a_0 a_{n-1} + a_1 a_{n-2} + \ldots + a_{n-1} a_0$, hence its generating function

$$f(x) = a_0 + a_1 x + a_2 x^2 + a_3 x^3 + \ldots$$

satisfies the differential equation $\dfrac{f'(x)}{1 + f^2(x)} = \dfrac{1}{2}$ with initial condition $f(0) = a_0 = 1$.

For further generalizations see L. Carlitz [6] and L. Carlitz, R. Scoville [7].

6.   Show that the number $c(n, k)$ of permutations of n elements that contain k cycles is equal to the coefficient of x^k in the polynomial $x(x + 1)(x+2) \ldots (x+n-1)$, i.e. to $|s(n, k)|$.

H i n t: Prove that these numbers satisfy the same recurrence relation.

7.   For $f \in S_m$ let $c(f)$ be the number of cycles (including those of length one) in the distinct cycle decomposition of the permutation f. Prove that for any positive integers m and n,

$$\frac{1}{m!} \sum_{f \in S_m} n^{c(f)} = \binom{n + m - 1}{m};$$

$$\frac{1}{m!} \sum_{f \in S_m} p(f)\, n^{c(f)} = \binom{n}{m}$$

where $p(f)$ is the signature of f.

(Marcus, 1970)

H i n t: Use the previous result.

8.   A circular permutation of the numbers $1, 2, \ldots, n$, denoted by $[a_1, a_2, \ldots, a_n]$, has a rise in a_i $(1 \leqslant i \leqslant n)$ if $a_i > a_{i-1}$ and $a_i > a_{i+1}$, where the indices are considered modulo n.

Prove that the number of circular permutations of the numbers $1, 2, \ldots, n$ with k rises, denoted $M(n, k)$, satisfies the following recurrence relation

$$M(n, k) = 2kM(n - 1, k) + (n + 1 - 2k)\, M(n - 1, k - 1).$$

9.   Prove that $\displaystyle\sum_{f \in S_n} I(f) = \frac{1}{2} n! \binom{n}{2}$.

H i n t: If $f_1 = \begin{pmatrix} 1 & 2 & \ldots & n \\ a_1 & a_2 & \ldots & a_n \end{pmatrix}$ and $f_2 = \begin{pmatrix} 1 & 2 & \ldots & n \\ a_n & a_{n-1} & \ldots & a_1 \end{pmatrix}$ then $I(f_1) + I(f_2) = \binom{n}{2}$.

10.   Let $p(n, k)$ denote the number of permutations $f \in S_n$ having $I(f) = k$ inversions.

Prove that:

i) $p(n + 1, k) = p(n, k) + p(n, k - 1) + \ldots + p(n, k - n)$ where $p(n, i) = 0$ for $i > \binom{n}{2}$ or $i < 0$, and $p(n, 0) = 1$;

ii) $p(n, k) = p(n, k - 1) + p(n - 1, k)$ for $k < n$;

iii) $p(n, k) = p\left(n, \binom{n}{2} - k\right)$;

iv) $\displaystyle\sum_{0 \leqslant k \leqslant \binom{n}{2}} p(n, k) x^k = (1 + x)(1 + x + x^2) \ldots (1 + x + x^2 + \ldots + x^{n-1})$.

11.   Let $X = \{1, 2, \ldots, n\}$ and let $D(n)$ denote the number of permutations of X with no fixed points (derangements). Let $E(n)$ denote the number of even permutations of X with no fixed points. Show that

$$E(n) = \frac{1}{2} \left[D(n) - (-1)^n (n - 1) \right].$$

H i n t: Use the principle of inclusion and exclusion and formula (3.8) for $D(n)$.

12.   Prove that the number of elementary cycles of the complete graph K_n is equal to

$$\frac{1}{2} \sum_{k=3}^{n} \frac{[n]_k}{k}.$$

13.   Prove Cauchy's identity

$$\sum_{\substack{c_1 + 2c_2 + \ldots = n \\ c_i \geqslant 0}} \frac{1}{c_1! \, c_2! \, \ldots \, 1^{c_1} 2^{c_2} \ldots} = 1.$$

14.   Prove that the expression $d(f, g) = \max\limits_{i=1,\ldots,n} |f(i) - g(i)|$ where f, g are two permutations of the set $\{1, 2, \ldots, n\}$, defines a distance in the set S_n. If $f(n, r)$ stands for the number of permutations f with the property $d(e, f) \leqslant r$, where e is the identity permutation, show that $f(n, 1) = F_n$ (the Fibonacci number).

15.   Let P_n be the number of permutations $f \in S_n$ such that $f^2 = e$. Show that:

i) $P_n = P_{n-1} + (n - 1)P_{n-2}$; $P_0 = P_1 = 1$;

ii) $\displaystyle\sum_{n \geqslant 0} P_n \frac{t^n}{n!} = \exp\left(t + \frac{t^2}{2}\right)$;

iii) $P_n = n! \displaystyle\sum_{i + 2j = n} (i! \, j! \, 2^j)^{-1}$.

If $P_n(m)$ denotes the number of permutations $f \in S_n$ with the property $f^m = e$, show that

$$\sum_{n=0}^{\infty} P_n(m) \frac{t^n}{n!} = \exp\left(\sum_{k \mid m} \frac{t^k}{k}\right).$$

(Jacobstahl, 1949; Chowla, Herstein, Scott, 1952; Moser, Wyman, 1955)

16.   Prove that the number of distinct ways of labelling the vertices of a graph G with n vertices is equal to $\dfrac{n!}{|\text{Aut }(G)|}$, where Aut (G) is the group of automorphisms of G.

H i n t: Use Burnside's lemma.

17.   Let $d(n, k)$ be the number of permutations $f \in S_n$ with no fixed points which contain k cycles. Prove that:

i) $d(n + 1, k) = n(d(n, k) + d(n - 1, k - 1))$ where $d(0, 0) = 1$;

(Appell, 1880)

ii) $d(n, k) = \displaystyle\sum_{j=0}^{n} (-1)^j \binom{n}{j} c(n - j, k - j)$

where $c(n, k) = |s(n, k)| = (-1)^{n+k} s(n, k)$.

(Kaucky, 1971)

H i n t: Use Problem 6 and the principle of inclusion and exclusion.

BIBLIOGRAPHY

1. André, D., *Développement de* sec x *et* tg x, C. R. Acad. Sci. Paris, **88**, 1879, 965−967.
2. André, D., *Sur les permutations alternées*, J. Math. Pures Appl., **7**, 1881, 167−184.
3. Appell, P., *Développement en série entière de* $(1 + ax)^{1/x}$, Arch. Math. Phys. (ed: Grunert), **65**, 1880, 171−175.
4. Berge, C., *Principes de combinatoire*, Dunod, Paris, 1968.
   English edition: *Principles of combinatorics*, Academic Press, New York, 1971.
5. Burnside, W., *Theory of groups of finite order*, 2nd ed., Cambridge, 1911; reprinted Dover, New York, 1955.
6. Carlitz, L., *Enumeration of up-down sequences*, Discrete Math., **4**, 1973, 273−286.
7. Carlitz, L., Scoville, R., *Up-down sequences*, Duke Math. J., **39**, 1972, 583−598.
8. Chowla, S. D., Herstein, I. N., Scott, W. R., *The solutions of* $x^d = 1$ *in symmetric groups*, Norske Vid. Selsk. Forh. (Trondheim), **25**, 1952, 29−31.
9. Davis, R. L., *The number of structures of finite relations*, Proc. Amer. Math. Soc., **4**, 1953, 486−495.
10. Davis, R. L., *Structures of dominance relations*, Bull. Math. Biophys., **16**, 1954, 131−140.
11. Dénes, J., *The representation of a permutation as the product of a minimal number of transpositions, and its connection with the theory of graphs*, Magyar. Tud. Akad. Mat. Kutató Int. Közl., **4**, 1959, 63−71.
12. Dénes, J., *Some combinatorial properties of transformations and their connections with the theory of graphs*, J. Combinatorial Theory, **9**, 1970, 108−116.

13. Entringer, R. C., *A note on enumeration of permutations of* $(1, \ldots, n)$ *by number of maxima*, Gac. Mat. (Madrid), (1) **23**, 1971, 67—69.
14. Freese, R., *Solution to advanced problem* N° 5751, Amer. Math. Monthly, **78**, 1971, 1028—1029.
15. Harary, F., *The number of oriented graphs*, Michigan Math. J., **4**, 1957, 221—224.
16. Jacobstahl, E., *Sur le nombre d'éléments de* S_n *dont l'ordre est un nombre premier*, Norske Vid. Selsk. Forh. (Trondheim), **21**, 1949, 49—51.
17. Kauckẏ, J., *A contribution to the theory of permutations*, Mat Časopis Sloven. Akad. Vied, **21**, 1971, 82—86.
18. Kazuaki, H., *Generation of rosary permutations expressed in Hamiltonian circuits*, Comm. ACM, **14**, 1971, 373—379.
19. Moon, J. W., *Topics on tournaments*, Holt, Rinehart and Winston, New York, 1968.
20. Moser, L., Wyman, *On solutions of* $x^d = 1$ *in symmetric groups*, Canad. J. Math., **7**, 1955, 159—168.
21. Netto, E., *Lehrbuch der Combinatorik*, Teubner, Leipzig, 1927.
22. Riordan, J., *An introduction to combinatorial analysis*, Wiley, New York, 1958.

CHAPTER 8

The Pólya—De Bruijn Enumeration Method

8.1. COUNTING SCHEMES WITH RESPECT TO A PERMUTATION GROUP

Let X be a set of objects which we denote by $1, 2, ..., n$ and A a set of colours which we denote by $a_1, a_2, ..., a_m$.

A function $f: X \to A$ defines a colouring of the objects from the set X, the object i being coloured with colour $f(i)$ for $i = 1, 2, ..., n$. The number of colourings is equal to the number of functions from X to A, that is, to m^n. In the first three Chapters we have counted the various colourings generated by injective, surjective and bijective functions f. We now adopt another point of view, considering a group G of permutations of the set X. If $g \in G$, the mapping $fg: X \to A$ is also a colouring obtained by composition of the two mappings $g: X \to X$ and $f: X \to A$.

We say that two colourings f_1 and f_2 belong to the same scheme and we write $f_1 \sim f_2$, if there is a permutation $g \in G$ such that $f_1 g = f_2$.

The relation thus defined is an equivalence, with respect to the permutation group G, in the set of colourings because it is:

— reflexive: $f \sim f$ for the identity permutation $e \in G$ and $f = fe$;

— symmetric: $f_1 \sim f_2$ implies the existence of a permutation $g \in G$ such that $f_1 g = f_2$, hence $f_2 g^{-1} = f_1$ by right composing on both sides with $g^{-1}: X \to X$, therefore $f_2 \sim f_1$ since $g^{-1} \in G$;

— transitive: $f_1 \sim f_2$ and $f_2 \sim f_3$ imply $f_1 \sim f_3$, because $f_1 g = f_2$ and $f_2 h = f_3$ with $g, h \in G$ imply $(f_1 g) h = f_1(gh) = f_3$, where $gh \in G$, since G is a permutation group.

In other words, two colourings are equivalent if there exists a permutation in G which transforms each into the other. This equivalence relation generates a partition of the set of colourings into equivalence classes, or schemes with respect to the group G; their number will be determined by calculating the numerical value of a polynomial known as the cycle index of the group G. If the permutation $g \in G$, denote by $\lambda_i(g)$ the number of cycles of length i of the permutation g and let the polynomial

$$P(G; x_1, x_2, ..., x_n) = \frac{1}{|G|} \sum_{g \in G} x_1^{\lambda_1(g)} x_2^{\lambda_2(g)} \cdots x_n^{\lambda_n(g)} \qquad (8.1)$$

be called the *cycle index* of the group G, where n is the number of objects in the set X.

PROPOSITION 1 *(Pólya).* The number of colouring schemes of a set of n objects with m colours, with respect to a group G of permutations of the object set, is equal to the numerical value $P(G; m, m, ..., m)$ of the cycle index when all the variables are equal to m.

If we denote by F the set of the m^n colourings $f: X \to A$, then for every permutation $g \in S_n$ the mapping $f \mapsto \bar{g}(f) = fg$ is an injection of F into F. For $f_1 \neq f_2$ implies the existence of an object $i \in X$ such that $f_1(i) \neq f_2(i)$. Set $j = g^{-1}(i)$. Then $f_1 g(j) = f_1(i) \neq f_2(i) = f_2 g(j)$ so that $f_1 g \neq f_2 g$. Therefore the mapping $\bar{g}: F \to F$ is injective and as the set F is finite, it follows that $\bar{g}$ is also a surjection, and hence a bijection. In other words $\bar{g} \in \bar{S}$, where $\bar{S}$ stands for the set of permutations of the set F. We have thus defined an $\bar{S}$-valued mapping φ of the permutation group G, by associating with each permutation $g \in G \subset S_n$, a permutation $\bar{g}$ of the colourings.

If $m \geqslant 2$ we show that the mapping $\varphi: G \to \bar{S}$ is injective. Now $g_1 \neq g_2$ implies the existence of a $k \in X$ such that $g_1(k) \neq g_2(k)$, so we must show that the functions $\bar{g}_1: F \to F$ and $\bar{g}_2: F \to F$ are different, that is, there exists $f \in F$ such that $fg_1 \neq fg_2$ (taking into account the definition of the function $\bar{g}$). If $m \geqslant 2$, then since F is the set of all functions $f: X \to A$, there exists a colouring f such that the colours of the distinct elements $g_1(k)$ and $g_2(k)$ are distinct, i.e. $f(g_1(k)) \neq f(g_2(k))$ or, equivalently, $(fg_1)(k) \neq (fg_2)(k)$, thus implying that the functions fg_1 and fg_2 are distinct, hence φ is an injection. If the number of colours is $m = 1$, then φ is not injective, but in this case the number of colourings is obviously equal to 1, i.e. all objects have the same colour, while $P(G; 1, 1, ..., 1) = 1$ from (8.1), so that the proposition is verified in this case.

If we denote the set $\varphi(G)$ by $\bar{G} = \{\bar{g} | g \in G\}$, then since φ is an injection it follows that there exists a bijection $\varphi_1: G \to \bar{G} \subset \bar{S}$ defined by $\varphi_1(g) = \varphi(g)$ for every $g \in G$, therefore $|G| = |\bar{G}|$. But the set $\bar{G}$ is a subgroup of the group $\bar{S}$ of permutations of the set F of colourings, because $\bar{g}_1 \in \bar{G}$ and $\bar{g}_2 \in \bar{G}$ imply $\bar{g}_1 \bar{g}_2 \in \bar{G}$, for

$$\bar{g}_1 \bar{g}_2(f) = \bar{g}_1(fg_2) = (fg_2)g_1 = f(g_2 g_1) = \overline{g_2 g_1}(f),$$

hence the product $\bar{g}_1 \bar{g}_2$ of two elements, $\bar{g}_1$ and $\bar{g}_2$, of $\bar{G}$, is also an element of $\bar{G}$ and corresponds to the product $g_2 g_1 \in G$. According to the definition, two colourings f_1 and f_2 are equivalent if there exists $g \in G$ such that $f_1 g = f_2$ or $\bar{g}(f_1) = f_2$, therefore they belong to the same orbit of the subgroup $\bar{G}$ of the group $\bar{S}$. Since, the colouring schemes coincide with the orbits of the group $\bar{G}$, the number of schemes is equal to the number of orbits which, according to Burnside's theorem (7.3), is

$$|C_{\bar{G}}| = \frac{1}{|G|} \sum_{g \in G} \lambda_1(\bar{g}),$$

where $\lambda_1(\bar{g})$ represents the number of fixed points of the permutation $\bar{g}$, that is, the number of colourings f such that $\bar{g}(f) = f$, i.e. $fg = f$. But $fg = f$ implies that f is constant for every cycle of the permutation f, since otherwise $fg \neq f$. Hence there exist as many colourings f with the property $fg = f$, as mappings from the set of cycles, which contains $\lambda_1(g) + \lambda_2(g) + \ldots + \lambda_n(g)$ elements, into the set of m colours; so this number is $m^{\lambda_1(g)+\cdots+\lambda_n(g)}$. Also, since $|G| = |\bar{G}|$, we obtain

$$|C_{\bar{G}}| = \frac{1}{|G|} \sum_{g \in G} m^{\lambda_1(g)+\cdots+\lambda_n(g)} = P(G; m, m, \ldots, m).$$

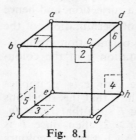

Fig. 8.1

Example 1. Colouring the cube. Consider a set X consisting of the six faces of a cube, denoted by $1, 2, \ldots, 6$ as in Fig. 8.1; the problem is to colour them with two colours, white and black, hence $A = \{a_1, a_2\}$. The vertices of the cube will be denoted by a, b, c, d, e, f, g, h. Let us find the number of colouring schemes, two coloured cubes being considered equivalent if they coincide by a certain motion.

As a matter of fact, in this case it is not too difficult to enumerate all possibilities. One finds 10 colouring schemes of the faces of the cube with two colours, namely: one scheme with white faces only, one scheme with only one black face, two schemes with two black faces, two schemes with three black faces, two schemes with four black faces, one scheme with five black faces, one scheme with black faces only. The two schemes with two black faces are obtained by taking as black faces the faces $(1, 2)$ and $(1, 3)$, respectively, the two schemes with three black faces are obtained by taking as black faces the faces $(1, 2, 5)$ and $(1, 3, 5)$, respectively, while the two schemes with four black faces are obtained from the two schemes with two black faces by interchanging the roles of the colours black and white.

In order to apply Pólya's method for determining the number of equivalence classes of colourings, i.e. of mappings $f: X \to A$, we determine the group G of rotations of the cube, which are:

— about the axis $abcd-efgh$: $[2, 6, 4, 5]$; $[2, 4]$ $[6, 5]$; $[2, 5, 4, 6]$;
— about the axis $bcfg-adhe$: $[1, 5, 3, 6]$; $[1, 3]$ $[5, 6]$; $[1, 6, 3, 5]$;
— about the axis $abfe-dcgh$: $[1, 2, 3, 4]$; $[1, 3]$ $[2, 4]$; $[1, 4, 3, 2]$;
— about the axis $a-g$: $[1, 4, 5]$ $[6, 3, 2]$; $[1, 5, 4]$ $[6, 2, 3]$;
— about the axis $b-h$: $[1, 5, 2]$ $[6, 4, 3]$; $[1, 2, 5]$ $[6, 3, 4]$;
— about the axis $c-e$: $[1, 2, 6]$ $[3, 4, 5]$; $[1, 6, 2]$ $[3, 5, 4]$;
— about the axis $d-f$: $[1, 6, 4]$ $[3, 5, 2]$; $[1, 4, 6]$ $[3, 2, 5]$;
— about the axis $ab-hg$: $[1, 5]$ $[3, 6]$ $[2, 4]$;
— about the axis $bc-eh$: $[1, 2]$ $[3, 4]$ $[5, 6]$;
— about the axis $cd-ef$: $[1, 6]$ $[3, 5]$ $[2, 4]$;
— about the axis $ad-fg$: $[1, 4]$ $[2, 3]$ $[5, 6]$;
— about the axis $bf-dh$: $[2, 5]$ $[6, 4]$ $[1,3]$;
— about the axis $cg-de$: $[2, 6]$ $[5, 4]$ $[1, 3]$.

The rotation about the axis $bcfg-adhe$ means, in fact, the rotation about the axis determined by the centre of the squares $bcfg$ and $adhe$, the rotation about the axis $ab-hg$ is the rotation about the axis determined by the mid-points of the edges ab and hg, etc. Since cycles of length 1 are usually omitted, the permutation $[2, 6, 4, 5]$ must be read $[1]$ $[3]$ $[2, 6, 4, 5]$, the permutation $[2, 4]$ $[6, 5]$ must be interpreted as $[1]$ $[3]$ $[2, 4]$ $[6, 5]$, etc.

The above permutations together with the identity permutation $e = [1]$ $[2]$ $[3]$ $[4]$ $[5]$ $[6]$ form a group G of 24 permutations written as products of cycles, therefore the cycle index in the case of the group of rotations of the cube is

$$P(G; x_1, x_2, x_3, x_4, x_5, x_6) = \frac{1}{24} (x_1^6 + 3x_1^2 x_2^2 + 6x_1^2 x_4 + 6x_2^3 + 8x_3^2). \qquad (8.2)$$

By using Pólya's theorem, we find that the number of colouring schemes of the faces of the cube with two colours is equal to

$$P(G;\ 2, 2, 2, 2, 2, 2) = \frac{1}{24}\ (2^6 + 3 \cdot 2^2 \cdot 2^2 + 6 \cdot 2^2 \cdot 2 + 6 \cdot 2^3 + 8 \cdot 2^2) = 10.$$

After the determination of the cycle index of the group G of permutations of the objects, the number of colouring schemes with no matter how many colours is obtained by a simple algebraic calculus; this is not the case for direct enumeration, when an appropriate reasoning must be applied to each case separately.

Example 2. The arrangements of a collection of objects that may contain identical objects. Consider a multiset $X = \{1, 2, 3, 4\}$, the objects 1 and 2 being identical spheres B, while 3 and 4 are identical cubes C, which we want to locate in two identical cells a and b, empty cells being allowed. An arrangement is a mapping $f: X \to \{a, b\}$. The order of the objects in a cell is immaterial. We find by direct enumeration the following nine possible schemes of arrangements, where we have denoted the empty set by $\varnothing$:

$$BB|CC \qquad BBCC|\varnothing \qquad BBC|C$$
$$BC|BC \qquad \varnothing|BBCC \qquad BCC|B$$
$$C|BBC \qquad B|BCC \qquad CC|BB.$$

Since the objects 1 and 2 are identical and the objects 3 and 4 are also identical, it follows that the group G consists of four permutations, namely: [1] [2] [3] [4], [1, 2] [3] [4], [1] [2] [3, 4] and [1, 2] [3, 4]. The cycle index is thus

$$P(G;\ x_1,\ x_2,\ x_3,\ x_4) = \frac{1}{4}\ (x_1^4 + 2x_1^2 x_2 + x_2^2).$$

The number of distinct arrangements in cells is equal to

$$P(G;\ 2, 2, 2, 2) = \frac{1}{4}\ (2^4 + 2 \cdot 2^2 \cdot 2 + 2^2) = 9,$$

as we obtained by direct enumeration.

8.2. Determination of the Weights of Schemes Invariant
with Respect to a Permutation
of the Colours

If $f: X \to A$ is a colouring of the objects from X with the colours from A, and G is a group of permutations of the objects in X, denote by $\bar{f}$ the colouring scheme of f with respect to the group G, i.e. the set of those colourings $f_1: X \to A$ for which there exists a permutation $g \in G$ such that $f_1 g = f$. If h is a permutation of the colours, that is, a bijection $h: A \to A$, then the function obtained by composing h with $f: X \to A$ is a colouring as well, since $hf: X \to A$. We say that a colouring $f_1 \in \bar{f}$ is invariant with respect to the permutation h of the colours, if the colouring $hf_1 \in \bar{f}$, i.e. if the colouring obtained from f_1 by the permutation h of the colours remains in the same equivalence class or colouring scheme as f_1. If $f_1 \in \bar{f}$ is left invariant by the permutation h of the colours, it follows that any other colouring $f_2 \in \bar{f}$ is invariant with respect to the permutation h of the colours. For, if $hf_1 \sim f$, there exists a permutation $g_1 \in G$ such that $hf_1g_1 = f$ and since $f_1 \sim f_2$, there exists $g_2 \in G$ such that $f_1 = f_2g_2$, hence we obtain $hf_2(g_2g_1) = f$ by substitution. Since G is a group, it follows that $g_2g_1 \in G$, hence $hf_2 \sim f$, that is, $hf_2 \in \bar{f}$, therefore f_2 is invariant with respect to the permutation h of the colours.

Therefore if in the colouring scheme $\bar{f}$ there is a colouring left invariant by the permutation h of the colours, then all the colourings of the scheme are invariant with respect to that permutation of the colours and we will say that $\bar{f}$ is invariant with respect to the permutation h.

With each colour a_i associate a weight $w(a_i) > 0$ $(i = 1, 2, ..., m)$. With a colouring f in which colour a_i appears r_i times $(i = 1, 2, ..., m)$ associate the multiplicative weight

$$W(f) = w(a_1)^{r_1} w(a_2)^{r_2} ... w(a_m)^{r_m}.$$

Clearly all the colourings of the same scheme $\bar{f}$, differing only by a permutation of the objects and not of the colours, will have the same weight. For this reason we define the weight of a colouring scheme $\bar{f}$ as being $W(\bar{f}) = W(f_1)$, where $f_1 \in \bar{f}$. We shall evaluate the sum of the weights of all colouring schemes that are left invariant by a permutation h of the colours. If we take $w(a_i) = 1$ for $i = 1, ..., m$, this sum will be precisely the number of schemes invariant with respect to the permutation h, for in this case $W(\bar{f}) = 1$ for every scheme $\bar{f}$.

PROPOSITION 2 (de Bruijn). *The sum of the weights of all colouring schemes left invariant by a permutation h of the colours is equal to* $P(G; p_1, p_2, ..., p_n)$, *where the polynomial*

$$P(G; x_1, x_2, ..., x_n) = \frac{1}{|G|} \sum_{g \in G} x_1^{\lambda_1(g)} x_2^{\lambda_2(g)} ... x_n^{\lambda_n(g)}$$

is the cycle index of the group G of permutations of the objects, while

$$p_k = \sum_{\substack{a \in A \\ h^k a = a}} w(a) \, w(ha) \, w(h^2 a) ... w(h^{k-1} a) \qquad (8.4)$$

for $k = 1, 2, ..., n$ and $p_k = 0$ if the set $\{a | h^k a = a\}$ is empty.

The sum of the weights of the colouring schemes $\bar{f}$ invariant with respect to the permutation h of the colours will be denoted by $\sum_{\bar{f}} W(\bar{f}) = \sum_{\bar{f}} \sum_{f \in \bar{f}} \frac{W(f)}{|\bar{f}|}$, the first summation being performed with respect to all schemes $\bar{f}$ left invariant by the permutation h of the colours, because all the colourings of an equivalence class have the same weight. But the equivalence class $\bar{f}$ of the colouring f with respect to the group G of permutations of the objects is in fact the orbit of the element $f \in F$ with respect to the group $\bar{G}$, where F is the set of colourings of X and $\bar{G}$ consists of the permutations $\bar{g}$ defined by $\bar{g}(f) = fg \in F$, with the notation from the proof of proposition 1. Using formula (7.2), we obtain

$$|\bar{f}| = |O_f| = \frac{|\bar{G}|}{|\bar{G}_f|} = \frac{|G|}{|\bar{G}_f|},$$

where $\bar{G}_f$ is the set of permutations $\bar{g} \in \bar{G}$ that leave f invariant, i.e. $\bar{g}(f) = f$; we have also used the property $|\bar{G}| = |G|$, which we know from the proof of proposition 1. We thus obtain

$$\sum_{\bar{f}} W(\bar{f}) = \frac{1}{|G|} \sum_{\bar{f}} \sum_{f \in \bar{f}} W(f) \, |\bar{G}_f|.$$

As we have seen, $|\bar{G}_f|$ represents the number of permutations $\bar{g} \in \bar{G}$ such that $\bar{g}(f) = f$; taking into account the definition of the permutation $\bar{g}$ and the bijection between G and $\bar{G}$, we see that this number is equal to the number of permutations $g \in G$ such that $fg = f$, since $\bar{g}(f) = fg$.

We now prove that the number of permutations $g \in G$ such that $fg = f$ is equal to the number of permutations $g \in G$ such that $hfg = f$, if the scheme $\bar{f}$ of the colouring f is invariant with respect to the permutation h of the colours. In this case there is a permutation $g_1 \in G$ such that $hf = fg_1$, since $hf \sim f$ and the relation $hfg = f$ becomes $f(g_1 g) = f$. We have thus established a mapping φ from the set of permutations $g \in G$ such that $hfg = f$ to the set of permutations $g \in G$ with the property $fg = f$, defined by $\varphi(g) = g_1 g$ if $hf \sim f$, since $g_1 g \in G$, G being a group.

This mapping is an injection, because $g' \neq g''$ and $hfg' = hfg'' = f$ imply $\varphi(g') \neq \varphi(g'')$, that is, $g_1 g' \neq g_1 g''$. For the contrary assumption $g_1 g' = g_1 g''$ would imply, by left multiplication with g_1^{-1}, that $g' = g''$, thus contradicting the hypothesis. The mapping φ is also a surjection, for every permutation $g \in G$ with $fg = f$ is the image by the function φ of the permutation $g_1^{-1} g \in G$ with the property $hf(g_1^{-1} g) = f$. Indeed, $hf(g_1^{-1} g) = fg_1 g_1^{-1} g = fg = f$ and $\varphi(g_1^{-1} g) = g_1(g_1^{-1} g) = g$.

Therefore in formula (8.4) we can take $|G_{h,f}|$ instead of $|\bar{G}_f|$, where $G_{h,f}$ stands for the set of permutations $g \in G$ such that $hfg = f$. But

$$W(f)| \, G_{h,f}| = \sum_{g \in G_{h,f}} W(f),$$

hence (8.4) becomes

$$\sum_{\bar{f}} W(\bar{f}) = \frac{1}{|G|} \sum_{\bar{f}} \sum_{f \in \bar{f}} \sum_{g \in G_{h,f}} W(f),$$

since $g \in G_{h,f}$ implies that the colouring scheme $\bar{f}$ is invariant with respect to the permutation h, for the existence of a permutation $g \in G$ such that $g \in G_{h,f}$, i.e. $hfg = f$, implies $hf = fg^{-1}$ with $g^{-1} \in G$, that is, $hf \in \bar{f}$. By changing the summation order, we obtain

$$\sum_{\bar{f}} W(f) = \frac{1}{|G|} \sum_{g \in G} \sum_{\substack{f \\ hfg = f}} W(f). \tag{8.5}$$

To evaluate the last sum, notice that the cycles of permutation g form a partition $(X_1, X_2, ..., X_s)$ of the set X of objects and choose s elements $x_i \in X_i$ for $i = 1, 2, ..., s$.

If $hfg = f$ and $k(i)$ is the length of that cycle of the permutation g which contains x_i, then setting $f(x_i) = a$, the colours of the set X_i are $f(x_i) = a$, $fg^{-1}(x_i) = hf(x_i) = h(a)$, $fg^{-2}(x_i) = h^2 f(x_i) = h^2(a)$, ..., $h^{k(i)-1}(a)$, respectively, while $fg^{-k(i)}(x_i) = h^{k(i)}(a) = a$, since $g^{-k(i)}(x_i) = x_i$. We have taken into account that $hfg = f$, hence $fg^{-1} = hf$, $fg^{-2} = fg^{-1}g^{-1} = hfg^{-1} = hhf = h^2 f$, etc. It follows that the weight of the colouring f is equal to

$$W(f) = \prod_{i=1}^{s} p_{k(i)}(f(x_i)),$$

where we have set $p_k(a) = w(a) w(ha) w(h^2 a) \ldots w(h^{k-1} a)$ if $a = h^k a$ and $p_k(a) = 0$ if $a \neq h^k a$. It we denote by A_k the set of colours $\{a | a \in A, h^k a = a\} = \{a_k^1, a_k^2, ..., a_k^{p_k}\}$; then $hfg = f$ implies that the colouring f is completely determined by the colours of the elements x_i for $i = 1, 2, ..., s$, i.e. it is completely determined by the s-tuple $(j_1, j_2, ..., j_s)$ through the equation $f(x_i) = a_{k(i)}^{j_i}$ for $i = 1, 2, ..., s$. We can therefore write

$$\sum_{hfg = f} W(f) = \sum_{(j_1, ..., j_s)} \sum_{x_i} p_{k(1)}(a_{k(1)}^{j_1}) p_{k(2)}(a_{k(2)}^{j_2}) \ldots p_{k(s)}(a_{k(s)}^{j_s}).$$

But this sum of products is the result of calculating the product of sums

$$\prod_{i=1}^{s} \prod_{x_i} [p_{k(i)}(a_{k(i)}^1) + p_{k(i)}(a_{k(i)}^2) + \ldots]$$

$$= \prod_{i=1}^{s} \prod_{x_i} \sum_{a \in A_{k(i)}} w(a) w(ha) \ldots w(h^{k(i)-1} a) = \prod_{i=1}^{s} \prod_{x_i} p_{k(i)} = p_1^{\lambda_1} p_2^{\lambda_2} \ldots p_n^{\lambda_n},$$

because the product is completed by factors equal to 1 for the cycles with $\lambda_i = 0$, for in this case $p_i^{\lambda_i} = 1$. We have taken into account that the product over x_i is evaluated over the $k(i)$ possibilities for choosing x_i in the cycle of length k_i of the permutation g. By introducing this value into (8.5), we obtain the statement of the proposition.

For other generalizations see [6], [31], [42].

Example 3. In the case of the colouring of the cube, the cycle index is given by formula (8.2).

The number of schemes with two symmetrically distributed colours a_1 and a_2, i.e. invariant with respect to the transposition $[a_1, a_2]$, is obtained by taking $h = [a_1, a_2]$ and $w(a_1) = w(a_2) = 1$. Since $h^2 = e$, one gets $h^2a = h^4a = h^6a = a$ for every $a \in \{a_1, a_2\}$, whereas $h^{2k+1}a \neq a$ for any $a \in \{a_1, a_2\}$ and $k = 0, 1, 2$, therefore $p_1 = p_3 = p_5 = 0$ and $p_2 = p_4 = p_6 = 2$. Therefore the number of colourings with two symmetrically distributed colours is $P(G; 0, 2, 0, 2, 0, 2) = \frac{1}{24}(6 \cdot 2^3) = 2$. This can be verified by direct enumeration of the cases when three faces of the cube are of colour a_1 and the other three faces of colour a_2.

Example 4. Colour the faces of the cube with six distinct colours $a_1, a_1', a_2, a_2', a_3, a_3'$ such that the permutation $h = [a_1, a_1'][a_2, a_2'][a_3, a_3']$ leaves invariant the colouring schemes relative to the group of rotations of the three-dimensional cube.

Setting $w(a_i) = w_i$ and $w(a_i') = w_i'$ for $i = 1, 2, 3$, we shall calculate the numbers p_k with the aid of formula (8.3):

$$p_1 = 0; \ p_2 = 2w_1w_1' + 2w_2w_2' + 2w_3w_3'; \ p_3 = p_5 = 0,$$

since h is a product of transpositions with no common elements. The numbers p_4 and p_6 do not appear in this calculation, because $p_1 = p_3 = p_5$ and

$$P(G; \ p_1, \ p_2, \ p_3, \ p_4, \ p_5, \ p_6) = \frac{1}{24} \cdot 6p_2^3 = 2(w_1w_1' + w_2w_2' + w_3w_3')^3$$

$$= 2 \sum_{i+j+k=3} \frac{3!}{i!j!k!} \ (w_1 \ w_1')^i (w_2 \ w_2')^j (w_3 \ w_3')^k.$$

Since the sum of the weights of the schemes invariant with respect to the permutation h has been obtained before, the required number of invariant schemes that utilize each colour $a_1, a_1', a_2, a_2', a_3, a_3'$, once only is precisely the coefficient of $w_1w_1' \ w_2w_2' \ w_3w_3'$ in this sum, which is obtained when $i = j = k = 1$; that is, 12 invariant schemes. Each invariant scheme which makes use of all 6 of the colours, has a weight equal to $w_1w_1' \ w_2w_2' \ w_3w_3'$ and the coefficient of this product will be precisely the number of invariant schemes which utilize each of the six colours once only. The number of invariant schemes that utilize each of the colours a_1 and a_1' twice and each of the colours a_2 and a_2' once for colouring the faces of the cube, is equal to the coefficient of $w_1^2w_1'^2 \ w_2w_2'$, which is obtained when $i = 2$, $j = 1$ and $k = 0$, that is, 6 schemes. The number of schemes invariant with respect to h, which utilize each of the colours a_1 and a_1' three times, is equal to the coefficient of $w_1^3w_1'^3$, i.e. to 2, as was seen in the previous example.

PROPOSITION 3 (*Pólya*). *Setting* $w_i = w(a_i)$, *the sum of the weights of all schemes relative to a group G of permutations of the objects, is equal to*

$$P\left(G; \sum_{i=1}^{m} w_i, \ \sum_{i=1}^{m} w_i^2, ..., \ \sum_{i=1}^{m} w_i^n\right), \tag{8.6}$$

where $P(G; x_1, x_2, ..., x_n)$ *is the cycle index of the group G, while the number of colouring schemes f which utilize the colour* a_i k_i *times, for* $i = 1, 2, ..., m$, *is the coefficient of* $w_1^{k_1} w_2^{k_2} ... w_m^{k_m}$ *from the expansion of the polynomial* (8.6).

If the permutation h of the colours is the identity permutation, then every colouring scheme relative to the permutation group G is invariant with respect to h and the sum of weights of all schemes with respect to G is given by the previous proposition. Since h is the identity permutation, $h^i a = a$ for every $a \in A$ and positive integer i, therefore $p_k = \sum_{i=1}^{m} w_i^k$ and the result is deduced from de Bruijn's theorem.

Every colouring scheme with respect to G that utilizes the colour a_i k_i times for $i = 1, 2, \dots, m$ has the weight $w_1^{k_1} w_2^{k_2} \dots w_m^{k_m}$ and since this correspondence between the schemes that utilize each colour a certain number of times and the products of variables $w_1, \dots, w_m$ raised to various powers, is injective, it follows that the number of schemes that utilize the colour a_i $(i = 1, \dots, m)$ k_i times is the coefficient of $w_1^{k_1} \dots w_m^{k_m}$ from the expansion of the polynomial (8.6).

We have thus obtained an efficient algorithm for the determination of the number of schemes which utilize certain colours a certain number of times; this algorithm starts from the cycle index of the group G and reduces to a calculation of polynomials and of certain coefficients.

Example 5 *(Pólya).* Given six spheres, two of colour a, one of colour b and three of colour c, find the number of ways in which we can assign them to the vertices of a free regular octahedron (that can be submitted to motions).

Let 1, 2, 3, 4, 5, 6 be the vertices of the octahedron. The motions of the octahedron, that is the rotations about certain axes, correspond to certain permutations of the set of its vertices.

For example, the permutation $g = [4, 5, 1, 6]$ [2] [3] corresponds to a rotation of 90° about the axis $2-3$ (Fig. 8.2).

The cycle index of the group $G \subset S_6$ of rotations of the octahedron is

$$P(G; x_1, x_2, x_3, x_4, x_5, x_6) = \frac{1}{24} (x_1^6 + 6x_1^2 x_4 + 3x_1^2 x_2^2 + 6x_2^3 + 8x_3^2). \qquad (8.7)$$

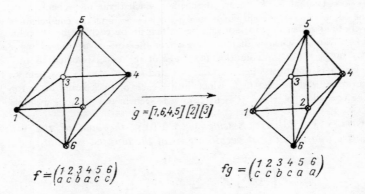

Fig. 8.2

Setting $w(a) = w_1$, $w(b) = w_2$ and $w(c) = w_3$, and substituting

$$x_1 = w_1 + w_2 + w_3, \quad x_2 = w_1^2 + w_2^2 + w_3^2, \quad x_3 = w_1^3 + w_2^3 + w_3^3 \text{ and } x_4 = w_1^4 + w_2^4 + w_3^4,$$

the required number of colourings will be the coefficient of $w_1^2 w_2 w_3^3$ in the polynomial

$$\frac{1}{24} [(w_1 + w_2 + w_3)^6 + 6(w_1 + w_2 + w_3)^2 (w_1^4 + w_2^4 + w_3^4)$$

$$+ 3(w_1 + w_2 + w_3)^2 (w_1^2 + w_2^2 + w_3^2)^2 + 6(w_1^2 + w_2^2 + w_3^2)^3 + 8(w_1^3 + w_2^3 + w_3^3)^2].$$

By using the multinomial formula (2.7), we find that this coefficient is equal to

$$\frac{1}{24} \left(\frac{6!}{2!1!3!} + 3 \frac{2!}{0!1!1!} \cdot \frac{2!}{1!0!1!} \right) = \frac{1}{24} (60 + 12) = 3.$$

This method enables us, for example, to count the various isomers of substances, isomers being compounds with the same chemical composition but with a different arrangement of the atoms, which results in different properties of the compounds.

It is known, for example, that the four valences of the carbon atom can be satisfied by two chlorine atoms and by CH_3 (methyl) and C_2H_5 (ethyl), located on the vertices of a regular tetrahedron, the carbon atom being situated in the centre of the tetrahedron.

8.3. DETERMINATION OF THE CYCLE INDEX

We present below without proofs the formulae from [18] which enable the calculation of the cycle index in certain cases:

1) The identity group $E_n = \{e\}$ contains only the identity permutation of n objects and $P(E_n; x_1) = x_1^n$.

2) The symmetric group S_n of degree n contains $n!$ permutations. Using Cauchy's formula (7.1) which gives the number of permutations of type $1^{\lambda_1} 2^{\lambda_2} \ldots n^{\lambda_n}$, we obtain

$$P(S_n; x_1, \ldots, x_n) = \sum_{\lambda_1 + 2\lambda_2 + \ldots + n\lambda_n = n} \frac{1}{1^{\lambda_1}\lambda_1! \, 2^{\lambda_2} \lambda_2! \ldots n^{\lambda_n} \lambda_n!} x_1^{\lambda_1} x_2^{\lambda_2} \ldots x_n^{\lambda_n}.$$

3) The alternating group A_n consists of the $\frac{1}{2} n!$ even permutations of n objects. Since the parity of a permutation of type $1^{\lambda_1} 2^{\lambda_2} \ldots n^{\lambda_n}$ is the same as the parity of the sum $\lambda_2 + \lambda_4 + \ldots$, we obtain

$$P(A_n; x_1, \ldots, x_n) =$$

$$= \sum_{\lambda_1 + 2\lambda_2 + \ldots + n\lambda_n = n} \frac{1 + (-1)^{\lambda_2 + \lambda_4 + \lambda_6 + \ldots}}{1^{\lambda_1} \lambda_1! \, 2^{\lambda_2} \lambda_2! \ldots n^{\lambda_n}\lambda_n!} x_1^{\lambda_1} x_2^{\lambda_2} \ldots x_n^{\lambda_n}.$$

4) The cyclic group C_n has order n and is generated by the cyclic permutation $[1, 2, \ldots, n]$.

$$P(C_n; x_1, \ldots, x_n) = \frac{1}{n} \sum_{k|n} \varphi(k) \, (x_k)^{\frac{n}{k}},$$

where the summation is over all divisors k of n, while $\varphi(k)$ denotes Euler's totient function (3.7) which gives the number of positive integers smaller than and prime to k.

5) The simple product $G \times H$ of the permutation groups G and H of orders r and s, respectively, is defined as follows: if G is a group of permutations of the set X and H a group of permutations of the set Y, the simple product $G \times H$ is a group of permutations of the set $X \cup Y$, defined by $(g, h)z = gz$ if $z \in X$ and $(g, h)z = hz$ if $z \in Y$, where $g \in G$ and $h \in H$. It is assumed that X and Y are disjoint sets and $|X| = n$, $|Y| = m$.

The order of the group $G \times H$ is rs and $P(G \times H; x_1, ..., x_n, y_1, ..., y_m) =$
$= P(G; x_1, ..., x_n) P(H; y_1, ..., y_m)$.

6) The Cartesian product $G \otimes H$ of the permutation groups G and H of orders r and s, respectively, is defined as follows: if G is a permutation group acting on the n-element set X and H is a permutation group acting on the m-element set Y, then the Cartesian product $G \otimes H$ is a permutation group acting on the Cartesian product $X \times Y$ as follows: $(g, h)(x, y) = (gx, hy)$, where $g \in G$, $h \in H$, $x \in X$ and $y \in Y$. Its order is rs and the cycle index is

$$P(G \otimes H; x_1, x_2, ..., x_{mn}) = \frac{1}{|G| |H|} \sum_{\substack{g \in G \\ h \in H}} \prod_{k,l \geqslant 1} x_{[k, l]}^{(k, l)\lambda_k(g)\lambda_l(h)},$$

where we have denoted by $[k, l]$ and by (k, l) the least common multiple and the greatest common divisor of the numbers k and l, respectively. We have also denoted by $\lambda_k(g)$ the number of cycles of length k of the permutation g and by $\lambda_l(h)$ the number of cycles of length l of the permutation h. This is proved as follows: if $x \in X$ is situated on a circuit of length k of the graph of the permutation g and $y \in Y$ is situated on a circuit of length l of the graph of the permutation h, then the pair $(x, y) \in X \times Y$ is situated on a circuit of length $[k, l]$ of the permutation (g, h), for $(g, h)^p(x, y) = (g^p x, h^p y)$ and the least p such that $(g, h)^p (x, y) = (x, y)$ is $p = [k, l]$. The number of these circuits of length $[k, l]$ that result from a circuit of length k of the graph G_g and from a circuit of length l of the graph G_h is equal to (k, l). For each circuit of the graph $G_{(g,h)}$ is determined from one of its vertices by successive applications of the permutation (g, h), thus generating the other vertices, it follows that there exist kl pairs of vertices, one from the circuit of length k and the other from the circuit of length l, that generate circuits of length $[k, l]$ in the graph $G_{(g,h)}$. But a circuit of length $[k, l]$ is generated in exactly $[k, l]$ different ways, taking as initial vertex any of its $[k, l]$ vertices. Hence the number of distinct circuits (corresponding to distinct cycles of the permutation (g, h)) from the graph $G_{(g,h)}$, made up of pairs of vertices from the cycle of length k of g and the cycle of length l of h is equal to $\dfrac{kl}{[k, l]} = (k, l)$.

Now the expression of the cycle index of the group $G \otimes H$ follows immediately, taking into account definition (8.1).

7) The composition (or wreath) product $G[H]$ of the permutation groups G and H, of orders r and s, respectively, is defined as follows: if G is a group of permutations acting on the set X with $|X| = n$ and H is a group of permutations acting on the set Y with $|Y| = m$, the composition product $G[H]$ is a group of permutations acting on the Cartesian product $X \times Y$ that contains the permutations $(g; h_1, h_2, ..., h_n)$ defined by $(g; h_1, h_2, ..., h_n)(x_i, y) = (gx_i, h_i y)$, where $g \in G$, $h_i \in H$ for $i = 1, ..., n$; $x_i \in X$ for $i = 1, ..., n$ and $y \in Y$. This group is of order rs^n. Pólya has shown that

$$P(G[H]; z_1, ..., z_{mn}) = P(G; p_1(H), p_2(H), ..., p_n(H)),$$

where

$$p_k(H) = P(H; z_k, z_{2k}, z_{3k}, ..., z_{mk}).$$

8) The power group H^G, where G is a group of permutations acting on the set X with $|X| = n$ and H is a group of permutations acting on the set Y with $|Y| = m$, is a group of permutations acting on the set Y^X of all functions $f: X \to Y$. Its elements are pairs (g, h) with $g \in G$, $h \in H$, which act as follows:

$$(g, h) f(x) = hfg(x).$$

This group is of order $|G| \cdot |H|$ and its cycle index is

$$P(H^G; x_1, ..., x_{m^n}) = \frac{1}{|G| |H|} \sum_{\substack{g \in G \\ h \in H}} \prod_{k=1}^{m^n} x_k^{\lambda_k(g,h)},$$

where

$$\lambda_1(g, h) = \prod_{k \geqslant 1} \sum_{r|k} [r\lambda_r(h)]^{\lambda_k(g)}$$

and for $k > 1$,

$$\lambda_k(g, h) = \frac{1}{k} \sum_{r|k} \mu(r, k) \lambda_1(g^r, h^r),$$

where $\mu(r, k)$ is the Möbius function of the lattice of divisors, to be presented in the next section.

Application. A Boolean function $f(x_1, ..., x_n)$ of n variables is a function $f: \{0, 1\}^n \to \{0, 1\}$. This function can be realized by means of a circuit with contacts, the conductibility of which is precisely the given function [28]. By performing permutations and negations of the letters that correspond to the contacts which appear in an optimal circuit for the function $f(x_1, ..., x_n)$, i.e. a circuit containing a minimum number of contacts, we obtain optimal circuits for other Boolean functions that are obtained from the given function by the same transformations upon the variables.

We thus come to the problem of classifying Boolean functions, multi-terminals etc., with respect to the hyperoctahedral group (which is the product of the group C_2^n of negations of the variables and the group S_n of permutations of the variables), or with respect to certain subgroups of this group.

The group C_2 is the set $\{0, 1\}$ endowed with addition modulo 2, while $C_2^n = \underbrace{C_2 \times C_2 \times \ldots \times C_2}_{n \text{ factors}}$,

the product being the above defined direct product of groups. The hyperoctahedral group is $G_n = \{(i, \sigma) \mid i \in C_2^n \text{ and } \sigma \in S_n\}$, where (i, σ) acts as follows on a Boolean function of n variables:

$$(i, \sigma) f(x_1, \ldots, x_n) = f(x_{\sigma^{-1}(1)}^{i_1}, \ldots, x_{\sigma^{-1}(n)}^{i_n}),$$

where $i = (i_1, i_2, \ldots, i_n)$ with $i_j \in \{0, 1\}$ for $j = 1, \ldots, n$, and $x^1 = x$, $x^0 = \bar{x}$.

The number of equivalence classes of the 2^{2^n} Boolean functions of n variables with respect to the hyperoctahedral group is given in Table 8.1.

TABLE 8.1

n	Number of classes with respect to G_n	2^{2^n}	$2^{2^n} - E(2, 2, n)$
1	3	4	2
2	6	16	6
3	22	256	38
4	402	65 536	942
5	1 228 158	4 294 967 296	325 262
6	400 507 806 843 728		

The last column contains the number of degenerate Boolean functions of n variables, where $E(2, 2, n)$ is obtained from (3.10).

Details of the calculation of the cycle index of the group G_n, as well as the table obtained by the Computation Laboratory of Harvard University, containing for each of the 402 types of Boolean functions of four variables, the corresponding minimal circuit, are given in M. Harrison's book [27].

For $n > 4$ the number of types of Boolean functions with respect to the group G_n increases rapidly, so that a catalogue of representatives of the equivalence classes cannot be made with present computational means.

The study of the classification of Boolean functions was initiated by Slepian [41]; the Romanian school of finite automata theory has obtained numerous results concerning the classification of various classes of Boolean functions, of systems of Boolean functions and of relay-contact circuits with respect to the hyperoctahedral group or various subgroups of it (P. Constantinescu [8]—[12], T. Gaşpar [13]—[15], Gr. C. Moisil [28], R. Popescu [34], [35], S. Rudeanu [39], Al. Şchiop [15], [40]).

8.4. COUNTING GRAPHS WITH NON-LABELLED VERTICES

Every permutation group G acting on the set X induces a permutation group $G^{(2)}$ acting on the set of pairs $\{x, y\}$ with $x, y \in X$ and $x \neq y$, in the following way: a permutation $g \in G$ defines a permutation $\bar{g}$ of the set of pairs, such that $\bar{g}\{x, y\} = \{g(x), g(y)\}$ with $g(x) \neq g(y)$ as g is an injection. It is easy to see that the mapping

g thus defined is a permutation of the set of (unordered) pairs $\{x, y\}$ with $x \neq y$ and the set $G^{(2)} = \{\bar{g} | g \in G\}$ is a permutation group.

Two graphs (X, U) and (X, V) with the sets of edges U and V, respectively, are said to be *isomorphic* if there is a bijection $g \colon X \to X$, i.e. a permutation g of the vertices, such that $\{x, y\} \in U$ if and only if $\bar{g}\{x, y\} = \{g(x), g(y)\} \in V$, where $\{x, y\}$ stands for an edge joining the vertices x and y. The isomorphism relation between undirected graphs is an equivalence relation which yields a partition of unoriented graphs into equivalence classes, called graphs with non-labelled vertices. In other words, if we denote by $\mathscr{P}_2(X)$ the set of $\dfrac{n(n-1)}{2}$ edges of the complete graph with vertices $x_1, x_2, \ldots, x_n$ (without loops or multiple edges), two graphs are isomorphic if the edge sets U and V are equivalent with respect to the group $S_n^{(2)}$ of permutations of the edges from $\mathscr{P}_2(X)$, where S_n is the group of all permutations of n objects. In this case we say that the edge sets U and V are equivalent modulo $S_n^{(2)}$ if there is a bijection $g \colon X \to X$ which induces a bijection $\bar{g} \colon U \to V$. To every mapping f of the edge set $\mathscr{P}_2(X)$ into a two-colour set $A = \{a_1, a_2\}$ corresponds an unoriented graph (X, U) defined by

$$U = \{u | u \in \mathscr{P}_2(X), \ f(u) = a_1\}. \tag{8.8}$$

Since this correspondence between the unoriented graphs with n vertices and the colourings with two colours of the set $\mathscr{P}_2(X)$ is a bijection, it follows that the number of non-labelled graphs with n vertices is equal to the number of colouring schemes with two colours of the elements of $\mathscr{P}_2(X)$ with respect to the permutation group $S_n^{(2)}$. By applying proposition 1 we find that the number of graphs with n non-labelled vertices is equal to $P(S_n^{(2)}; 2, 2, 2, \ldots)$, the maximum number of variables that can occur in the cycle index in this case being $\dbinom{n}{2}$. If we introduce the weights $w(a_1) = w$ and $w(a_2) = 1$, then the inverse of the correspondence (8.8) associates a graph with m edges with a two colour colouring scheme of the set $\mathscr{P}_2(X)$ under the action of the group $S_n^{(2)}$. The weight of the graph is w^m; therefore, by applying proposition 3, we find that the number of non-labelled graphs with n vertices and m edges is the coefficient of w^m in the expression

$$P(S_n^{(2)}; 1 + w, 1 + w^2, 1 + w^3, \ldots).$$

The *complementary graph* of the graph (X, U) is the graph $(X, \bar{U})$, where $\bar{U}$ is the complement of the set U with respect to the set of $\dbinom{n}{2}$ edges of the complete graph with n vertices.

The correspondence (8.8) induces a bijection between the non-labelled graphs isomorphic to their complements and the colouring schemes with two colours a_1, a_2 of the set $\mathscr{P}_2(X)$, that are invariant under the transposition $h = [a_1, a_2]$ of the colours. Using proposition 2, we find that the number of non-labelled graphs with n vertices, isomorphic to their complements, is equal to $P(S_n^{(2)}; 0, 2, 0, 2, ...)$.

If we choose the weights $w(a_1) = w$ and $w(a_2) = 1$, a non-labelled graph with n vertices and m edges isomorphic to its complement will correspond to a two colour colouring scheme of the set $\mathscr{P}_2(X)$, with respect to $S_n^{(2)}$, invariant under the transposition $[a_1, a_2]$, and of weight w^m. Therefore, the number of non-labelled graphs with n vertices and m edges isomorphic to their complements, is the coefficient of w^m in the expansion of the polynomial $P(S_n^{(2)}; 0, 2w, 0, 2w^2, 0, 2w^3, ...)$. To prove this, compute the numbers $p_1 = 0$, $p_2 = 2w$, $p_3 = 0$, $p_4 = 2w^2$, ... and apply proposition 2. Obviously a graph isomorphic to its complement has $\dfrac{1}{2}\dbinom{n}{2} = \dfrac{n(n-1)}{4}$ edges, if this number is an integer. Therefore, the single non-null term in the development of the polynomial $P(S_n^{(2)}; 0, 2w, 0, 2w^2, ...)$ is the term in w^m with $m = \dfrac{n(n-1)}{4}$ when this number is an integer (in the contrary case all terms are null) and its coefficient is equal to the number of non-labelled graphs with n vertices isomorphic to their complements, i.e. to $P(S_n^{(2)}; 0, 2, 0, 2, ...)$.

A formula for the computation of the polynomial $P(S_n^{(2)}; x_1, x_2, ...)$, as well as for the polynomial that occurs in counting oriented graphs, is given in the paper [21].

If we count oriented graphs with non-labelled vertices, the edges $\{x, y\}$ will be replaced by arcs $(x, y) \in X \times X$.

For example, for $n = 3$ the polynomial $P(S_3^{(2)}; x_1, x_2, x_3)$ is calculated as follows: set $a = \{1, 2\}$, $b = \{1, 3\}$, $c = \{2, 3\}$ and denote the permutations of S_3 by $g_1 = [1][2][3]$, $g_2 = [1][2, 3]$, $g_3 = [2][1, 3]$, $g_4 = [3][1, 2]$, $g_5 = [1, 2, 3]$ and $g_6 = [1, 3, 2]$. Then $\bar{g}_1(a) = a$, $\bar{g}_1(b) = b$ and $\bar{g}_1(c) = c$; $\bar{g}_2(a) = b$, $\bar{g}_2(b) = a$ and $\bar{g}_2(c) = c$; $\bar{g}_3(a) = c$, $\bar{g}_3(b) = b$ and $\bar{g}_3(c) = a$; $\bar{g}_4(a) = a$, $\bar{g}_4(b) = c$ and $\bar{g}_4(c) = b$; $\bar{g}_5(a) = c$, $\bar{g}_5(b) = a$ and $\bar{g}_5(c) = b$; $\bar{g}_6(a) = b$, $\bar{g}_6(b) = c$ and $\bar{g}_6(c) = a$. Therefore $\bar{g}_1 = [a][b][c]$, $\bar{g}_2 = [c][a, b]$, $\bar{g}_3 = [b][a, c]$, $\bar{g}_4 = [a][b, c]$, $\bar{g}_5 = [a, c, b]$ and $\bar{g}_6 = [a, b, c]$, whence we get

$$P(S_3^{(2)}; x_1, x_2, x_3) = \frac{1}{6}(x_1^3 + 3x_1 x_2 + 2x_3).$$

By performing such calculations for $n = 2, 3, 4$, we obtain Table 8.2.

In the case $n = 4$ the only graph isomorphic to its complement is the three-edge graph drawn in the last row of Table 8.2.

The counting of directed graphs with n labelled vertices is much simpler, their number being 2^{n^2} (or $2^{n^2 - n}$ if we only take into account loop-free graphs), since the Boolean adjacency matrix of the graph has n^2 elements each of which can be choosen to be 0 or 1. The number of loop-free undirected graphs with n vertices is equal to $2^{\binom{n}{2}}$.

In the paper [1], A. Balaban counted all possible valence-isomers of cyclopolyenes with the formula $C_{2p}H_{2p}$ for $p = 2, 3, 4$ and 5.

From the topological point of view, the problem requires the determination of all possible ways of joining $2p$ units CH by: a) simple lines or b) simple lines and

TABLE 8.2

n	Non-labelled graphs	$P(S_n^{(2)};\ x_1, x_2, \ldots)$	$P(S_n^{(2)};\ 2, 2, \ldots)$	$P(S_n^{(2)};\ 0, 2, 0, 2, \ldots)$	$P(S_n^{(2)};\ 1+w,\ 1+w^2,\ \ldots)$
2		x_1	2	0	$1 + w$
3		$\dfrac{1}{6}\,(x_1^3 + 3x_1 x_2 + 2x_3)$	4	0	$1 + w + w^2 + w^3$
4		$\dfrac{1}{24}\,(x_1^6 + 9x_1^2 x_2^2 + 8x_3^2 + 6x_2 x_4)$	11	1	$1 + w + 2w^2 + 3w^3 + 2w^4 + w^5 + w^6$

double lines such that three simple lines or one simple line and one double line come from each unit CH, corresponding to the valency 3 of it.

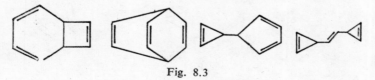

Fig. 8.3

Fig. 8.3 depicts four isomers of the cyclooctatetraene C_8H_8 each of which contains three double lines and six simple lines; the units CH, situated at the vertices of the polygons, have not been represented. These isomers can be transformed into one another by the action of various external factors and the enumeration of all possible isomers can direct research towards obtaining them in the laboratory. Thus, Fig. 8.4 represents two isomers of the cyclooctatetraene C_8H_8 which, under

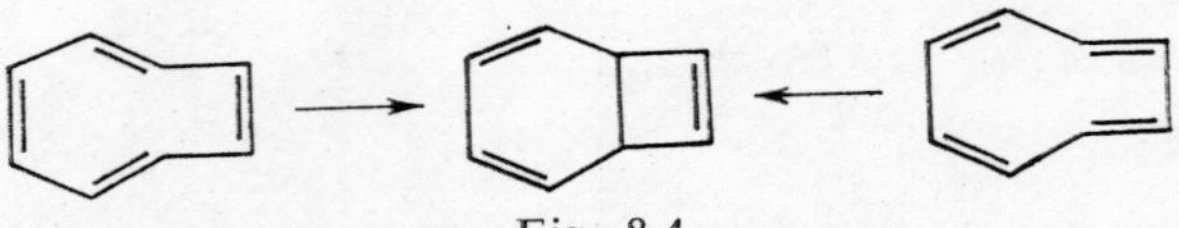

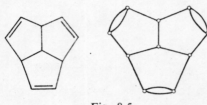

Fig. 8.4

the influence of heat and of ultra-violet rays, respectively, are transformed into bicyclooctatriene. These graphical representations of chemical links can be regarded as graphs or as multigraphs (when several edges exist joining the same two vertices) with non-labelled vertices, as in Fig. 8.5 for the representation of triquinacene $(10 - -3-14)$, which is an isomer of the cyclopolyene $C_{10}H_{10}$. Thus the enumeration of all possible isomers of cyclopolyenes reduces to: a) the enumeration of all regular graphs of degree 3 (every vertex incident to three edges) with $2p$ non-labelled vertices and b) the enumeration of all regular multigraphs of degree 3, with $2p$ non-labelled vertices and simple and/or double edges. As these problems are not yet solved in the general case, in [1] all possible solutions are enumerated for small values of p. Thus, for $2p = 4$ one gets an isomer with simple edges and an isomer with two double edges, for $2p = 6$ one obtains two isomers with simple edges, one isomer with a double edge, two isomers with two double edges each, and one isomer with three double edges (benzene) etc., these hydrocarbons being represented as graphs or as multigraphs in Fig. 8.6.

Fig. 8.5

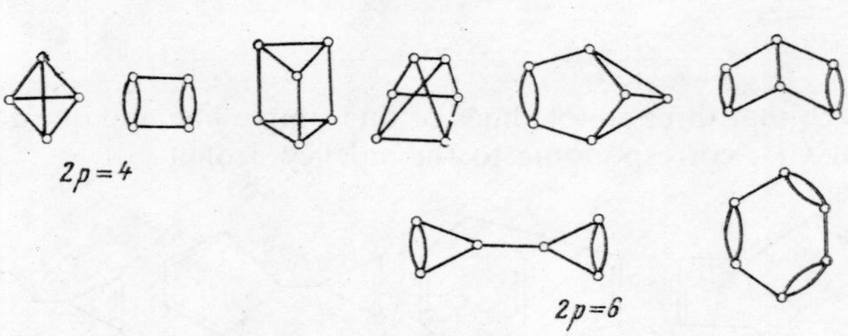

$2p = 4$

$2p = 6$

Fig. 8.6

In the paper [2], proposition 3 (Pólya) is used to calculate the number of isomers of a monocyclic aromatic compound $X_x Y_y Z_z$ with $x + y + z = m$, by replacing each variable y_k in the cycle index of the dihedral group D_m (which has order $2m$

and is generated by the permutations $[1, 2, ..., m]$ and $[1, m] [2, m - 1] [3, m - 2]...$),
by $r^k + s^k + t^k$. We mention that the dihedral group D_m has the cycle index

$$P(D_m; x_1, ..., x_m) = \frac{1}{2} [P(C_m; x_1, ..., x_m) + x_1 x_2^{(m-1)/2}] \quad \text{for} \quad \text{odd} \quad m \quad \text{and}$$

$$P(D_m; x_1, ..., x_m) = \frac{1}{2} [P(C_m; x_1, ..., x_m) + \frac{1}{2} (x_2^{m/2} + x_1^2 x_2^{(m-2)/2})] \text{ for even } m,$$

where $P(C_m; x_1, ..., x_m)$ is the cycle index of the cyclic group C_m.

Problems

1.   Prove that the number of nonisomorphic multigraphs with n vertices and m edges equals the coefficient of x^m from

$$P\left(S_n^{(2)}, \frac{1}{1 - x}, \frac{1}{1 - x^2}, \frac{1}{1 - x^3}, ...\right)$$

(Pólya, 1937)

H i n t: A multigraph is a mapping $f: \mathscr{P}_2(X) \to \{0, 1, 2, ...\}$. Apply Pólya's theorem
by choosing $w(i) = x^i$ and using the property $\frac{1}{1 - x^r} = 1 + x^r + x^{2r} + ...$

2.   In how many ways can six balls (two red, two yellow, one blue and one white) be arranged at the vertices of a regular hexagon?
H i n t: Use Proposition 3 with the cycle index of the dihedral group D_6. One finds 16 possibilities.

3.   Prove that for $p \leqslant m$ the coefficient of x^p in the expansion of

$P(S_n; 1 + x + x^2 + ... + x^m, \ 1 + x^2 + x^4 + ... x^{2m}, \ 1 + x^3 + x^6 + ... + x^{3m}, ...)$

equals the number of partitions of p into at most n parts.
H i n t: Apply Pólya's theorem with $X = \{1, 2, ..., n\}$, $A = \{a_0, a_1, ..., a_m\}$, $w_i = x^i$
for $i = 0, 1, ..., m$. If $p = k_1 + 2k_2 + ...$, write:

$$p = \underbrace{1 + 1 + ... + 1}_{k_1} + \underbrace{2 + ... + 2}_{k_2} + ...$$

4.   Let f be a permutation of m objects which has order r and $P(f; x_1, x_2, ...)$ denote the cycle index of the cyclic group generated by f. Prove that

$$P(f; x_1, x_2, ...) = \frac{1}{r} \sum_{i=1}^{r} \prod_{k=1}^{r} \frac{x_k^{(k, i) \lambda_k(f)}}{(k, i)}.$$

(Redfield, 1927)

H i n t: The cyclic group generated by f is $\{f, f^2, ..., f^r = e\}$ and the permutation c^i where c is a cycle of length k has (k, i) cycles, each of length $\dfrac{k}{(k, i)}$.

5.   A bicoloured graph is a pair (G, f) where G is a graph and f is a function from the set of vertices of G onto the set $\{1, 2\}$ such that if a and b are adjacent vertices, then $f(a) \neq f(b)$. Two bicoloured graphs are chromatically isomorphic if there is a colour preserving isomorphism between them.

Thus (G_1, f_1) is chromatically isomorphic with (G_2, f_2) if there is an isomorphism $\theta: X_1 \to X_2$ and a permutation $\sigma \in S_2$ such that $\sigma(f_1(a)) = f_2(\theta(a))$ for every vertex a in X_1, where $G_1 = (X_1, U_1)$ and $G_2 = (X_2, U_2)$.

A bicoloured graph with p vertices may be considered as a spanning subgraph of the bipartite graph $K_{m,n}$ for which $m + n = p$.

Let $g_{mn, p}$ be the number of chromatically nonisomorphic spanning subgraphs of $K_{m,n}$ having q lines, and let

$$g_{mn}(x) = \sum_{q=0}^{mn} g_{mn, q} \, x^q.$$

Prove that the counting polynomial $g_{mn}(x)$ for bicoloured graphs is given by:

$g_{mn}(x) = P(S_m \times S_n; \ 1 + x, \ 1 + x^2, 1 + x^3, ...)$ when $m \neq n$;

$g_{mn}(x) = P(S_2[S_n]; \ \ 1 + x, \ 1 + x^2, 1 + x^3, ...)$ when $m = n$.

(Harary, 1958)

H i n t: The automorphism group of $K_{m,n}$ is $S_m \times S_n$ when $m \neq n$ and it is $S_2[S_n]$ when $m = n$. Apply Proposition 3.

6.   Prove that the number of unoriented nonisomorphic multigraphs having three vertices and m edges is equal to $\left[\dfrac{(m + 3)^2}{12} \right]$.

H i n t: This number is equal to $P(m, 1) + P(m, 2) + P(m, 3) = 1 + \left[\dfrac{m}{2} \right] +$

$+ 1 + P(m - 3, 2) + P(m - 3, 3) = \left[\dfrac{(m + 3)^2}{12} \right]$ using induction.

7.   If we set $Q_n(x) = P(S_n^{(2)}; \ 1 + x, \ 1 + x^2, \ 1 + x^3, ...)$ prove that $Q_n(x) = x^{\binom{n}{2}} Q_n\left(\dfrac{1}{x} \right)$.

8.   Show that $P(S_n, 1 + x, 1 + x^2, ..., 1 + x^n) = 1 + x + x^2 + ... + x^n$.
H i n t: Use Pólya's theorem.

9. In how many ways can five balls (three red and two white) be arranged in three distinct boxes so that the numbers of the balls are respectively 2, 2 and 1?

H i n t: $P(S_2 \times S_3; x_1, x_2, x_3) = P(S_2; x_1, x_2) P(S_3; x_1, x_2, x_3)$

$$= \frac{1}{12} (x_1^5 + 4x_1^3 x_2 + 3x_1 x_2^2 + 2x_1^2 x_3 + 2x_2 x_3).$$

Applying Pólya's theorem we find 5 possibilities.

10. In how many geometrically different ways is it possible to paint the faces of a regular tetrahedron using four different colours so that any two faces have different colours?

H i n t: The cycle index of the group of rotations of the faces (or vertices) of a regular

tetrahedron is equal to $\dfrac{1}{12}(x_1^4 + 8x_1 x_3 + 3x_2^2)$.

If we substitute $x_1 = w_1 + w_2 + w_3 + w_4$, $x_2 = w_1^2 + w_2^2 + w_3^2 + w_4^2$, $x_3 = w_1^3 + w_2^3 + w_3^3 + w_4^3$ we find the value 2 for the coefficient of $w_1 w_2 w_3 w_4$.

11. Let $C_2 = \langle \{0, 1\}, \oplus \rangle$ where $\oplus$ is addition modulo 2 and let $C_2^n = \underbrace{C_2 \times C_2 \times ... \times C_2}_{n \text{ times}}$ be the direct product of n copies of C_2. Prove that the cycle

index of C_2^n is equal to $P(C_2^n; x_1, x_2) = \dfrac{1}{2^n} (x_1^{2^n} + (2^n - 1) x_2^{2^{n-1}})$.

H i n t: Use induction on n.

12. The direct product C_2^n induces a group of permutations on the set B_n of Boolean functions of n variables in the following way: for any $f \in B_n$ and $i = (i_1, i_2, ..., i_n) \in C_2^n$, we define $if = f(x_1^{i_1}, x_2^{i_2}, ..., x_n^{i_n})$ where $x_j^{i_j} = x_j$ if $i_j = 1$ and $x_j^{i_j} = \bar{x}_j$, if $i_j = 0$, for $j = 1, ..., n$.

Prove that the number of equivalence classes of Boolean functions of n variables with the property that $|f^{-1}(1)| = k$ relative to C_2^n is equal to

$$\frac{1}{2^n} \binom{2^n}{k} \text{ for } k \equiv 1 \pmod 2;$$

$$\frac{1}{2^n} \left[\binom{2^n}{k} + (2^n - 1) \binom{2^{n-1}}{k/2} \right] \text{ for } k \equiv 0 \pmod 2$$

and the total number of equivalence classes of Boolean functions of n variables

relative to the group C_2^n is equal to $\dfrac{1}{2^n} (2^{2^n} + (2^n - 1) 2^{2^{n-1}})$.

H i n t: Apply Pólya's theorem and find the coefficient of w^k in the expansion of $P(C_2^n; 1 + w, 1 + w^2)$, respectively $P(C_2^n; 2, 2)$.

13.   Show that the cycle index of the group of permutations of the 8 vertices of the cube, induced by the group of rotations of the cube, is equal to

$$\frac{1}{24}(x_1^8 + 9x_2^4 + 6x_4^2 + 8x_1^2 x_3^2),$$

while in the case of the 12 edges of the cube, the cycle index is

$$\frac{1}{24}(x_1^{12} + 3x_2^6 + 6x_4^3 + 6x_1^2 x_5^2 + 8x_3^4).$$

BIBLIOGRAPHY

1. Balaban, A. T., *Valence-isomerism of cyclopolyenes*, Rev. Roumaine Chim., **11**, 1966, 1097—1116; erratum **12**, 1967, 103.
2. Balaban, A. T., Harary, F., *Chemical graphs IV: Dihedral groups and monocyclic aromatic compounds*, Rev. Roumaine Chim., **12**, 1967, 1511—1515.
3. Beckenbach, E. F. (ed.), *Applied combinatorial mathematics*, Wiley, New York, 1964.
4. Berge, C., *Principes de combinatoire*, Dunod, Paris, 1968.
   English edition: *Principles of combinatorics*, Academic Press, New York, 1971.
5. de Bruijn, N. G., *Generalization of Pólya's fundamental theorem in enumerative combinatorial analysis*, Nederl. Akad. Wetensch. Proc. Ser. A, **62** (Indag. Math., **21**), 1959, 59—69.
6. de Bruijn, N. G., *Enumeration of mapping patterns*, J. Combinatorial Theory (A), **12**, 1972, 14—20.
7. de Bruijn, N. G., Klarner, D. A., *Enumeration of generalized graphs*, Nederl. Akad. Wetensch. Proc. Ser. A, **72** (Indag. Math., **31**), 1969, 1—9.
8. Constantinescu, P., *On the classification of symmetric Boolean functions* (in Romanian), Stud. Cerc. Mat., **11**, 1960, 193—206.
9. Constantinescu, P., *On the number of the types of Boolean functions with respect to some subgroups of the hyperoctahedral group*, Bull. Math. Soc. Sci. Math. Phys. R. P. Roumaine, **4** (52), 1960, 1—16.
10. Constantinescu, P., *On the number of synonimy types of 1-k multipoles* (in Romanian), An. Univ. Bucureşti, Ser. Şti. Natur., Mat.-Fiz., **10**, 1961, 137—144.
11. Constantinescu, P., *On the number of types of generalized symmetric Boolean functions* (in Romanian), Stud. Cerc. Mat., **12**, 1961, 449—455.
12. Constantinescu, P., *On the classification of some codes with respect to the symmetric group*, Bull. Math. Soc. Sci. Math. Phys. R. P. Roumaine, **6** (54), 1962, 133—138.
13. Gaşpar, T., *The number of types of 1-k multipoles with respect to the group of negations* (in Romanian), An. Univ. Bucureşti, Ser. Şti. Natur., Mat.-Fiz., **10**, 1961, 167—169.
14. Gaşpar, T., *The number of synonimy types of the classes of functionally equivalent dipoles with rectifiers* (in Romanian), Stud. Cerc. Mat., **16**, 1964, 1153—1163.
15. Gaşpar, T., Şchiop, A., *Nombres de types de multipoles (1-k) d'après le groupe hyper-octa-édrique*, Rev. Roumaine Math. Pures Appl., **10**, 1965, 1431—1436.
16. Harary, F., *The number of linear, directed, rooted and connected graphs*, Trans. Amer. Math. Soc., **78**, 1955, 445—463.
17. Harary, F., *On the number of bicolored graphs*, Pacific J. Math., **8**, 1958, 743—755.
18. Harary, F., *Graphical enumeration problems*, Applied Combinatorial Mathematics (ed: E. F. Beckenbach), Wiley, New York, 1966.
19. Harary, F., *A proof of Pólya's enumeration theorem*, A Seminar on Graph Theory (ed: F. Harary), Holt, Rinehart and Winston, New York, 1967, 21—24.

20. Harary, F., *Applications of Pólya's theorem to permutation groups*, A Seminar on Graph Theory (ed: F. Harary), Holt, Rinehart and Winston, New York, 1967, 25—33.
21. Harary, F., *Enumeration of graphs and digraphs*, A Seminar on Graph Theory (ed: F. Harary), Holt, Rinehart and Winston, New York, 1967, 34—41.
22. Harary, F., *Graph theory*, Addison-Wesley, Reading, Mass., 1969.
23. Harary, F., *Enumeration under group action: unsolved graphical enumeration problems* IV, J. Combinatorial Theory, **8**, 1970, 1—11.
24. Harary, F., Palmer, E., *Enumeration of finite automata*, Information and Control, **10**, 1967, 499—508.
25. Harary, F., Palmer, E., *The power group enumeration theorem*, J. Combinatorial Theory, **1**, 1966, 157—173.
26. Harrison, M., *A census of finite automata*, Canad. J. Math., **17**, 1965, 100—113.
27. Harrison, M., *Introduction to switching and automata theory*, McGraw-Hill, New York, 1965.
28. Moisil, G. C., *Algebraic theory of contact-relay networks* (in Romanian), Editura Tehnică, București, 1965.
29. Oberman, R. M., *Disciplines in combinatorial and sequential circuit design*, McGraw-Hill, New York, 1970.
30. Oberschelp, W., *Die Anzahl nicht-isomorphen m-Graphen*, Monatsh. Math., **72**, 1968, 220—223.
31. Parthasarathy, K.R., Sridharan, M.R., *On structure enumeration theory*, Nederl. Akad. Wetensch. Proc. Ser. A, **74** (Indag. Math., **33**), 1971, 327—339.
32. Palmer, E. M., *Methods for the enumeration of multigraphs*, The Many Facets of Graph Theory, Lecture Notes in Math. N° 110, Springer, Berlin, 1969, 251—261.
33. Pólya, G., *Kombinatorische Anzahlbestimmungen für Gruppen, Graphen und Chemische Verbindungen*, Acta Math., **68**, 1937, 145—254.
34. Popescu, R., *The number of types of (1-k) multipoles with respect to the product between the alternating subgroup and the subgroup of negations (subgroups of the hyperoctahedral group)* (in Romanian), Stud. Cerc. Mat. **18**, 1966, 889—894.
35. Popescu, R., *The number of types of Boolean functions with respect to the product between the alternating subgroup and the subgroup of negations (subgroups of the hyperoctahedral group)* (in Romanian), Stud. Cerc. Mat. **18**, 1966, 895—907.
36. Redfield, J. H., *The theory of group-reduced distributions*, Amer. J. Math., **49**, 1927, 433—455.
37. Riordan, J., *An introduction to combinatorial analysis*, Wiley, New York, 1958.
38. Robinson, R. W., *Enumeration of colored graphs*, J. Combinatorial Theory, **4**, 1968, 181—190.
39. Rudeanu, S., *The use of Galois fields in the theory of switching circuits X: Classification of functions of two variables in GF* (2^2) (in Romanian). Stud. Cerc. Mat., **9**, 1958, 217—287.
40. Șchiop, A., *The number of types of (1-k) multipoles with respect to the alternating group* (in Romanian), Stud. Cerc. Mat., **17**, 1965, 451—456.
41. Slepian, D., *On the number of symmetry types of Boolean functions of n variables*, Canad. J. Math., **5**, 1953, 185—193.
42. Williamson, S. G., *Operator theoretic invariants and the enumeration theory of Pólya and de Bruijn*, J. Combinatorial Theory, **8**, 1970, 162—169.

CHAPTER 9

Inversion Formulae

9.1. THE FIRST INVERSION FORMULA AND APPLICATIONS

In certain counting problems, the unknown functions appear in sums which we can evaluate. Under certain conditions, specified in the sequel, these formulae can be solved with respect to the unknown functions, which are thus obtained as sums the coefficients of which are precisely the sums already evaluated.

Consider two families of polynomials $\{p_k(x)\}_{k=0,1,...,n}$ and $\{q_k(x)\}_{k=0,1,...,n}$, both polynomials p_k and q_k being of degree k, for $k = 0, 1, ..., n$.

Assume that these polynomials are connected by the relations

$$p_k(x) = \sum_{i=0}^{k} \alpha_{i,k} q_i(x)$$

and

$$q_k(x) = \sum_{i=0}^{k} \beta_{i,k} p_i(x) \tag{9.1}$$

for every $k = 0, 1, ..., n$.

PROPOSITION 1. *If relations* (9.1) *hold, then for every* $a_0, a_1, a_2, ..., a_n$ *and* $b_0, b_1, ..., b_n$ *such that* $a_k = \sum_{i=0}^{k} \alpha_{i,k} b_i$ *for* $k = 0, 1, ..., n$, *the relations* $b_k = \sum_{i=0}^{k} \beta_{i,k} a_i$ *also hold for* $k = 0, 1, ..., n$.

Setting $\alpha_{i,k} = 0$ for $i = k + 1, ..., n$ and $\beta_{i,k} = 0$ for $i = k + 1, ..., n$, the numbers $\alpha_{i,k}$ and $\beta_{i,k}$ form two $(n + 1) \times (n + 1)$ matrices A and B, respectively. Denoting by a the column vector which has the $n + 1$ components $a_0, a_1, ..., a_n$ and by b the column vector which has the $n + 1$ components $b_0, b_1, ..., b_n$, the proposition states that $a = Ab$ implies $b = Ba$. But if we show that the matrices A and B are inverse to each other, i.e. $BA = I_{n+1} = $ the identity matrix with $n + 1$ rows and $n + 1$ columns, then by left multiplying the equation $a = Ab$ with B, we obtain $Ba = (BA)b = I_{n+1}b = b$, as required.

To prove that A and B are inverse to each other, we make use of relations (9.1), by completing the sums up to n with zero terms:

$$p_k(x) = \sum_{i=0}^{n} \alpha_{i,k} q_i(x) = \sum_{i=0}^{n} \alpha_{i,k} \sum_{j=0}^{n} \beta_{j,i} p_j(x) = \sum_{j=0}^{n} \left(\sum_{i=0}^{n} \beta_{j,i} \alpha_{i,k} \right) p_j(x).$$

Since the polynomials $p_k(x)$ for $k = 0, ..., n$ are of different degrees, it follows that they are linearly independent, therefore the above representation is unique. So $\sum_{i=0}^{n} \beta_{j,i}\alpha_{i,k} = \delta_{j,k} =$ the *Kronecker symbol* defined by $\delta_{j,k} = 1$ for $j = k$ and $\delta_{j,k} = 0$ for $j \neq k$; therefore $BA = I_{n+1}$.

PROPOSITION 2. *If the numbers* $a_0, a_1, a_2, ..., a_n$ *and* $b_0, b_1, b_2, ..., b_n$ *satisfy the relations* $a_k = \sum_{i=0}^{k} \binom{k}{i} b_i$ *for* $k = 0, 1, ..., n$, *then the numbers* $b_0, b_1, ..., b_n$ *are given by the relations*

$$b_k = \sum_{i=0}^{k} \binom{k}{i} (-1)^{k-i} a_i.$$

These relations are known as the *inverse binomial formulae.*

The proof follows from Newton's binomial formula

$$x^k = (x - 1 + 1)^k = \sum_{i=0}^{k} \binom{k}{i} (x - 1)^i$$

and

$$(x - 1)^k = \sum_{i=0}^{k} \binom{k}{i} (-1)^{k-i} x^i.$$

By taking the family of polynomials $p_k(x) = x^k$ and $q_k(x) = (x - 1)^k$ for $k = 0, 1, ..., n$, proposition 1 yields the desired result.

We have seen that the Stirling numbers of the second kind satisfy the relation

$$k^p = \sum_{i=0}^{k} \binom{k}{i} i! \, S(p, i),$$

where $S(p, 0) = 0$. It follows by inversion of this relation that

$$k! \, S(p, k) = \sum_{i=0}^{k} (-1)^{k-i} \binom{k}{i} i^p,$$

which is just formula (4.2).

PROPOSITION 3. *If the numbers* $a_1, a_2, ..., a_n$ *and* $b_1, b_2, ..., b_n$ *satisfy the relations* $a_k = \sum_{i=1}^{k} s(k,i) b_i$ *for* $k = 1, 2, ..., n$, *then* $b_k = \sum_{i=1}^{k} S(k,i) a_i$ *for* $k = 1, ..., n$.
These formulae are known as the *inverse Stirling formulae.*

We have defined the Stirling numbers $s(k, i)$ of the first kind by the relation $[x]_k = \sum_{i=1}^{k} s(k, i)x^i$. We have shown in Chap. 4 that the following formula (4.3) holds:

$$x^k = \sum_{i=1}^{k} S(k, i)\,[x]_i.$$

By taking the families of polynomials $p_k(x) = [x]_k$ and $q_k(x) = x^k$ for $k = 1, 2, \ldots, n$, the proof follows from proposition 1.

Application. A r r a n g e m e n t s   i n t o   c e l l s   o f   s e t s   w h i c h   c o n t a i n   i d e n t i c a l   o b j e c t s. Given a set X with n objects, we shall identify certain objects which will be said to be of the same kind.

We say that X is a collection of objects of type $1^{\lambda_1} 2^{\lambda_2} \ldots n^{\lambda_n}$ if there is a partition of X that contains λ_1 classes with one element, λ_2 classes with two elements, $\ldots$, λ_n classes with n elements, and we identify the objects that belong to the same class of the partition. For example, if X is a set consisting of the following 12 objects: one white ball, one black ball, three red balls, three blue balls and four green balls, then X is a collection of objects of type $1^2 3^2 4^1$. Obviously $\lambda_1 + 2\lambda_2 + \ldots + n\lambda_n = n$.

Let us arrange the objects in cells $a_1, a_2, \ldots, a_m$, some of which are identical so that they form a collection of cells of type $1^{\mu_1} 2^{\mu_2} \ldots m^{\mu_m}$ with $\mu_1 + 2\mu_2 + \ldots + m\mu_m = m$.

An arrangement of the objects into cells is a mapping $f : X \to A$; if $f(x) = a_i$, the object $x \in X$ will be in the cell a_i. We say that two *arrangements* are *equivalent* if they are obtained from one another by a permutation of the objects of the same kind or of the cells of the same kind. The classes of this equivalence are called *arrangement schemes* of the objects into cells. The methods presented in the previous Chapter yield an approach to the counting of arrangement schemes. If we denote by G the group of the $(1!)^{\lambda_1} (2!)^{\lambda_2} \ldots (n!)^{\lambda_n}$ permutations for which the objects of X of the same type are left invariant, and by H the group of the $(1!)^{\mu_1} (2!)^{\mu_2} \ldots (m!)^{\mu_m}$ permutations for which the cells of A of the same type are invariant, then the power group H^G has order $|G|\,|H|$ and consists of permutations (g, h) that act as follows on an arrangement $f : X \to A$:

$$(g, h)f = hfg : X \to A, \text{ where } g \in G \text{ and } h \in H.$$

The number of arrangement schemes of the objects into cells is equal to the number of equivalence classes or orbits of the set A^X of arrangements with respect to the permutation group H^G.

Other computational methods are based on the consequences of Pólya's method (see previous Chapter). However, in certain particular cases the number of arrangements can be obtained by simple formulae.

Denote by $A_{\mathscr{C}}(1^{\lambda_1} 2^{\lambda_2} \ldots n^{\lambda_n}; 1^{\mu_1} 2^{\mu_2} \ldots m^{\mu_m})$ the number of arrangement schemes of a collection of objects of type $1^{\lambda_1} 2^{\lambda_2} \ldots n^{\lambda_n}$ into a collection of cells of type $1^{\mu_1} 2^{\mu_2} \ldots m^{\mu_m}$ and by $A(1^{\lambda_1} 2^{\lambda_2} \ldots n^{\lambda_n}; 1^{\mu_1} 2^{\mu_2} \ldots m^{\mu_m})$ the number of arrangements for which no cell is empty. Thus $A(1^n; m) = S(n, m) =$ the Stirling number of the second kind and $A_{\mathscr{C}}(1^n; m) = S(n, 1) + S(n, 2) + \ldots + S(n, m)$. For, in this case the n objects are pairwise distinct and the m cells are identical, hence the number of arrangements with no empty cells is equal to the number of partitions of n objects into m classes and $A_{\varnothing}(1^n; m) = \sum_{k=1}^{m} S(n, k)$, because we can leave $m-1$ $m-2, \ldots, 0$ empty cells.

Similarly $A(1^n; 1^m) = s_{n,m} = m!S(n, m)$, because if the objects and the cells are pairwise different, the number of arrangements with no empty cells is equal to the number of surjections from an n-element set onto an m-element set. If no condition is imposed upon the arrangements, the number of them is equal to the number of functions $f: X \to A$, hence $A_\emptyset(1^n, 1^m) = m^n$.

The number $A(n, m) = P(n, m) =$ the number of partitions of n into m parts, because if all objects and all cells are identical, then a bijection can be established between the set of partitions of n into m parts, $n = p_1 + p_2 + \ldots + p_m$ with $p_1 \geqslant p_2 \geqslant \ldots \geqslant p_m \geqslant 1$ and the set of arrangements of the set X into cells of A, such that each cell contains at least one object; namely, p_1 objects are located in a cell, p_2 objects in another cell,..., p_n objects in the last cell. If cells without objects are permitted, then $A_\emptyset(n; m) = P(n, 1) + P(n, 2) + \ldots + P(n, m)$, because $m - 1$, $m - 2$,..., 0 cells without objects may exist and the corresponding arrangements form disjoint sets.

$A(n; 1^m) = \binom{n-1}{m-1}$, because the objects are identical whereas the cells are pairwise distinct, so that $A(n; 1^m)$ is the number of ways of writing n as a sum of m natural numbers: $n = u_1 + u_2 + \ldots + u_m$, two sums being distinct if they differ in the nature of their terms or in their order. We have seen in Chap. 2 that this number equals $\binom{n-1}{m-1}$. If the existence of empty cells is admitted, then this is covered by the possibility that certain terms of the sum vanish, therefore $A_\emptyset(n; 1^m) = \binom{n+m-1}{n}$, this being the number of ways in which n can be written as a sum of m numbers: $n = u_1 + u_2 + \ldots + u_m$ with $u_i \geqslant 0$ for $i = 1,\ldots, m$, two sums differing either in the nature of their terms, or in their order.

As a matter of fact, in each case the existence of a bijection between the set of arrangements with the given properties and the set of representations of n as a sum of m integers is understood.

PROPOSITION 4 (MacMahon). *The following relations hold:*

$$A_\emptyset(1^{\lambda_1} 2^{\lambda_2} \ldots n^{\lambda_n}; 1^m) = \binom{m}{1}^{\lambda_1} \binom{m+1}{2}^{\lambda_2} \ldots \binom{m+n-1}{n}^{\lambda_n};$$

$$A(1^{\lambda_1} 2^{\lambda_2} \ldots n^{\lambda_n}; 1^m) = \sum_{k=1}^{m} (-1)^{m-k} \binom{m}{k} \binom{k}{1}^{\lambda_1} \binom{k+1}{2}^{\lambda_2} \ldots \binom{k+n-1}{n}^{\lambda_n}. \quad (9.2)$$

If the existence of cells without objects is permitted, we can first arrange the n_1 objects of the first kind, then the n_2 objects of the second kind, ..., then the n_k objects of the last kind. But we have seen that

$$A_\emptyset(n_p; 1^m) = \binom{n_p + m - 1}{n_p},$$

therefore we obtain

$$A_\emptyset(n_1 n_2 \ldots n_k; 1^m) = A_\emptyset(n_1; 1^m) \ldots A_\emptyset(n_k; 1^m) =$$
$$= \binom{n_1 + m - 1}{n_1} \ldots \binom{n_k + m - 1}{n_k},$$

whence we obtain the first formula, grouping together the factors of the product that are equal.

Denote by $A(k)$ the number of arrangements of the n objects into a subset $K \subset A$ of pairwise distinct cells with $|K| = k$, such that no cell is empty, and by $A_\varnothing(k)$ the number of arrangements when empty cells are permitted. The numbers $A(k)$ do not depend on the choice of the subset $K \subset A$ with k elements and we obtain

$$A_\varnothing(m) = \sum_{k=1}^{m} \sum_{\substack{K \subset A \\ |K|=k}} A(k) = \sum_{k=1}^{m} \binom{m}{k} A(k).$$

By applying the inverse binomial formulae (proposition 2), we get

$$A(m) = \sum_{k=1}^{m} (-1)^{m-k} \binom{m}{k} A_\varnothing(k),$$

where $A_\varnothing(k) = A_\varnothing(1^{\lambda_1} 2^{\lambda_2} \ldots n^{\lambda_n}; 1^k)$, i.e. the second formula (9.2). The summation starts with $k = 1$ because for $k = 0$ the corresponding term in each sum is zero.

For example, let us determine the number of arrangements of one red ball, one green ball and two blue balls in three cells such that no cell is empty. In this case the objects are of type $1^2 2^1$, hence $n = 4$, $m = 3$, $\lambda_1 = 2$, $\lambda_2 = 1$ and $\lambda_3 = \lambda_4 = 0$. According to (9.2) we obtain

$$A(1^2 2^1; 1^3) = \sum_{k=1}^{3} (-1)^{3-k} \binom{3}{k} \binom{k}{1}^2 \binom{k+1}{2}$$

$$= \binom{3}{1}\binom{1}{1}^2 \binom{2}{2} - \binom{3}{2}\binom{2}{1}^2 \binom{3}{2} + \binom{3}{3}\binom{3}{1}^2 \binom{4}{2} = 21$$

possibilities. If the existence of empty cells is permitted, then there exist $A_\varnothing(1^2 2^1; 1^3) = \binom{3}{1}^2 \binom{4}{2} = 54$ possible arrangements. In the case when no cell is empty, the 21 possible arrangements are the following:

$(R)(G)(B, B);$ $(B)(G)(B, R);$ $(R)(B)(B, G);$ $(R)(B, B)(G);$

$(B)(R, B)(G);$ $(R)(B, G)(B);$ $(G)(R)(B, B);$ $(B)(R)(G, B);$

$(G)(B)(R, B);$ $(G)(B, B)(R);$ $(B)(G, B)(R);$ $(G)(B, R)(B);$

$(B, B)(R)(G);$ $(B, R)(B)(G);$ $(B, G)(R)(B);$ $(B, B)(G)(R);$

$(B, R)(G)(B);$ $(B, G)(B)(R);$ $(B)(B)(R, G);$ $(B)(R, G)(B);$

and $(R, G)(B)(B),$

where R stands for the red ball, G for the green ball and B for the blue ball.

The number $A(1^{\lambda_1} 2^{\lambda_2} \ldots n^{\lambda_n}; 1^m)$ also represents the number of possibilities of writing a natural number p as a product of m factors (where two products differ if they differ either in the nature or in the order of their factors), if the decomposition of p into distinct prime factors contains λ_1 factors with exponent 1, λ_2 factors with exponent 2, ..., λ_n factors with exponent n, where $\lambda_1 + 2\lambda_2 + \ldots + n\lambda_n = n =$ the number of factors in the product obtained by writing each factor with exponent 1.

Take, for example, the number $150 = 2.3.5^2$. It has $A(1^2 2^1; 1^3) = 21$ decompositions as a product of three natural numbers. These products are obtained from the above arrangements of balls if G is replaced by 2, R is replaced by 3 and B is replaced by 5.

9.2. The Möbius Function

The general theory of Möbius function, due to Gian-Carlo Rota [8], has an increasing importance in combinatorics, occurring in various problems. For instance, in the counting of cyclic words (C. Moreau), the determination of types of functions depending on multi-valued arguments [4] (which occur in the study of the real operation of relay-contact switching circuits [6]), the study of the chromatic number of a graph [11], and in the proof of the Ryser formula for the permanent of a matrix [2]. In this way connections are established between apparently disparate results, thus yielding a unified and systematic theory.

We shall define the Möbius function for a locally finite partially ordered set. Consider a set X, partially ordered by a relation denoted by $\leqslant$ (reflexive, antisymmetric and transitive). We can assume that X has a least element denoted by 0 (which can always be added to the original set X), i.e. $0 \leqslant x$ for every $x \in X$.

If for every $x, y \in X$, the interval $[x, y] = \{u | u \in X, u \geqslant x \text{ and } u \leqslant y\}$, i.e. the set of elements between x and y, is a finite set, then the partially ordered set X is called a *locally finite ordered set*.

Examples of locally finite ordered sets are: the set of non-negative integers endowed with the usual ordering, and the set of integers greater than or equal to one, endowed with the following order relation: $x \leqslant y$ if x divides y, which we write $x|y$. In the former case the least element is the number zero, while in the latter case the least element is the number one, which divides any integer.

Given a locally finite ordered set X, denote by A the set of all functions $f: X \times X \to R$, that is, the set of real-valued functions of two variables $x, y \in X$ (more generally, we can consider functions with values in an arbitrary field) having the following property: $f(x, x) \neq 0$ for every $x \in X$ and $f(x, y) = 0$ for $x \nleqslant y$, i.e. if $x > y$ or if x and y are incomparable with respect to the partial order.

We define a product $*$ on the set A by the equations:

$$f * g(x, y) = \sum_{x \leqslant u \leqslant y} f(x, u)g(u, y) \text{ if } x \leqslant y;$$

$$f * g(x, y) = 0 \text{ if } x \nleqslant y. \tag{9.3}$$

PROPOSITION 5. *The set A, endowed with the operation $*$, is a group, called the group of arithmetic functions.*

Now

$$* g : X \times X \to R \text{ and } f * g(x, x) = f(x, x)g(x, x) \neq 0,$$

while for $x \nleqslant y$, $f * g(x, y) = 0$ by definition, therefore $f * g \in A$. Notice that the sum in (9.3) has a meaning because of the finiteness of the interval $[x, y]$. Let us prove the associativity of the above defined product, that is,

$$(f * g) * h = f * (g * h).$$

Let $x \leqslant z$. Then

$$[(f * g) * h](x, z) = \sum_{x \leqslant y \leqslant z} (f * g)(x, y)h(y, z)$$

$$= \sum_{x \leqslant y \leqslant z} (\sum_{x \leqslant u \leqslant y} f(x, u)g(u, y))h(y, z)$$

$$= \sum_{x \leqslant u \leqslant z} (f(x, u) \sum_{u \leqslant y \leqslant z} g(u, y)h(y, z)) = [f * (g * h)](x, z).$$

For $x \nleqslant z$ both sides are equal to zero, hence $(f * g) * h = f * (g * h)$. The unit element for this product is the Kronecker function: $\delta(x, y) = 1$ for $x = y$ and $\delta(x, y) = 0$ for $x \neq y$. This follows since, $\delta \in A$ and $(f * \delta)(x, y) = \sum_{x \leqslant u \leqslant y} f(x, u)\delta(u, y) = $ $= f(x, y)$, because $\delta(u, y) = 0$ for every $u \neq y$. Therefore $f * \delta = f$ for every $f \in A$ and similarly $\delta * f = f$.

Given $f \in A$, let us define a left inverse $f^{-1} \in A$, such that $f^{-1} * f = \delta$. For a fixed $x \in X$, $f^{-1}(x, y)$ is defined by recurrence as follows:

1) if $y = x$, $f^{-1}(x, y) = \dfrac{1}{f(x, x)}$;

2) if $y > x$, $f^{-1}(x, y) = \dfrac{-1}{f(y, y)} \sum_{x \leqslant u < y} f^{-1}(x, u)f(u, y)$;

3) if $x \nleqslant y$, $f^{-1}(x, y) = 0$.

As X is a locally finite ordered set, the sum in 2) has a meaning, for the interval $[x, y) = [x, y] \setminus \{y\}$ is finite and $f \in A$ implies $f(x, x) \neq 0$ and $f(y, y) \neq 0$.

To define the function f^{-1} at the point y, we have assumed it has been defined on the interval $[x, y)$. This is possible, because if the interval $[x, y)$ contains the single element x, then $f^{-1}(x, y)$ follows from 2), the sum containing a single term. If the function $f^{-1}(x, y)$ has been defined for all the elements y such that the interval $[x, y)$ contains $1, 2, ..., n-1$ elements, then $f^{-1}(x, y)$ can be defined by formula 2) for all the elements y such that the interval $[x, y)$ contains n elements, because the values $f^{-1}(x, u)$ have been defined, the intervals $[x, u)$ with $x \leqslant u < y$ containing at most $n - 1$ elements. Thus f^{-1} will be defined for every pair (x, y), because the interval $[x, y]$ is finite. Let us prove that $f^{-1}(x, y)$ is actually a left inverse for the function $f(x, y)$:

$$(f^{-1} * f)(x, x) = f^{-1}(x, x)f(x, x) = 1,$$

which follows from relation 1). If $x < y$, then definition 2) implies

$$(f^{-1} * f)(x, y) = \sum_{x \leqslant u \leqslant y} f^{-1}(x, u)f(u, y)$$

$$= \sum_{x \leqslant u < y} f^{-1}(x, u)f(u, y) + f^{-1}(x, y)f(y, y) = 0.$$

If $x \not\leqslant y$, then obviously $(f^{-1} * f)(x, y) = 0$, hence $f^{-1} * f = \delta$, the unit element with respect to multiplication.

Let us prove that the function f^{-1} defined above is also a right inverse of the function f. Setting $f * f^{-1} = g$, we obtain $f * f^{-1} = g = \delta * g = g^{-1} * g * g = g^{-1} * f * f^{-1} * f * f^{-1}$.

We have made use of the fact that every function in A has a left inverse, therefore this is in particular the case for the function g, hence $\delta = g^{-1} * g$. By replacing $f^{-1} * f$ by δ, we get

$$f * f^{-1} = g^{-1} * f * \delta * f^{-1} = g^{-1} * f * f^{-1} = g^{-1} * g = \delta,$$

hence f^{-1} is also a right inverse for $f(x, y)$, i.e. it is the unique inverse of the function $f \in A$ with respect to the operation $*$. Summarizing: $(A, *)$ is a group.

The following property will be of a particular interest to us: every function of A having an inverse, from any relation of the form $f_1 = g_1 * \alpha$ one deduces, by right multiplication with α^{-1}, that

$$g_1 = f_1 * \alpha^{-1}.$$

If we set $f(x) = f_1(0, x)$ and $g(x) = g_1(0, x)$ for every $x \in X$, it follows that every function $\alpha(x, y)$ such that $\alpha(x, x) \neq 0$ and $\alpha(x, y) = 0$ for $x \not\leqslant y$, can be associated with a function $\beta(x, y)$ having the same properties, which is the inverse of the function $\alpha(x, y)$ with respect to the multiplication $*$. Furthermore

$$f(x) = \sum_{0 \leqslant u \leqslant x} \alpha(u, x)g(u),$$

implies (9.4)

$$g(x) = \sum_{0 \leqslant u \leqslant x} \beta(u, x)f(u).$$

If X is a locally finite ordered set, the *Riemann function* is a function in the group A of arithmetic functions over X, defined as follows: $\zeta(x, y) = 1$ if $x \leqslant y$ and $\zeta(x, y) = 0$ for $x \not\leqslant y$. To obtain the inverse of this function, we shall use the relations 1), 2) and 3). We thus obtain the *Möbius function* $\mu(x, y)$:

$$\mu(x, x) = 1;$$

$$\mu(x, y) = - \sum_{x \leqslant u < y} \mu(x, u) \text{ for } x < y;$$

$$\mu(x, y) = 0 \text{ for } x \not\leqslant y.$$

If we take the Riemann function $\zeta(x, y)$ as the function α and the Möbius function $\mu(x, y)$ as the function β, then relations (9.4) imply:

THE MÖBIUS INVERSION THEOREM. *If X is a locally finite ordered set, while $f(x)$ and $g(x)$ are functions defined on X such that*

$$f(x) = \sum_{0 \leqslant u \leqslant x} g(u) \text{ for every } x \in X,$$

then

$$g(x) = \sum_{0 \leqslant u \leqslant x} \mu(u, x) f(u) \text{ for every } x \in X.$$

We present in the sequel several examples illustrating the computation of the Möbius function for various locally finite ordered sets. The partially ordered sets referred to below are, in fact, lattices, i.e. partially ordered sets such that for every two elements $a, b \in X$ the greatest lower bound and the least upper bound of the elements a and b exist.

A finite partially ordered set X can be visualized in the plane by a graph the vertices of which represent the elements of the set X, an oriented arrow from vertex x to vertex y meaning that $x, y \in X$ and x covers y, that is, $x > y$ and there is no z such that $x > z > y$. Such a graph does not contain circuits and is called the *Hasse diagram* of the partially ordered set X. Sometimes arrows are omitted, the relation "x covers y" being indicated by the fact that x and y are joined by a line (without arrow), the element x being situated above the element y.

Example 1. Let X be the set of nonnegative integers $\{0, 1, 2, \ldots\}$ endowed with the usual ordering. Then X is a locally finite ordered set, for there is a finite number of integers t eween any two integers a and b with $b > a \geqslant 0$, namely $|[a, b]| = b - a + 1$. This set is a lattice, the least upper bound of the elements a and b with $a \leqslant b$ being b, which is also denoted

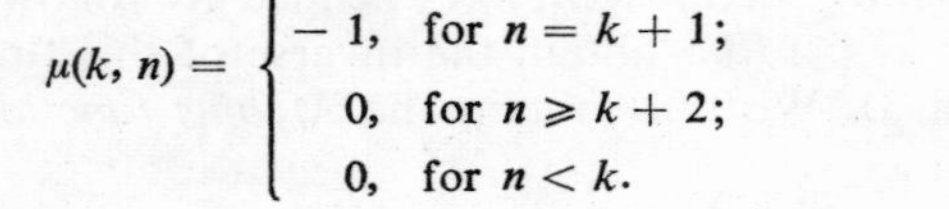

Fig. 9.1

by max $\{a, b\}$, the greatest lower bound being a, which is also denoted min $\{a, b\}$. The Möbius function $\mu(k, n)$ is defined (Fig. 9.1) for every $n \geqslant k$ by the equalities

$$\mu(k, n) = \begin{cases} 1, & \text{for } n = k; \\ -1, & \text{for } n = k + 1; \\ 0, & \text{for } n \geqslant k + 2; \\ 0, & \text{for } n < k. \end{cases}$$

By taking the inverse of the formula $f(n) = \sum_{k=0}^{n} g(k)$, we obtain

$$g(n) = f(n) - f(n - 1) \text{ for } n \geqslant 1 \text{ and } g(0) = f(0).$$

Example 2. Let X be the set of natural numbers $\{1, 2, \ldots\}$ endowed with the order relation $a \leqslant b$ if a divides b (which is written $a|b$). This set is partially ordered, because the divisibility relation is reflexive, symmetric and transitive, the least element being the number 1 which divides every number. This ordered set is locally finite and is a lattice, because for every two natural numbers the least upper bound with respect to divisibility is the least common multiple of the two numbers, while the greatest lower bound is the greatest common divisor of the two numbers. The Möbius function $\mu(d, n)$ of the lattice of divisors is equal to 1 for $d = n$, $\mu(d, n) = (-1)^k$ if $n = p_1 p_2 \ldots p_k d$ with p_i pairwise distinct prime numbers (the number 1 not being considered a prime number), and $\mu(d, n) = 0$ in all other cases, i.e. if d is not a divisor of n or if in the decomposition of n/d at least two prime numbers p_i and p_j are equal.

This function was introduced by Möbius (1832) with a view to studying the distribution of prime numbers.

An example of the computation of the Möbius function starting from the definition is given in Fig. 9.2, where a and b are distinct prime numbers.

By taking the inverse of the formula

$$f(n) = \sum_{d \mid n} g(d),$$

we obtain

$$g(n) = \sum_{d \mid n} \mu(d, n) f(d),$$

both formulae being valid for every natural number n.

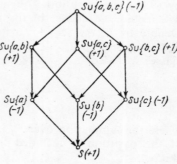

Fig. 9.2

Example 3. If M is a finite set, the family of subsets of M, denoted by $\mathscr{P}(M)$, is ordered by the inclusion relation, and this has the properties of a partial order relation. The least element of the family $\mathscr{P}(M)$ is the empty set $\varnothing$, while the greatest element is the set M.

$\mathscr{P}(M)$ is a locally finite set, because the number of sets Y with the property $B \subset Y \subset C$ is finite for every $B, C \subset M$. $\mathscr{P}(M)$ is a lattice as well, the least upper bound of the sets B and C with respect to the inclusion relation being their union $B \cup C$, while the greatest lower bound is their meet $B \cap C$. It can be shown that the Möbius function $\mu(S, B) = (-1)^{|B| - |S|}$ for $S \subset B \subset M$ and $\mu(S, B) = 0$ for $S \not\subset B$, as can be checked in Fig. 9.3, where $a, b, c \notin S$.

By taking the inverse of the formula $f(B) = \sum_{S \subset B} g(S)$,

we obtain

$$g(B) = \sum_{S \subset B} (-1)^{|B| - |S|} f(S),$$

both formulae being valid for every $B \subset M$.

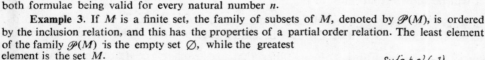

Fig. 9.3

Example 4. Consider a finite set X and a partition $\mathscr{A} = (A_1, A_2, \ldots, A_k)$ of it, i.e. $\bigcup_{i=1}^{k} A_i = X$, $A_i \neq \varnothing$ for $i = 1, \ldots, k$ and $A_i \cap A_j = \varnothing$ for every $i, j = 1, \ldots, k$ and $i \neq j$. Introduce the following partial ordering on the set of all partitions of X: the partition $\mathscr{A} = (A_1, \ldots, A_k)$ is said to be finer than the partition $\mathscr{B} = (B_1, \ldots, B_m)$, if for every two indices $i \in \{1, \ldots, k\}$ and $j \in \{1, \ldots, m\}$, the fact that $A_i \cap B_j \neq \varnothing$ implies that $A_i \subset B_j$. In other words, $\mathscr{A} < \mathscr{B}$ if every class of the partition $\mathscr{B}$ is the union of some classes of the partition $\mathscr{A}$.

For example, the partition $(\{1\}, \{2, 3\}, \{4\}, \{5\})$ is finer than the partition $(\{1, 2, 3\}, \{4, 5\})$.

This relation is a partial ordering on the set of partitions of X, for which the least element is the partition with $|X|$ classes, each of them consisting of a single element, while the greatest element is the partition with one class, namely the set X itself. The set of partitions of X is locally finite with respect to this order relation.

The set of partitions of X is also a lattice with respect to this partial order relation, the least upper bound of the partitions $\mathscr{A}$ and $\mathscr{B}$ being the partition $\mathscr{C}$ constructed as follows: two elements $i, j \in X$ will belong to the same class of $\mathscr{C}$ if and only if they belong to the same class of $\mathscr{A}$ or to the same class of $\mathscr{B}$. The greatest lower bound of the partitions $\mathscr{A}$ and $\mathscr{B}$ is the partition $\mathscr{D}$ for which two elements $i, j \in X$ belong to the same class if and only if they belong to the same class both in $\mathscr{A}$ and in $\mathscr{B}$.

For example, the least upper bound of the partitions $(\{1\}, \{2, 3\}, \{4, 5\})$ and $(\{1, 2\}, \{3, 4\}, \{5\})$ is the one-class partition $(\{1, 2, 3, 4, 5\})$, while the greatest lower bound is the five-class partition $(\{1\}, \{2\}, \{3\}, \{4\}, \{5\})$.

PROPOSITION 6 *(Schützenberger). If the partition $\mathscr{A}$ of the set X consists of the p classes $A_1, A_2, ..., A_p$ and if the partition $\mathscr{B} < \mathscr{A}$ has the property that every class A_i of the partition $\mathscr{A}$ contains n_i classes of the partition $\mathscr{B}$, then*

$$\mu(\mathscr{B}, \mathscr{A}) = (-1)^{p + n_1 + n_2 + \dots + n_p}(n_1 - 1)!\,(n_2 - 1)! \dots (n_p - 1)!. \qquad (9.6)$$

The inductive proof will proceed as follows: assuming that the expression (9.6) for $\mu(\mathscr{B}, \mathscr{X})$ is already established for every partition $\mathscr{X}$ such that $\mathscr{B} \leqslant \mathscr{X} < \mathscr{A}$, it will be shown that $\mu(\mathscr{B}, \mathscr{A}) = -\sum\limits_{\mathscr{B} \leqslant \mathscr{X} < \mathscr{A}} \mu(\mathscr{B}, \mathscr{X})$, where the terms $\mu(\mathscr{B}, \mathscr{X})$ and $\mu(\mathscr{B}, \mathscr{A})$ are given by formula (9.6), since (9.6) holds for $\mathscr{A} = \mathscr{B}$. In other words, let us show that $\sum\limits_{\mathscr{B} \leqslant \mathscr{X} \leqslant \mathscr{A}} \mu(\mathscr{B}, \mathscr{X}) = 0$ where all the terms $\mu(\mathscr{B}, \mathscr{X})$ have been determined by (9.6). The partition $\mathscr{X}$ being finer than $\mathscr{A}$, each class A_i of the partition $\mathscr{A}$ will include m_i classes of the partition $\mathscr{X}$, where $n_i \geqslant m_i$ for $\mathscr{B}$ is finer than $\mathscr{X}$. Denote by P_i the set of partitions $\mathscr{X}$ with the property that the classes included in the classes $A_1, A_2, ..., A_{i-1}, A_{i+1}, ..., A_p$ of the partition $\mathscr{A}$ are fixed, while the classes included in A_i vary in all possible ways with $\mathscr{B} \leqslant \mathscr{X} \leqslant \mathscr{A}$. Then

$$\sum\limits_{\mathscr{B} \leqslant \mathscr{X} \leqslant \mathscr{A}} \mu(\mathscr{B}, \mathscr{X}) = \sum\limits_{P_i} \sum\limits_{\mathscr{X} \in P_i} \mu(\mathscr{B}, \mathscr{X}),$$

where the summation over P_i is performed over all possible choices of the classes of the partition $\mathscr{X}$ included in $A_1, A_2, ..., A_{i-1}, A_{i+1}, ..., A_p$, with $\mathscr{B} \leqslant \mathscr{X} \leqslant \mathscr{A}$. But

$$\sum\limits_{\mathscr{X} \in P_i} \mu(\mathscr{B}, \mathscr{X}) = \sum\limits_{k=1}^{n_i} \sum\limits_{\lambda_1 + 2\lambda_2 + \dots + k\lambda_k = n_i} \frac{(-1)^{n_i + (\lambda_1 + \dots + \lambda_k)}\, n_i!}{1^{\lambda_1} 2^{\lambda_2} \dots k^{\lambda_k} \lambda_1! \dots \lambda_k!}.$$

Since the classes of the partition $\mathscr{X}$ included in A_i vary in all possible ways provided that they include the n_i classes of $\mathscr{B}$, the value of the Möbius function $\mu(\mathscr{B}, \mathscr{X})$ is the same for all the partitions of A_i that have λ_1 classes each of which includes one class of $\mathscr{B}$, ..., λ_k classes each of which include k classes of the partition $\mathscr{B}$. If the classes of $\mathscr{X}$ included in A_i are obtained in this way, $\mu(\mathscr{B}, \mathscr{X})$ is calculated, according to the inductive hypothesis, by (9.6) and we get $\mu(\mathscr{B}, \mathscr{X}) = (-1)^{\lambda_1 + \lambda_2 + \cdots + \lambda_k + n_i} (0!)^{\lambda_1} (1!)^{\lambda_2} \ldots ((k-1)!)^{\lambda_k}.$

Since the number of partitions of type $1^{\lambda_1} 2^{\lambda_2} \ldots k^{\lambda_k}$ is equal to

$$\frac{n_i!}{(1!)^{\lambda_1} (2!)^{\lambda_2} \ldots (k!)^{\lambda_k} \lambda_1! \lambda_2! \ldots \lambda_k!}$$

by (4.6), the above equality follows by multiplication and summation. But

$$n_i + (\lambda_1 + \lambda_2 + \ldots + \lambda_k) = 2\lambda_1 + 3\lambda_2 + 4\lambda_3 + \ldots,$$

and this sum has the parity of $\lambda_2 + \lambda_4 + \ldots$. From the Cauchy formula (7.1), the expression

$$\frac{n_i!}{1^{\lambda_1} 2^{\lambda_2} \ldots k^{\lambda_k} \lambda_1! \ldots \lambda_k!}$$

gives precisely the number of permutations which contain λ_1 cycles with one element, ..., λ_k cycles with k elements. The parity of such a permutation is the same as the parity of the sum $\lambda_2 + \lambda_4 + \ldots$, so that by separating the double sum according to the signs of the terms, we obtain

$$\sum_{\mathscr{X} \in P_i} \mu(\mathscr{B}, \mathscr{X}) = \frac{1}{2} n_i! - \frac{1}{2} n_i! = 0,$$

because the number of even permutations is equal to that of odd permutations. Hence $\sum_{\mathscr{B} \leqslant \mathscr{X} \leqslant \mathscr{A}} \mu(\mathscr{B}, \mathscr{X}) = 0$. Since $p + n_1 + \ldots + n_p$ has the same parity as $(n_1 - 1) + \ldots + (n_p - 1)$, it follows that $\mu(\mathscr{B}, \mathscr{A}) = \mu(\mathscr{B}_{n_1}, A_1) \ldots \mu(\mathscr{B}_{n_p}, A_p)$, where we have denoted by $\mathscr{B}_{n_i}$ the subpartition of $\mathscr{B}$ that contains the n_i classes included in A_i, while $\mu(\mathscr{B}_{n_i}, A_i)$ stands for the Möbius function of the lattice of partitions of A_i, for $i = 1, \ldots, p$. This relation can be generalized to the case of the product of lattices [8].

An example of the Möbius function for the lattice of partitions of a set with three elements a, b, c, is given in Fig. 9.4. By taking the inverse of the formula $f(\mathscr{A}) = \sum_{\mathscr{X} \leqslant \mathscr{A}} g(\mathscr{X})$, we obtain

$$g(\mathscr{A}) = \sum_{\mathscr{X} \leqslant \mathscr{A}} \mu(\mathscr{X}, \mathscr{A}) f(\mathscr{X}),$$

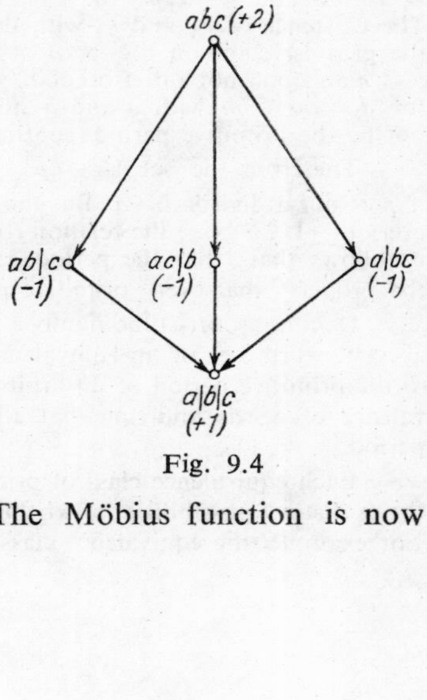

Fig. 9.4

both equations being valid for every partition $\mathscr{A}$ of the set X, while the summation starts with the finest partition of X, the classes of which are singletons. The Möbius function is now calculated by means of (9.6).

Application. *Determination of the number of cyclic words.* Consider an alphabet A with m distinct symbols, called letters and denoted by $a_1, a_2, \ldots, a_m$. Every mapping $f: X \to A$, where $X = \{1, 2, \ldots, n\}$, can be viewed as a word of length n, namely the word $f(1)f(2) \ldots f(n)$. We have seen in Chap. 2 that this correspondence between the set of functions from X into A and the set of all words of length n made up of letters from the alphabet A, is a bijection. As is usually done, we shall identify a word with the function associated with it by this bijection.

The following equivalence relation is introduced on the set of words of length n made up of letters from A: we say that the words f_1 and f_2 are equivalent, or constitute the same cyclic word, if there is a number $p \in \{0, 1, \ldots, n-1\}$ such that $f_2(i) = f_1(i + p)$ for every $i = 1, 2, \ldots, n$, the sums being taken modulo n (i.e. in the case of sums greater than n, only the remainder after division by n is taken into consideration). We then write $f_1 \sim f_2$. This relation is an equivalence relation, since it is:

1) reflexive: $f \sim f$ for every $f \in A^X$, choosing $p = 0$;

2) symmetric: $f_1 \sim f_2$ means by definition the existence of p such that $0 \leqslant p \leqslant n-1$ and $f_2(i) = f_1(i + p)$ for every $i = 1, 2, \ldots, n$. Hence $f_1(i + p) = f_2(i + n)$, where the sums are understood modulo n, therefore $f_1(i) = f_2(i + n - p)$ for every $i = 1, \ldots, n$, that is, $f_2 \sim f_1$;

3) transitive: $f_1 \sim f_2$ and $f_2 \sim f_3$ implies the existence of two numbers p_1 and p_2 in $\{0, 1, \ldots, n-1\}$ such that $f_2(i) = f_1(i + p_1)$ for $i \geqslant 1$ and $f_3(i) = f_2(i + p_2)$ for every $i \geqslant 1$. Therefore $f_3(i) = f_2(i + p_2) = f_1(i + p_1 + p_2)$ for every $i \geqslant 1$, that is, setting $p = p_1 + p_2$ (modulo n), $f_3(i) = f_1(i + p)$ for every $i \geqslant 1$, which means $f_1 \sim f_3$.

The classes of this equivalence will be called cyclic words with n letters. We say that p is a period of the word f if $1 \leqslant p \leqslant n$ and the word f satisfies the equation $f(i + p) = f(i)$ for every $i \geqslant 1$, the sums being taken modulo n. Obviously, one of the periods of every word of length n is the number n itself.

The least period p of the word f is called the *primitive period* of f. For example, the word *abababab* has the primitive period $p = 2$, every multiple of the primitive period which is less than or equal to n being a period, too. We shall prove that the primitive period p of the word f is a divisor of its length n. In order to proceed by *reductio ad absurdum,* assume that the primitive period p does not divide n. In this case we shall prove that for every $i, j \in \{1, 2, \ldots, n\}$ the relation $f(i) = f(j)$ holds. That is the word f has period 1 (which is a divisor of n), hence $1 \neq p$ and so there exists a period smaller than p. By iterating the relation $f(i) = f(i + p)$, we obtain $f(i) = f(i + kp)$ for every integer $k \geqslant 1$.

Assume that the numbers of the form $i + kp$, where $k = 1, 2, \ldots, n$, are pairwise distinct modulo n, that is, there are no numbers $k_1, k_2 \in \{1, 2, \ldots, n\}$, with $k_1 \neq k_2$, such that $i + k_1 p = a + rn$ and $i + k_2 p = a + sn$, where r and s are nonnegative integers and $a \in \{1, 2, \ldots, n\}$. The existence of k_1 and k_2 with the above properties would imply $(k_1 - k_2)p = (r - s)n$, hence the greatest common divisor d of p and n would satisfy $1 < d < p$, because we have assumed $p < n$ and p is not a divisor of n, while $k_1 - k_2 \leqslant n - 1$. In this case d is a period of the word f, for $d = ap + bn$ with a and b integers and $f(i) = f(i + ap + bn) = f(i + d)$, so that p would not be the primitive period, contrary to the hypothesis.

Therefore the set $\{i + kp\}_{k=1, \ldots, n}$ contains only pairwise distinct numbers modulo n if p is not a divisor of n. But the pairwise distinct numbers modulo n are $1, 2, \ldots, n$, hence for every $j \in \{1, 2, \ldots, n\}$ the relation $f(i) = f(j)$ holds. Since all the letters of the word f are identical, it follows that f has the period $p = 1$, which divides n, thus proving by *reductio ad absurdum* the property that every primitive period p divides n.

Denote by $M(p)$ the number of cyclic words of primitive period p. The primitive period of a cyclic word, i.e. of an equivalence class with respect to the above defined equivalence, is equal to the primitive period of an arbitrary word of that equivalence class, the definition of the equivalence of words implying that all the words of an equivalence class have the same primitive period.

Each equivalence class of primitive period p contains exactly p words, which correspond to the p distinct possibilities of writing a cycle of length p taking as initial letter each of its p letters. For example, the equivalence class of the word *abc* of primitive period $p = 3$ contains also the words *bca* and *cab*. Therefore the $M(p)$ equivalence classes of primitive period p contain $pM(p)$ words from A^X. Since the equivalence classes form a partition of the set of m^n words of length n made up of letters from A, it follows that

$$m^n = \sum_{p|n} pM(p).$$

By taking the inverse of this formula, we obtain $pM(p) = \sum_{q|p} \mu(q, p)m^q$, where $\mu(q, p)$ is the Möbius function of the lattice of divisors (example 2), defined as follows: $\mu(q, p) = (-1)^k$ if $p = p_1 p_2 \ldots p_k q$ with p_i pairwise distinct prime numbers, and $\mu(q, p) = 0$ otherwise. The total number of cyclic words of length n made up of m letters is equal to the number of equivalence classes, that is,

$$C(n, m) = \sum_{p|n} M(p) = \sum_{p|n} \frac{1}{p} \sum_{q|p} \mu(q, p)m^q. \qquad (9.7)$$

For example, for $n = 3$ and $m = 3$ one obtains $C(3, 3) = \sum_{p|3} \frac{1}{p} \sum_{q|p} \mu(q, p)3^q =$

$$= \sum_{q|1} \mu(q, 1)3^7 + \frac{1}{3} \sum_{q|3} \mu(q, 3)3^q = 3\mu(1, 1) + \frac{1}{3}(\mu(1, 3)3^1 + \mu(3, 3)3^3) = 3 + \frac{1}{3}(-3 + 27) =$$

$= 3 + 8 = 11$ cyclic words. If the alphabet $A = \{a, b, c\}$, then these cyclic words (taking one representative from each class) are the following:

$$aaa, \; bbb, \; ccc, \; abc, \; acb, \; aab, \; aac, \; bbc, \; acc, \; abb, \; bcc.$$

Assuming that these words are codified messages written around a circle and we do not know the beginning of each message, but the direction of reading is fixed, say clockwise sense, then $C(n, m)$ represents the maximum number of those messages of length n made up of m letters that can be correctly decodified. Indeed, by choosing one message in each equivalence class and knowing which message has been chosen in each class, then, wherever we begin reading the message, in the fixed direction of reading, we obtain a message from the equivalence class of the original message written around the circle. Therefore the determination of the equivalence class results in the decoding of the transmitted message.

Fig. 9.5 represents the equivalence class of the cyclic word *abc* of primitive period $p = 3$, which thus contains three words.

A generalization of this problem in information theory is the problem of constructing a comma free dictionary with a maximum number of words. A *comma free dictionary* is a set of words made up of n letters with the following property: for every two words of the dictionary, there is no integer k with $1 \leqslant k \leqslant n - 1$ such that the last $n - k$ letters of the first word followed by the first k letters of the second word form a word of the same dictionary; this ensures the uniqueness of decoding.

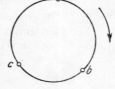

Fig. 9.5

We mention that these problems occur within the context of genetic codes, which are comma free codes written on circles or on spatial spirals.

In music the problem of determining modal richness reduces to the problem of finding the number of cyclic words that contain certain letters a certain number of times; this can be solved by means of Pólya's method, by utilizing the cycle index of the cyclic group in n variables.

Problems

1.   Let $\mathscr{A} = (A_1, \ldots, A_k)$ and $\mathscr{B} = (B_1, \ldots, B_m)$ be two partitions of the set X. We define a bipartite nonoriented graph as follows: the vertices are denoted by $a_1, \ldots, a_k, b_1, \ldots, b_m$ and the edges are those pairs $[a_i, b_j]$ for which $A_i \cap B_j \neq \varnothing$.

Prove that the greatest lower bound of the partitions $\mathscr{A}$ and $\mathscr{B}$ has its blocks of the form: $A_i \cap B_j$ where $[a_i,\ b_j]$ is an edge of the graph, and that the least upper bound of these partitions has its blocks of the form $\bigcup A_i$, where the join is over all vertices a_i belonging to the same connected component of the graph.

2.   Prove the following generalization of proposition 2: If $a_0, a_1, ..., a_n$ and $b_0, b_1, ..., b_n$ are related by the equations:

$$a_k = \sum_{i=0}^{k} \begin{bmatrix} k \\ i \end{bmatrix}_q b_i, \text{ then } b_k = \sum_{i=0}^{k} (-1)^i q^{\binom{i}{2}} \begin{bmatrix} k \\ i \end{bmatrix}_q a_i.$$

(Goldman, Rota, 1970)

3.   By using the Möbius function of the lattice of partitions of an n-set, show that the number of nonoriented connected graphs with n labelled vertices is equal to

$$\sum_{(k_1,...,k_n)} (-1)^{\sum_{i=1}^{n} k_i - 1} 2^{\sum_{i=2}^{n} k_i \binom{i}{2}} \left(\sum_{i=1}^{n} k_i - 1 \right)! \frac{n!}{(1!)^{k_1} k_1! ... (n!)^{k_n} k_n!}.$$

The sum is over all solutions in nonnegative integers of the equation:

$$k_1 + 2k_2 + ... + nk_n = n.$$

(Rota, Crapo, 1971)

H i n t: We put $V = \{x_1, x_2, ..., x_n\}$, for any graph G with the vertex-set V and we define a partition $\pi(G)$ of V, a block of this partition being the set of vertices of a connected component of G. The graph is connected if $\pi(G) = 1$, where 1 is the partition with a single block V.

Let $C(\pi)$ enumerate the graphs G with $\pi(G) = \pi$ and let $D(\pi)$ enumerate the graphs G with $\pi(G) \leqslant \pi$.

If π is of the type $1^{k_1} 2^{k_2} ... n^{k_n}$, then we have: $D(\pi) = 2^{\sum_{i=2}^{n} k_i \binom{i}{2}} = \sum_{\sigma \leqslant \pi} C(\sigma)$. Applying the Möbius inversion formula we get $C(1) = \sum_{\sigma \leqslant 1} D(\sigma) \mu(\sigma, 1)$.

4.   A chain of length p of the lattice of partitions of an n-element set X is a sequence of distinct partitions such that:

$$\mathscr{P}_0 < \mathscr{P}_1 < \mathscr{P}_2 < ... < \mathscr{P}_p.$$

If $p = n$ then obviously $\mathscr{P}_0$ has n blocks and $\mathscr{P}_p$ has a unique block, equal to X.

This chain is connected if $\mathscr{P}_{i+1}$ covers $\mathscr{P}_i$, i.e. there exists no partition $\mathscr{Q}$ different from $\mathscr{P}_i$ and $\mathscr{P}_{i+1}$ such that $\mathscr{P}_i < \mathscr{Q} < \mathscr{P}_{i+1}$ for $i = 0, ..., p - 1$.

Prove that the number of connected chains of length n of a finite set X having n elements is equal to

$$\frac{(n-1)!\, n!}{2^{n-1}}.$$

H i n t: If $\mathscr{P}_0, \ldots, \mathscr{P}_{k-1}$ are fixed partitions, then $\mathscr{P}_k$ may be chosen in $\binom{n-k+1}{2}$ ways in order to cover $\mathscr{P}_{k-1}$, because $\mathscr{P}_k$ is obtained from $\mathscr{P}_{k-1}$ by the join of two blocks of $\mathscr{P}_{k-1}$.

Hence the number of connected chains of length n is equal to $\prod\limits_{k=1}^{n-1}\binom{n-k+1}{2}$.

5.   Let us define the numbers $L(n, k)$ (called Lah numbers) by

$$[-x]_n = \sum_{k=1}^{n} L(n, k)[x]_k.$$

Prove that:

i) If $a_k = \sum\limits_{i=1}^{k} L(k, i)b_i$ for $k = 1, 2, \ldots, n$, then

$$b_k = \sum_{i=1}^{k} L(k, i)a_i \quad \text{for} \quad k = 1, 2, \ldots, n;$$

ii) $L(n + 1, k) = -(n + k)L(n, k) - L(n, k - 1);$

iii) $L(n, k) = (-1)^n \dfrac{n!}{k!}\dbinom{n-1}{k-1}.$

6.   Prove that:

$$A(1^p\, q;\, m) = \binom{p}{m} + \binom{p}{m-1} + \ldots + \binom{p}{m-q};$$

$$A(n_1 n_2 \ldots n_k;\, m) = A(n_2 \ldots n_k;\, m) + A(n_2 \ldots n_k;\, m - 1) + \ldots$$

$$+ A(n_2 \ldots n_k;\, m - n_1).$$

7.   If we set $L(n, m) = A(n;\, 1^m)$, prove that

$$L(n, m) + \binom{m}{1} L(n, m - 1) + \ldots + \binom{n}{k} L(n, m - k) + \ldots + L(n, 0)$$

$$= \binom{n + m - 1}{n}.$$

By applying proposition 2 derive the identity

$$\binom{n-1}{m-1} = \sum_{k=0}^{m-1} (-1)^k \binom{m}{k}\binom{m+n-k-1}{n}.$$

8.   Prove that the number of circular words $C(n, m)$ is equal to $P(C_n; m, m, ..., m)$ where $P(C_n; x_1, ..., x_n)$ is the cycle index of the cyclic group C_n and the number of circular words which use the letter a_1 k_1 times, ..., letter a_m k_m times $\left(\sum_{i=1}^{m} k_i = n \right)$ is equal to the coefficient

$$\frac{1}{n} \sum_{d|(k_1, ..., k_m)} \varphi(d) \frac{(n/d)!}{(k_1/d)! \, ... \, (k_m/d)!}$$

of $w_1^{k_1} ... w_p^{k_p}$ in the expansion of the polynomial

$$P\left(C_n; \sum_{i=1}^{p} w_i, \sum_{i=1}^{p} w_i^2, ..., \sum_{i=1}^{p} w_i^n \right)$$

where $(k_1, ..., k_m)$ is the greatest common divisor of $k_1, ..., k_m$.
H i n t: Apply Pólya's theorem.

9.   Prove that the Möbius function $\mu(S, B) = (-1)^{|B| - |S|}$ for $S \subset B$.

10.   Verify that in the lattice of divisors of n, $\mu(d, n) = 1$ for $d = n$, $\mu(d, n) = (-1)^k$ for $n = p_1 p_2 ... p_k d$ where p_i are pairwise distinct prime numbers and $\mu(d, n) = 0$ otherwise.

11.   Let us consider orderings of elements in which the order of the groups is significant, and also the order of the elements in the groups.

Prove that if n distinct things are ordered into m distinguishable groups, all groups nonempty, them the number of orderings is $n! \binom{n-1}{m-1}$, and if the groups are indistinguishable, then the number of orderings is $\dfrac{n!}{m!} \binom{n-1}{m-1}$.

H i n t: Use the property which asserts that $A(n; 1^m) = \binom{n-1}{m-1}$.

BIBLIOGRAPHY

1. Berge, C., *Principes de combinatoire*, Dunod, Paris, 1968.
   English edition: *Principles of combinatorics*, Academic Press, New York, 1971.
2. Crapo, H., *Permanents by Möbius inversion*, J. Combinatorial Theory, **4**, 1968, 198–200.

3. Goldman, J., Rota, G.-C., *On the foundations of combinatorial theory IV: Finite vector spaces and Eulerian generating functions*, Studies in Appl. Math., **49**, 1970, 239—258.
4. Harrison, M., *Sur la classification des fonctions logiques à plusieurs valeurs*, Bull. Math. Soc. Sci. Math. R. S. Roumanie, **13** (61), 1969, 41—54.
5. MacMahon, P. A., *Combinatory Analysis*, Vol. I, Cambridge Univ. Press, Cambridge, 1915, reprinted Chelsea, New York, 1960.
6. Moisil, G. C., *Théorie structurelle des automates finis*, Gauthier-Villars, Paris, 1967.
7. Möbius, A. F., *Über eine besondere Art von Umkehrung der Reihen*, J. Reine Angew. Math., **9**, 1832, 105—123.
8. Rota, G.-C., *On the foundations of combinatorial theory I: Theory of Möbius functions*, Z. Wahrscheinlichkeitstheorie und Verw. Gebiete, **2**, 1964, 340—368.
9. Rota, G.-C., Crapo, H., *Combinatorial geometries*, M.I.T. Press, Cambridge, Mass., 1971.
10. Schützenberger, M. P., *Contributions aux applications statistiques de la théorie de l'information*, Publ. Inst. Statist. Univ. Paris, **3**, 1954, 5—117.
11. Wilf, H. S., *Hadamard determinants, Möbius functions and the chromatic number of a graph*, Bull. Amer. Math. Soc., **74**, 1968, 960—964.

CHAPTER 10

Systems of Distinct Representatives

10.1. THE EXISTENCE THEOREM

The problem of the existence of a system of distinct representatives (SDR), as well as the counting of them is a combinatorial problem which occurs for example in the study of matchings of bipartite graphs.

Let S be an arbitrary set and $\mathscr{P}(S)$ the family of its subsets. Let $M(S) = (S_1, S_2, ..., S_m)$ with $S_i \in \mathscr{P}(M)$. The m-tuple $(a_1, a_2, ..., a_m)$ is an SDR for $M(S)$, if $a_i \in S_i$ for $i = 1, 2, ..., m$ and $a_i \neq a_j$ for any $i, j \in \{1, ..., m\}$ with $i \neq j$.

For example, let $S = \{1, 2, 3, 4, 5, 6\}$ and $S_1 = \{2, 6\}$, $S_2 = \{2, 4\}$, $S_3 = \{2, 4\}$, $S_4 = \{1, 2, 5, 6\}$, $S_5 = \{3, 6\}$. In this case the 5-tuple $(6, 2, 4, 1, 3)$ will be an SDR for $(S_1, S_2, S_3, S_4, S_5)$. If S_4 is replaced by the set $\{4, 6\}$, then obviously these subsets have no SDR. The reason is that $\{2, 6\} \cup \{2, 4\} \cup \{2, 4\} \cup \{4, 6\} = \{2, 4, 6\}$; if an SDR were to exist, then the union of the four sets would contain at least four elements. Also, if the number of sets exceeds that of elements, $m > |S|$, then the existence of an SDR is impossible. These remarks are included in the following more comprehensive result:

PROPOSITION 1 *(Philip Hall)*. *The sets $S_1, S_2, ..., S_m$ have an SDR if and only if the set $S_{i_1} \cup S_{i_2} \cup ... \cup S_{i_k}$ contains at least k elements for every $k = 1, 2, ..., m$ and for every choice of pairwise distinct numbers $\{i_1, i_2, ..., i_k\} \subset \{1, ..., m\}$.*

To prove necessity, assume that the subsets $S_1, ..., S_m$ have an SDR, namely $(a_1, a_2, ..., a_m)$. Since $S_{i_1} \cup S_{i_2} \cup ... \cup S_{i_k} \supset \{a_{i_1}, a_{i_2}, ..., a_{i_k}\}$, it follows that $|S_{i_1} \cup S_{i_2} \cup ... \cup S_{i_k}| \geqslant k$ for every k and every choice of pairwise distinct numbers $i_1, i_2, ..., i_k$.

We shall now prove a result stronger than sufficiency, which will enable us to obtain a lower bound for the number of SDR's. Assume that the subsets $S_1, S_2, ..., S_m$ satisfy the necessary condition for the existence of an SDR and each of these subsets contains at least t elements. If $t \leqslant m$, there exist at least $t!$ SDR's, while if $t > m$ there exist at least $\dfrac{t!}{(t-m)!}$ SDR's (M. Hall [8]). The proof of this property is inductive. For $m = 1$, the subset S_1 contains t elements. If $t = 1$, there exists a unique SDR, while if $t > 1$ there exist $t = \dfrac{t!}{(t-1)!}$ SDR's, and the proposition holds.

We now assume the proposition to be true for every $m' < m$ and we prove it for $M(S) = (S_1, S_2, ..., S_m)$. Two cases may occur:

1) The set $S_{i_1} \cup S_{i_2} \cup \ldots \cup S_{i_k}$ contains at least $k + 1$ elements for $k = 1, 2, \ldots, m - 1$ and for every choice of distinct numbers $\{i_1, i_2, \ldots, i_k\} \subset \{1, \ldots, m\}$.

Let $a_1 \in S_1$. Remove a_1 from the other sets of $M(S)$, wherever it appears, and denote the resulting sets by $S_2', S_3', \ldots, S_m'$. Then $M'(S) = (S_2', S_3', \ldots, S_m')$ satisfies the necessary and sufficient condition for the existence of an SDR, because the set $S_{i_1} \cup S_{i_2} \cup \ldots \cup S_{i_k}$ contains at least $k + 1$ elements and we have deleted a single element a_1. If $t \leqslant m$, then $t - 1 \leqslant m - 1$ and since every set from $M'(S)$ contains at least $t - 1$ elements, it follows from the inductive hypothesis that $M'(S)$ has at least $(t - 1)!$ SDR's. Indeed, if we are able to choose S_1 so that it is not a set with the minimum number of elements in $M(S)$, then $\min\limits_{2 \leqslant i \leqslant m} |S_i'| = t - 1$ or t. The only unfavourable situation would be that in which this minimum is equal to t and $t = m > m - 1$; but, considering the case $t > m - 1$, there exist $t! > (t - 1)!$ SDR's for $M'(S)$.

Similarly, if $t > m$, we obtain $t - 1 > m - 1$ and $M'(S)$ has at least $\dfrac{(t - 1)!}{(t - m)!}$ SDR's by the inductive hypothesis, because $\min\limits_{2 \leqslant i \leqslant m} |S_i'| \geqslant t - 1 > m - 1$. But the element a_1, which can be chosen in at least t distinct ways in the set S_1, together with an SDR for $M'(S)$ form an SDR for $M(S)$, in which a_1 represents S_1, because a_1 does not appear in the sets of $M'(S)$. It follows that if $t \leqslant m$, there are at least $t!$ SDR's, while if $t > m$ there are at least $\dfrac{t!}{(t - m)!}$ SDR's for $M(S)$.

2) If there is a $k \in \{1, \ldots, m - 1\}$ and a choice $\{i_1, i_2, \ldots, i_k\} \subset \{1, 2, \ldots, m\}$ such that $|S_{i_1} \cup S_{i_2} \cup \ldots \cup S_{i_k}| = k$, then we renumber the sets $S_1, S_2, \ldots, S_m$ so that S_{i_1} becomes S_1, S_{i_2} becomes S_2, ..., S_{i_k} becomes S_k.

In this case we have $t \leqslant k$, because $S_i \subset S_1 \cup S_2 \cup \ldots \cup S_k$ for every $i = 1, \ldots, k$.

According to the inductive hypothesis, $(S_1, S_2, \ldots, S_k)$ has at least $t!$ SDR's. Denote by $D^* = (a_1, a_2, \ldots, a_k)$ one of these SDR's and remove the elements of D^* wherever they appear in the sets $S_{k+1}, S_{k+2}, \ldots, S_m$. Denote the sets thus obtained by $S_{k+1}^*, S_{k+2}^*, \ldots, S_m^*$. The system of sets $M^*(S) = (S_{k+1}^*, \ldots, S_m^*)$ contains $m - k$ sets and satisfies the necessary and sufficient condition for the existence of an SDR. For, if $S_{i_1}^* \cup S_{i_2}^* \cup \ldots \cup S_{i_{k^*}}^*$ with $\{i_1, i_2, \ldots, i_{k^*}\} \subset \{k + 1, \ldots, m\}$ contains fewer than k^* elements, then $S_1 \cup S_2 \cup \ldots \cup S_k \cup S_{i_1}^* \cup \ldots \cup S_{i_{k^*}}^*$ would contain fewer than $k + k^*$ elements, because $|S_1 \cup \ldots \cup S_k| = k$, thus contradicting the hypothesis that $M(S)$ satisfies the necessary and sufficient condition for the existence of an SDR.

According to the inductive hypothesis, $M^*(S)$ has at least one SDR, which, together with D^* forms an SDR for $M(S)$, because $M^*(S)$ does not contain the elements of D^*. But $(S_1, \ldots, S_k)$ has at least $t!$ SDR's by the inductive hypothesis, hence $M(S)$ has at least $t!$ SDR's.

M. Hall [8] extended proposition 1 to the infinite case as follows: let I be an infinite set, and let $\mathfrak{A} = (A_i : i \in I)$ be a family of finite subsets of the non-empty set E. A subset F of E is called a transversal of $\mathfrak{A}$ (or an SDR of $\mathfrak{A}$) if there is a bijection $\theta : I \to F$ such that $\theta(i) \in A_i (i \in I)$.

M. Hall's theorem asserts that the family $\mathfrak{A}$ has a transversal if and only if, for every natural number k, the union of every k A's contains at least k (distinct) elements of E. Other generalizations are given, for example, in [24], [32]. We mention here the following generalization by R. Rado of Hall's theorem [20, p. 530]:

RADO'S THEOREM. *Let $(M_i : i \in I)$ be a family of subsets of E and assume that if I is infinite, then all M_i are finite. Furthermore, let r be a natural number. Then the family $(M_i : i \in I)$ possesses a system of representatives in which no element of E occurs more than r times if and only if, for each natural number $k \leqslant |I|$, the union of any k M_i's contains at least k/r elements.*

A *bipartite graph* is a graph $G = (X, Y, \Gamma)$ with $X \cap Y = \emptyset$, where $\Gamma : X \rightarrow \mathscr{P}(Y)$, that is, its set of vertices is $X \cup Y$ and the arcs go from vertices of X to vertices of Y. A *matching* C of the bipartite graph G is a set of arcs, such that two arbitrary arcs of the matching have no common extremities and from every vertex of X starts an arc of C. In other words, the arcs of a matching establish an injection $f : X \rightarrow Y$.

PROPOSITION 2 *(König-Hall). The bipartite graph $G = (X, Y, \Gamma)$ has a matching C if and only if for every subset $A \subset X$ the relation $|\Gamma A| \geqslant |A|$ holds.*
We recall the notation $\Gamma A = \bigcup_{x \in A} \Gamma(x)$ used in graph theory. If the set X consists of the m elements $x_1, x_2, \ldots, x_m$, set $S_i = \Gamma x_i$ for $i = 1, \ldots, m$.

There exists a bijection from the set of matchings of the graph G onto the set of SDR's of the set system $M(S) = (S_1, S_2, \ldots, S_m)$, with $S_i \subset Y$ for $i = 1, \ldots, n$. For every matching C is characterized by m arcs: $(x_1, y_1), \ldots, (x_m, y_m)$, no two of which have common extremities, hence the elements $y_1, y_2, \ldots, y_m$ are distinct and since $y_i \in S_i$ for $i = 1, \ldots, m$, the m-tuple $(y_1, \ldots, y_m)$ is an SDR for $M(S)$. This mapping from the set of matchings into the set of SDR's for the above defined sets is an injection, since to distinct mappings correspond distinct SDR's. It is a surjection as well, because an SDR denoted by $(y_1, y_2, \ldots, y_m)$ is the image of the matching $C = \{(x_1, y_1), \ldots, (x_m, y_m)\}$ by the above described mapping. Since $y_1, y_2, \ldots, y_m$ are distinct, it follows that C is a matching of the graph G, because we also have $y_i \in S_i = \Gamma x_i$ for $i = 1, \ldots, m$. Therefore the bipartite graph G has a matching if and only if the system of sets $M(S)$ has an SDR. Now it follows from proposition 1, that $M(S)$ has an SDR if and only if $|S_{i_1} \cup S_{i_2} \cup \ldots \cup S_{i_k}| \geqslant k$ for every $k = 1, \ldots, m$ and every choice of the pairwise distinct numbers $\{i_1, i_2, \ldots, i_k\} \subset \{1, 2, \ldots, m\}$. Setting

$$A = \{x_{i_1}, x_{i_2}, \ldots, x_{i_k}\}$$

we obtain

$$|S_{i_1} \cup \ldots \cup S_{i_k}| = \left| \bigcup_{j=1}^{k} \Gamma x_{i_j} \right| = |\Gamma A|$$

where $k = |A|$.

To count the matchings of a bipartite graph, as well as the SDR's, we shall use the permanent of a matrix, defined in a certain way.

Let us return to the previous example in which $S = \{1, 2, 3, 4, 5, 6\}$ and $S_1 = \{2, 6\}$, $S_2 = S_3 = \{2, 4\}$, $S_4 = \{1, 2, 5, 6\}$ and $S_5 = \{3, 6\}$, with the *SDR*: $(6, 2, 4, 1, 3)$. This corresponds to the bipartite graph and to the matching drawn in thin lines in Fig. 10.1. The sense of the arcs, from X to Y, has not been marked.

If certain costs are associated with the arcs of a bipartite graph, the problem of determining a matching for which the sum of the costs of arcs is a minimum is known as the *assignment problem* and it can be solved by means of combinatorial algorithms, for instance, by the Hungarian algorithm [11].

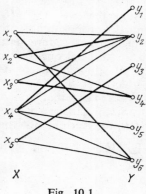

Fig. 10.1

An *independent set* of a graph G is a set of vertices that are not joined by any edge, while a *complete subgraph* or a *clique* is a set of vertices that are pairwise joined by edges in all possible ways. Another application of Hall's theorem is the following proposition, dealing with undirected bipartite graphs $G = (A, B, U)$, where U is the set of edges and $A \cap B = \varnothing$.

PROPOSITION 3 *(König). Every bipartite graph $G = (A, B, U)$ is perfect, i.e. the minimum number of complete subgraphs that constitute a partition of the set $A \cup B$ of vertices is equal to the maximum number of elements in an independent set.*

A complete subgraph of a bipartite graph consists either of a vertex, or of a pair of vertices joined by an edge.

Denote by $s(G)$ the maximum number of elements of an independent set, which is known as the *independence number*, or the *number of internal stability*, of the graph G. Also denote by $\omega(G)$ the minimum number of complete subgraphs that constitute a partition of $A \cup B$. Notice that $\omega(G) \geqslant s(G)$, because each class of a partition into $\omega(G)$ complete subgraphs contains at most one of the vertices of an independent set with $s(G)$ elements. We now prove the converse inequality by constructing a partition of the set $A \cup B$ made up of $s(G)$ complete subgraphs, thus implying $\omega(G) \leqslant s(G)$. Assume that $s(G) = |A|$, because in the contrary case we shall show that we can add to the set A sufficiently many vertices joined to all the vertices of B, which either does not modify the number $\omega(G)$ or yields the proof of the proposition.

We shall prove that, if the addition to the set A of a new vertex x joined to all vertices from B results in a graph G_1 such that $\omega(G_1) = \omega(G) + 1$, then $|A| = = s(G) = \omega(G)$. If $\omega(G_1) = \omega(G) + 1$, then a partition of $A \cup B \cup \{x\}$ into a minimum number of complete subgraphs can be obtained as follows: $C_1, C_2, ..., C_k$, $\{x\}$, where $k = \omega(G)$ and the vertex x is not joined to at least one vertex from each class $C_1, ..., C_k$, since otherwise $\omega(G_1) = \omega(G) = k$. The vertex x being joined to all vertices from B, it follows that each of these vertices belongs to one of the classes $C_1, ..., C_k$, together with a vertex of A, therefore $k = |A|$, since every class C_i contains one vertex of A. So $|A| = \omega(G)$. But the set A is independent, hence

$s(G) \geqslant |A| = \omega(G)$. Since the converse inequality $s(G) \leqslant \omega(G)$ also holds, it follows that $s(G) = \omega(G)$ and the proposition is established.

We now reason as follows: if the addition of a new vertex to the vertex set A results in a bipartite graph G_1 such that $s(G_1) = s(G) + 1$, we infer that $|A| = s(G)$ because the new vertex is joined to all vertices of B. If $s(G_1) = s(G)$, assume that $\omega(G_1) = \omega(G) + 1$. But we have shown that in this case $\omega(G) = s(G)$ and the proposition is proved. Finally, if $s(G_1) = s(G)$ and $\omega(G_1) = \omega(G)$, we continue this procedure until we obtain either the case $|A| = s(G)$ and $\omega(G)$ equal to the original one, or the proof of the proposition.

It still remains to be proved that if $|A| = s(G)$, then the bipartite graph is perfect. Let $B = \{b_1, b_2, ..., b_m\}$ and for each b_i denote by $S(b_i)$ the set of those vertices $a \in A$ that are joined by an edge to b_i. The set system $(S(b_1),\ S(b_2),\ ...$ $..., S(b_m))$ satisfies the necessary and sufficient condition for the existence of an SDR, for if it did not then some elements $b_{i_1}, b_{i_2}, ..., b_{i_l} \in B$ would exist such that

$$\left| \bigcup_{k=1}^{l} S(b_{i_k}) \right| < l. \quad \text{But in this case}$$

$$\left| A \setminus \bigcup_{k=1}^{l} S(b_{i_k}) \right| > |A| - l,$$

therefore the vertex set

$$\bigcup_{k=1}^{l} \{b_{i_k}\} \cup \left(A \setminus \bigcup_{k=1}^{l} S(b_{i_k}) \right)$$

is an independent set and contains more than $l + |A| - l = |A|$ elements, hence $s(G) > |A|$, thus contradicting the hypothesis. As the system $(S(b_1), ..., S(b_m))$ has an SDR, denote the latter by $(a_{i_1}, a_{i_2}, ..., a_{im})$ with $a_{i_k} \in A$ for $k = 1, ..., m$. Therefore we can construct a partition of the vertex set $A \cup B$ consisting of $s(G) = |A|$ classes which are complete subgraphs, in the following way: take the two-vertex classes $\{b_1, a_{i_1}\}$, $\{b_2, a_{i_2}\}$, ..., $\{b_m, a_{i_m}\}$, each of the remaining classes consisting of a single vertex from the set $A \setminus \{a_{i_1}, a_{i_2}, ..., a_{i_m}\}$, the total number of classes being equal to $s(G) = |A|$. We have thus proved the inequality $\omega(G) \leqslant s(G)$ and the proposition is established.

For example, in the case of the bipartite graph from Fig. 10.1, we have $s(G) = 6 = |Y|$, while a partition into 6 complete subgraphs of $X \cup Y$ consists of the 5 pairs of vertices which determine the edges of the matching represented by thick lines, together with the class consisting of the single vertex y_5.

The following types of graphs are other examples of classes of perfect graphs: trees, triangular graphs, unimodular graphs, transitive graphs and their complements [2].

10.2. LATIN RECTANGLES AND SQUARES

If S is a set with n elements, a *Latin rectangle* made up of elements from S is a matrix $A = \{a_{ij}\}_{\substack{i=1,...,r \\ j=1,...,s}}$ with r rows and s columns, where $a_{ij} \in S$, such that each row and each column of A consists of pairwise distinct elements. This condition implies $r \leqslant n$ and $s \leqslant n$. If the elements of S are denoted by $1, 2, ..., n$ and the number of columns is $s = n$, then every row of the Latin rectangle is a permutation of the numbers $1, 2, ..., n$ and these permutations must be chosen in such a way that no column contains the same element twice. An r by n Latin rectangle is said to be *normalized* if the first row is the identity permutation of the group S_n, i.e. if it is $1, 2, ..., n$.

Denoting by $L(r, n)$ the number of r by n Latin rectangles and by $K(r, n)$ the number of r by n normalized Latin rectangles, the relation $L(r, n) = n! \, K(r, n)$ holds. This relation is due to the fact that from a normalized Latin rectangle we obtain $n!$ distinct Latin rectangles by performing the same permutation $f \in S_n$ on all the elements of the rectangle and by iterating this operation for all $f \in S_n$, thus generating all $L(r, n)$ Latin rectangles with r rows and n columns.

The reader is warned that the term "permutation" means either a function $f \in S_n$ or an n-tuple $f(1) f(2) \dots f(n)$, the distinction resulting from the context.

The normalized Latin rectangles with two rows and n columns are, in fact, the permutations of n elements with no fixed elements, hence their number is $K(2, n) = D(n)$, where the numbers $D(n)$ are given by formula (3.8).

The number of normalized 3 by n Latin rectangles for which the first two rows are $1\,2\,3 \dots n$ and $n\,1\,2 \dots n-1$, respectively, is equal to the number $T(n)$ of solutions for Lucas' problem and so is given by formula (4.12).

The third row must contain a number different from 1 and n in the first column, a number different from 2 and 1 in the second column, a number different from 3 and 2 in the third column, etc., and these numbers form a permutation of the set $\{1, 2, ..., n\}$. Riordan has shown that an expression for $K(3, n)$ is given by

$$K(3, n) = \sum_{k=0}^{m} \binom{n}{k} D(n - k) \, D(k) \, T(n - 2k),$$

where $m = \left[\dfrac{n}{2} \right]$ and $T(0) = 1$. This relation enables the computation of the number of 3 by n Latin rectangles; however, in the general case the problem remains unsolved.

If $r = s = n$, the Latin rectangle is said to be a *Latin square* of order n. Such configurations can be considered as multiplication tables of general algebraic structures, in which the operations need not be associative, but have certain right or left simplification properties.

The multiplication table of a finite group (the *Cayley table*), which is constructed by numbering the elements of the group by the indices $1, 2, ..., n$ if the group is of order n and by locating on row i and column j the index of the product of the elements a_i and a_j taken in this order, is a Latin square. The numbers in each row and in each column are pairwise distinct, because $a_i \neq a_j$ implies $ba_i \neq ba_j$ and $a_i b \neq a_j b$ for any element b of the group. Assuming, for instance, that $ba_i = ba_j$, it follows by left multiplication by b^{-1}, that $a_i = a_j$, contradicting the hypothesis.

The Cayley table of the permutation group S_3, consisting of $g_1 = [1][2][3]$, $g_2 = [1][2, 3]$, $g_3 = [2], [1, 3]$, $g_4 = [3][1, 2]$, $g_5 = [1, 2, 3]$ and $g_6 = [1, 3, 2]$ is the 6 by 6 Latin square represented in Table 10.1.

TABLE 10.1

1	2	3	4	5	6
2	1	5	6	3	4
3	6	1	5	4	2
4	5	6	1	2	3
5	4	2	3	6	1
6	3	4	2	1	5

If l_n represents the number of Latin squares of order n in which the first row and the first column are written in the natural order $1, 2, ..., n$, then it can be shown that the total number of Latin squares of order n is

$$L(n, n) = n!(n - 1)! \, l_n.$$

By applying the theory of SDR's to Latin rectangles we shall show that Latin rectangles of arbitrary orders can be constructed and we shall obtain a lower bound for the numbers l_n. Consider an r by s Latin rectangle with elements from the set $\{1, 2, ..., n\}$. We say that this rectangle can be embedded into a Latin square of order n if $n - r$ rows and $n - s$ columns can be added such that the resulting configuration is a Latin square. We shall assume that the Latin square obtained in this way contains the r by s Latin rectangle in the upper left corner.

PROPOSITION 4. *Every Latin rectangle with r rows and n columns $(r < n)$ made up from the elements $1, 2, ..., n$, can embedded into a Latin square of order n.*

Let $S = \{1, 2, ..., n\}$ and let S_i be the set of elements of S which are not contained in column i of the Latin rectangle. For every $i = 1, 2, ..., n$, the set S_i has $n - r$ elements and $M(S) = (S_1, S_2, ..., S_n)$ is a collection of subsets of S. We shall prove that $M(S)$ satisfies the necessary and sufficient condition for the existence of an SDR.

Let $i \in S$. Since i occurs in each row of the Latin rectangle only once, it follows that i appears r times in r distinct columns, hence the element i is contained in exactly $n - r$ sets of $M(S)$ by the definition of the sets S_i.

Assume that $M(S)$ does not satisfy the condition for the existence of an SDR; this implies the existence of an index k and of a set $\{i_1, i_2, ..., i_k\} \subset \{1, 2, ..., n\}$ of pairwise distinct indices such that $|S_{i_1} \cup S_{i_2} \cup ... \cup S_{i_k}| = p \leqslant k - 1$. But each of the p elements of $S_{i_1} \cup ... \cup S_{i_k}$ can occur in this union at most $n - r$ times, therefore

$$\sum_{j=1}^{k} |S_{i_j}| \leqslant p(n - r) \leqslant (k - 1)(n - r).$$

On the other hand, since each set S_i contains $n - r$ elements, it follows that

$$\sum_{j=1}^{k} |S_{i_j}| = k(n - r)$$

which is a contradiction.

The necessary and sufficient condition for the existence of an *SDR*, given in proposition 1 is satisfied, so that $M(S)$ has an *SDR* which we denote by $(i_1, i_2, ..., i_n)$ and which is precisely a permutation of the numbers $1, 2, ..., n$, because the numbers $i_1, i_2, ..., i_n$ are pairwise distinct. By adding the row $i_1, i_2, ..., i_n$ to the r by n Latin rectangle, we obtain an $r + 1$ by n Latin rectangle, because the numbers of each row and of each column are distinct. This process can be continued until we obtain a Latin square of order n.

The above proposition implies the existence of Latin squares of any order n, since we can embed a 1 by n Latin rectangle, which is a permutation of the numbers $1, 2, ..., n$, into a Latin square of order n.

The next proposition indicates a lower bound for the number of r by n Latin rectangles, hence also for the Latin squares of order n.

PROPOSITION 5. *There exist at least* $n!(n - 1)! ... (n - r + 1)!$ *Latin rectangles with r rows and n columns, made up from the numbers* $1, 2, ..., n$.

Since there exist $n!$ permutations of the set $\{1, 2, ..., n\}$, there exist $n!$ Latin rectangles with one row and n columns, made up from the numbers $1, 2, ..., n$. It follows from the previous proposition that each of these Latin rectangles can be augmented by adding a new row which is an *SDR* for $M(S) = (S_1, S_2, ..., S_n)$, where each of the sets $S_1, S_2, ..., S_n$ contains $n - 1$ elements, Since $n - 1 < n$, proposition 1 implies the existence of at least $(n - 1)!$ *SDR*'s, therefore each 1 by n Latin rectangle will generate at least $(n - 1)!$ Latin rectangles with 2 rows and n columns. We continue this procedure until we get r rows, thus obtaining at least $n!(n - 1)! ... (n - r + 1)!$ Latin rectangles with r rows and n columns, with elements from the set $\{1, 2, ..., n\}$.

If $r = n$, we obtain at least $n!(n - 1)! ... 1!$ Latin squares of order n. Since $l_n = \dfrac{L(n, n)}{n!(n - 1)!}$, it follows that $l_n \geqslant (n - 2)! (n - 3)! ... 1!$ which, however, is not a satisfactory bound, as may be seen from Table 10.2, where all known values of l_n for $n \geqslant 3$ are given. Latin rectangles and squares can be generated with the aid of a computer [33].

Yamamoto, proving a conjecture of Erdös and Kaplansky [6], showed that the asymptotic value of the number $L(r, n)$ for $r < \sqrt[3]{n}$ is equal to $n!^r e^{-\binom{r}{2}}$ [34].

TABLE 10.2

n	3	4	5	6	7
l_n	1	4	56	9 408	16 942 080
$(n-2)! (n-3)! ... 1!$	1	2	12	288	34 560

10.3. The Permanent of a Matrix

Consider a matrix $A = \{a_{ij}\}_{\substack{i=1,\ldots,m \\ j=1,\ldots,n}}$ and assume $m \leqslant n$. If the elements of the matrix A are real numbers or, more generally, elements of a commutative field, the permanent of the matrix A is defined by the equality

$$\text{per}(A) = \sum_f a_{1f(1)} a_{2f(2)} \cdots a_{mf(m)}, \tag{10.1}$$

where f runs over all injective mappings

$$f: \{1, 2, \ldots, m\} \to \{1, 2, \ldots, n\}$$

the number of which is $[n]_m$, i.e. the sum of all products of m elements located on different rows and columns of the matrix A. It follows from this definition that the permanent of the matrix A is invariant with respect to permutations of the rows or of the columns of the matrix A, while the multiplication of a row of the matrix A by a scalar results in the multiplication of per (A) by that scalar, because each term of the sum (10.1) contains a single factor from a given row.

If A is a square matrix, then f runs through the symmetric group S_n, therefore per (A) can be obtained from the expansion of det (A) by replacing the minus sign which precedes the products corresponding to odd permutations, by a plus sign and then performing the sum of the $n!$ terms.

In the case of square matrices, the commutativity of multiplication implies

$$a_{f(1)1} a_{f(2)2} \cdots a_{f(m)m} = a_{1f^{-1}(1)} a_{2f^{-1}(2)} \cdots a_{mf^{-1}(m)}$$

where f^{-1} is a permutation of S_n as well, therefore per $(A) = $ per (A^T), where A^T is the transpose of the square matrix A, because the two sums consist of the same terms written in different orders. The *Laplace expansion* of a determinant has an analogue for the permanent of a square matrix, which is obtained by replacing all signs by plus; yet the property $\det(AB) = \det(A)\det(B)$ ceases to hold in the case of permanents.

Similarly, the addition of a row to another row multiplied by a constant does not leave per(A) invariant and the methods for computing the permanent of a matrix are rather difficult. An algorithm which computes the permanent of an arbitrary matrix without calculating all $[n]_m = n(n-1) \ldots (n-m+1)$ products which appear in (10.1) is given by:

Proposition 6 (*Ryser*). *Let A be an m by n matrix with $m \leqslant n$ and A_r a matrix obtained from A by replacing the elements of r columns of the matrix A by zeros. Denote by $S(A_r)$ the product of the sums of the elements in the rows of the matrix A_r and by $\sum S(A_r)$ the sum of these products with respect to all matrices*

A_r, with fixed r. Then the following relation holds:

$$\text{per}(A) = \sum S(A_{n-m}) - \binom{n-m+1}{1} \sum S(A_{n-m+1}) +$$

$$+ \binom{n-m+2}{2} \sum S(A_{n-m+2}) - \dots + (-1)^{m-1} \binom{n-1}{m-1} \sum S(A_{n-1}).$$

(10.2)

Consider the set of m-tuples

$$J = \{(j_1, j_2, \dots, j_m) | j_i \in \{1, 2, \dots, n\}\}$$

and denote by P_i the property "the m-tuple $(j_1, j_2, \dots, j_m)$ does not contain the integer i" for $i = 1, 2, \dots, n$. Set

$$X_A = \{(a_{1j_1}, a_{2j_2}, \dots, a_{mj_m}) | (j_1, j_2, \dots, j_m) \in J\}$$

and define the measure of an element from X_A by the equation $m(a_{1j_1}, \dots, a_{mj_m}) = a_{1j_1} a_{2j_2} \dots a_{mj_m}$ and the measure of a set $Y \subset X_A$ by $m(Y) = \sum_{x \in Y} m(x)$. But per (A) is the sum of the products of the elements that satisfy exactly $n - m$ of the properties P_i for $i = 1, 2, \dots, n$, by definition (10.1); therefore per (A) is the measure of the set of those elements from X_A which belong to exactly $n - m$ sets

$$A_i = \{(a_{1j_1}, a_{2j_2}, \dots, a_{mj_m}) | i \notin \{j_1, j_2, \dots, j_m\}\},$$

because only in this case are the numbers $j_1, \dots, j_m$ pairwise distinct. On the other hand, since only the additivity of the measure has been used in the proof of the sieve formula (3.6) (and not the property of being positive), we can apply the sieve formula, thus obtaining

$$\text{per}(A) = \sum_{k=n-m}^{n} (-1)^{k-n+m} \binom{k}{n-m} \sum_{\substack{K \subset N \\ |K|=k}} m(\bigcap_{i \in K} A_i),$$

where $N = \{1, 2, \dots, n\}$. But

$$m(\bigcap_{i \in K} A_i) = S(A_k) \quad \text{if} \quad K = \{i_1, i_2, \dots, i_k\},$$

because by replacing the elements of columns $i_1, i_2, \dots, i_k$ by zeros, by performing the sums of the elements in the rows of the matrix thus obtained and then by multiplying the m sums, we obtain a sum of products which contains all factors of the form $a_{1j_1} a_{2j_2} \dots a_{mj_m}$ with $j_1, j_2, \dots, j_m \in \{1, 2, \dots, n\} \setminus K$.

Notice that for $k = n$, $S(A_n) = 0$ since we have replaced all columns of the matrix A by zeros. Also, since $\binom{k}{n-m} = \binom{k}{k-n+m}$, we obtain

$$\text{per}(A) = \sum_{k=n-m}^{n-1} (-1)^{k-n+m} \binom{k}{k-n+m} \sum S(A_k)$$

$$= \sum S(A_{n-m}) - \binom{n-m+1}{1} \sum S(A_{n-m+1}) + \binom{n-m+2}{2} \sum S(A_{n-m+2}) - \cdots$$

$$+ (-1)^{m-1} \binom{n-1}{m-1} \sum S(A_{n-1}).$$

If A is a matrix with n rows and n columns, formula (10.2) becomes

$$\text{per}(A) = S(A) - \sum S(A_1) + \sum S(A_2) - \cdots + (-1)^{n-1} \sum S(A_{n-1}), \qquad (10.3)$$

the number of terms in $\sum S(A_k)$ being equal to $\binom{n}{k}$, because this is the number of possible choices for the columns that must be replaced by zeros.

By using the duality theorem from convex programming, L. Bregman recently proved [3] Minc's conjecture that asserts $\text{per}(A) \leqslant \prod_{i=1}^{n} (r_i!)^{\frac{1}{r_i}}$ where $r_i = \sum_{j=1}^{n} a_{ij}$ for any square (0,1)-matrix A of order n.

An extension of proposition 6 to a more general concept of permanent has been given by M. Abramson [1].

The concept of a permanent enables us to find the number of SDR's. Let $S = \{a_1, a_2, \ldots, a_n\}$ and $M(S) = (S_1, S_2, \ldots, S_m)$ with $S_i \in \mathscr{P}(S)$. Define the incidence matrix of the elements of S for the subsets $S_1, S_2, \ldots, S_m$ as being the matrix $A = \{a_{ij}\}_{\substack{i=1, \ldots, m \\ j=1, \ldots, n}}$ defined by $a_{ij} = 1$ if $a_j \in S_i$ and $a_{ij} = 0$ if $a_j \notin S_i$. It follows from this definition that the number of 1's in row i of the incidence matrix A represents the number of elements of S_i, while the number of 1's in column j represents the number of sets from $M(S)$ that contain the element a_j.

Conversely, given an m by n matrix A with elements 0 and 1, and an n-element set S, there exist sets $S_1, S_2, \ldots, S_m \in \mathscr{P}(S)$ such that the incidence matrix of the elements of S for the subsets $S_1, S_2, \ldots, S_m$ is precisely A: take $S_i = \{a_j | a_{ij} = 1\}$.

The choice of the elements 0 and 1 is due to their convenience for computations, the 0−1 matrices playing an important role in many combinatorial problems. The adjacency matrix of the vertices of a graph is also a 0−1 matrix; the determination of the transitive closure and of the algebraic closure of a graph reduce to operations with these 0−1 matrices [11].

PROPOSITION 7. *Let* $S_1, S_2, ..., S_m$ *be subsets of a set S with n elements,* $m \leqslant n$. *If A is the incidence matrix of the elements of S for these subsets, then the number of SDR's of* $M(S) = (S_1, S_2, ..., S_m)$ *is* per(A).

Notice first that if $m > n$, then $|S_1 \cup S_2 \cup ... \cup S_m| = n < m$, which implies, in view of proposition 1, that there is no *SDR* for *M(S)*. If $m \leqslant n$, then since the elements of the matrix A are 0 and 1, per(A) represents the number of products $a_{1f(1)} a_{2f(2)} \cdots a_{mf(m)}$ equal to 1, i.e. for which $a_{1f(1)} = a_{2f(2)} = ... = a_{mf(m)} = 1$. But in this case the subsets $S_1, S_2, ..., S_m$ contain the distinct elements $(a_{f(1)}, a_{f(2)}, ..., a_{f(m)})$, which constitute an *SDR* for *M(S)*. We thus obtain i rredundantly all *SDR*'s for *M(S)*.

For example, in the previous case when $S = \{1, 2, 3, 4, 5, 6\}$, $S_1 = \{2, 6\}$, $S_2 = S_3 = \{2, 4\}$, $S_4 = \{1, 2, 5, 6\}$ and $S_5 = \{3, 6\}$, the incidence matrix $\tilde{A}$ of the elements of S for these subsets is the following:

$$\tilde{A} = \begin{bmatrix} 0 & 1 & 0 & 0 & 0 & 1 \\ 0 & 1 & 0 & 1 & 0 & 0 \\ 0 & 1 & 0 & 1 & 0 & 0 \\ 1 & 1 & 0 & 0 & 1 & 1 \\ 0 & 0 & 1 & 0 & 0 & 1 \end{bmatrix}$$

while

$$\text{per}(\tilde{A}) = \Sigma S(\tilde{A}_1) - \binom{2}{1} \Sigma S(\tilde{A}_2) + \binom{3}{2} \Sigma S(\tilde{A}_3) - \binom{4}{3} \Sigma S(\tilde{A}_4) + \binom{5}{4} \Sigma S(\tilde{A}_5)$$

$$= 162 - 2 \cdot 142 + 3 \cdot 50 - 4 \cdot 6 = 4,$$

because $\Sigma S(\tilde{A}_5) = 0$, each column containing a zero. This result can be obtained by hand computation, by covering certain columns and by performing the sums of the elements of the rows, then the product of these sums.

The four *SDR*'s are the following:

$$(6, 2, 4, 5, 3); \quad (6, 4, 2, 5, 3); \quad (6, 2, 4, 1, 3) \text{ and } (6, 4, 2, 1, 3),$$

which correspond to the four matchings of the bipartite graph in Fig. 10.1:

$$C_1 = \{[x_1, y_6], [x_2, y_2], [x_3, y_4], [x_4, y_5], [x_5, y_3]\},$$

$$C_2 = \{[x_1, y_6], [x_2, y_4], [x_3, y_2], [x_4, y_5], [x_5, y_3]\},$$

$$C_3 = \{[x_1, y_6], [x_2, y_2], [x_3, y_4], [x_4, y_1], [x_5, y_3]\}$$

and

$$C_4 = \{[x_1, y_6], [x_2, y_4], [x_3, y_2], [x_4, y_1], [x_5, y_3]\}.$$

Consider now a matrix $A = \{a_{ij}\}_{\substack{i=1, ..., m \\ j=1, ..., n}}$ with $a_{ij} \in \{0, 1\}$. The *index* of the matrix A is by definition the maximum number of 1's with the property that no row and no column of A contains two 1's from the subset considered. The concept of index generalizes the concept of *SDR* for the subsets $S_1, S_2, ..., S_m$ of the n-element set S $(m \leqslant n)$. If A is the incidence matrix of the elements of S for these subsets, then these subsets have an *SDR* if and only if per $(A) \neq 0$ or, equivalently, if the index of the matrix A is equal to m.

In the case of the above matrix $\tilde{A}$ the index is 5, while the minimum number of rows and columns of $\tilde{A}$ which contain all 1's of $\tilde{A}$ is also equal to 5; for example, rows 4 and 5 and columns 2, 4 and 6. This property can be generalized as follows:

PROPOSITION 8 *(König). If A is a $0-1$ matrix, the minimum number of rows and columns of A which contain all the 1's of the matrix A is equal to the index of A.*

Let A be an m by n matrix, q the minimum number of rows and columns of A which contain all the 1's of A and p the index of the matrix A. Since no row or column of A can contain two 1's which contribute to the index p, it follows that $q \geqslant p$. Let a minimum covering of 1's by q rows and columns consist of e rows and f columns with $e + f = q$. Then the invariance of the numbers p and q with respect to permutations of the rows and columns of A implies that we can arrange the e rows and f columns as the first rows and columns of A. We thus obtain

$$A = \begin{bmatrix} A_1 & A_2 \\ A_3 & A_4 \end{bmatrix}$$

where A_1 is an e by f matrix and A_4 consists of 0's only.

Let us prove that A_2 has index e. Now A_2 can be regarded as the incidence matrix of the elements $f + 1, f + 2, ..., n$ for e subsets which we denote by $S_1, S_2, ..., S_e$. These subsets satisfy the necessary and sufficient condition for the existence of an *SDR*, because if not then the relation $|S_{i_1} \cup S_{i_2} \cup ... \cup S_{i_k}| = l < k$ would hold for some indices $i_1, i_2, ..., i_k$. In this case we could replace the rows $i_1, i_2, ..., i_k$ by l columns corresponding to the elements of $S_{i_1} \cup S_{i_2} \cup ... \cup S_{i_k}$, thus obtaining a covering of the 1's of the matrix A with less than $e + f$ rows and columns, which contradicts the minimality of q. Since the subsets $S_1, S_2, ..., S_e$ have an *SDR*, it follows that the index of the matrix A_2 is e. One proves similarly that the index of A_3 is f, because the transpose A_3^T can be viewed as an incidence matrix.

Since the rows and columns of the matrix A_2 are disjoint from the rows and columns of the matrix A_3, it follows that $p \geqslant e + f = q$. The double inequality implies $p = q$.

If, for a matrix A with coefficients in an arbitrary field, the non-zero elements are replaced by 1, one obtains the following result: the minimum number of rows and columns of the matrix A which contain all non-zero elements of A, is equal to the maximum number of non-zero elements with the property that no row and no column of A contains two non-zero elements from the subset considered.

An m by n *permutation matrix* $(m \leqslant n)$ is a matrix which contains one 1 in each row and at most one 1 in each column, the remaining elements being zero.

PROPOSITION 9. *Let A be a matrix, the elements of which are nonnegative real numbers such that the sum of elements of every row of the matrix A is equal to p, while the sum of elements of every column of A is equal to q. Then A can be written in the form*

$$A = c_1 P_1 + c_2 P_2 + \ldots + c_t P_t,$$

where P_i are permutation matrices and c_i are nonnegative real numbers, for $i = 1, 2, \ldots, t$.

Assume that A has m rows and n columns, $m \leqslant n$. If A is not a square matrix, add $n - m$ rows equal to $\left[\dfrac{p}{n}, \dfrac{p}{n}, \ldots, \dfrac{p}{n} \right]$ to the lower part of the matrix A, thus obtaining a new matrix A'. The sum of the elements of each row of A' is equal to p, while the sum of the elements of each column is equal to $(n - m) \dfrac{p}{n} + q = p$, because $mp = nq$ is the sum of the elements of the matrix A. Notice that if A is a square matrix, then $p = q$.

If A' is not the null matrix, there exist n positive elements which appear in different rows and columns of A', because otherwise the previous proposition would imply the possibility of covering the set of positive elements of the matrix A' by e rows and f columns such that $e + f < n$. But this would imply

$$\sum_{i,j=1}^{n} a'_{ij} = np \leqslant (e + f)p < np,$$

because the elements from the intersection of a row and a column are repeated in a sum. We have thus obtained a contradiction, hence there exist n positive elements which are situated in different rows and columns of the matrix A'.

Let P'_1 be the n by n permutation matrix in which the 1's occupy the same positions as the n positive elements of the matrix A' which are situated in different rows and columns. Let c_1 be the smallest of these n elements. The matrix $A' - c_1 P'_1$ is a matrix with nonnegative real elements, such that the sum of the elements of each row and of each column is equal to $p - c_1 \geqslant 0$; furthermore, the number of 0's in $A' - c_1 P'_1$ is greater than the number of 0's of A'. By repeating the above reasoning for the matrix $A' - c_1 P'_1$, we shall finally obtain the null matrix, because the matrix A' has at most n^2 elements different from 0, therefore we can write

$$A' = c_1 P'_1 + c_2 P'_2 + \ldots + c_t P'_t.$$

This decomposition of the matrix A' will generate a decomposition of the required form of the matrix A by deletion of the last $n - m$ rows of the permutation matrices P_i' with $i = 1, ..., t$.

If A is a $0-1$ matrix with n rows and n columns, in which the sum of the elements of each row and of each column is equal to $k > 0$, then $A = P_1 + P_2 + ... + P_h$, where the P_i's are permutation matrices. For, it follows from the previous construction that all coefficients $c_1, c_2, ..., c_t$ are equal to 1, hence $A = P_1 + P_2 + ... + P_t$. If we write down the sum of the elements of a row in the matrix A and in the matrix $P_1 + ... + P_t$, we obtain $t = k$, because each of the matrices $P_1, ..., P_t$ contributes one unit to the sum. Clearly a square permutation matrix contains one 1 in each row and in each column, the remainder of the elements being zero.

This decomposition property gives an affirmative answer to the following problem [28]: n boys and n girls attend a party; each boy has been introduced to k girls and each girl has been introduced to k boys. Can they dance in couples, each consisting of a boy and a girl who have already been introduced to one another?
Let $A = \{a_{ij}\}_{i,j=1, ..., n}$ be the $0-1$ square matrix of order n defined as follows: $a_{ij} = 1$ if the boy j has been introduced to the girl i and $a_{ij} = 0$ otherwise. Then the sum of the elements in each row and in each column of the matrix A is equal to k, hence A has a decomposition as a sum of permutation matrices. Each of these permutation matrices defines a solution of the above problem: if the element in row i and column j of such a permutation matrix is equal to 1, then the boy j will dance with the girl i.

A square matrix of order n is said to be *bistochastic* if its elements are nonnegative and the sum of elements in each row and in each column is equal to 1. It follows from proposition 9 that every bistochastic matrix A with n rows and n columns has a decomposition of the form $A = c_1P_1 + c_2P_2 + ... + c_t P_t$, where the P_i's are permutation matrices and $c_i > 0$ for $i = 1, ..., t$ and $\sum_{i=1}^{t} c_i = 1$. This result is known as the *Birkhoff-von Neumann theorem*. The relation $\sum_{i=1}^{t} c_i = 1$ is obtained by equating the sums of the elements of a row or of a column on the two sides of the relation $A = c_1P_1 + ... + c_t P_t$. Taking into account definition (10.1), the permanent of a bistochastic matrix A cannot be greater than the product of the sums of the elements of the rows of the matrix A, because the elements of A are nonnegative. Therefore $\operatorname{per}(A) \leqslant 1$, the equality holding in the case when A is a permutation matrix.

By using the decomposition $A = c_1 P_1 + ... + c_t P_t$, one infers that $\operatorname{per}(A) \geqslant \geqslant c_1 \operatorname{per}(P_1) = c_1 > 0$, hence the permanent of every bistochastic matrix is positive.

The problem of determining the least possible value for the permanent of a bistochastic matrix of order n has not yet been solved, but the *van der Waerden conjecture* states that $\operatorname{per}(A) \geqslant \dfrac{n!}{n^n}$, the equality being satisfied for example for a matrix A with all elements equal to $\dfrac{1}{n}$.

The van der Waerden conjecture has been proved for $n=2$ (an obvious case), 3, 4, 5, the latter case being recently solved by P. Eberlein [5].

Problems

1.   Prove that if per $(A) = 0$ then the graph whose adjacency matrix is A cannot have a Hamiltonian circuit.

2.   A balanced incomplete block design (*BIBD*) is a set B of v objects (sometimes called varieties) together with a family $\mathscr{B}$ of β subsets of B, called blocks, such that:

    i) each block contains exactly k objects;

    ii) each object belongs to exactly r blocks, and

    iii) each pair of distinct objects belongs to exactly λ blocks.

Prove that these parameters $(\beta, v, r, k, \lambda)$ satisfy the following equations:

$$\beta k = vr, \quad r(k - 1) = \lambda(v - 1).$$

H i n t: Count in two different ways objects and pairs of objects of B.

3.   For a bipartite graph $G = (X, Y, \Gamma)$ the number

$$\delta_0 = \max_{A \subset X} (|A| - |\Gamma A|)$$

is called its deficiency. If we set $\delta(A) = |A| - |\Gamma A|$ for any $A \subset X$, prove that the set

$$\mathscr{A} = \{A \mid \delta(A) = \delta_0\}$$

is a lattice relative to $\cup$ and $\cap$, i.e. $A_1, A_2 \in \mathscr{A} \Rightarrow A_1 \cup A_2 \in \mathscr{A}$ and $A_1 \cap A_2 \in \mathscr{A}$.

H i n t: Use the inequality:

$$\delta(A_1 \cup A_2) + \delta(A_1 \cap A_2) \geqslant \delta(A_1) + \delta(A_2)$$

which holds for any $A_1, A_2 \subset X$.

4.   Prove the König—Ore theorem:

In a bipartite graph $G = (X, Y, \Gamma)$, the maximal number of edges of a matching equals $|X| - \delta_0$.

H i n t: Verify that: a) if $\delta_0 > 0$, then $A_0 = \bigcap_{\delta(A) = \delta_0} A \neq \varnothing$;

b) if we delete a vertex $x \notin A_0$, the deficiency of the graph remains unchanged;

c) if we delete a vertex $x \in A_0$, then the deficiency decreases by a unit.

5.   Let X be a v-element set with $v \geqslant 3$.

A Steiner triple system of order v is a set of 3-subsets or triples of X such that each 2-subset of X is a subset of exactly one triple.

Prove that a necessary and sufficient condition for the existence of a Steiner triple system of order v is that $v \equiv 1$ or 3 (mod 6).

(Kirkman, 1847)

H i n t: To prove necessity observe that a Steiner triple system of order v is a balanced incomplete block design with $k = 3$ and $\lambda = 1$.

For sufficiency see a simplified proof in [10].

6.　　Let X be a v-set and let $X_1, X_2, ..., X_v$ be subsets of X. These subsets are called a (v, k, λ)-configuration provided that they satisfy the following requirements:

　i) $|X_i| = k$ for $i = 1, 2, ..., v$.

　ii) $|X_i \cap X_j| = \lambda$ for any $i \neq j$.

　iii) $0 < \lambda < k < v - 1$.

Prove that $AA^T = (k - \lambda) I + \lambda J$ if and only if A is the incidence matrix of a (v, k, λ)-configuration, where A^T denotes the transpose of A, the matrix J is the matrix of 1's of order v and I is the identity matrix of order v.

7.　　Prove that a (v, k, λ)-configuration is a balanced incomplete block design with the parameters (v, v, k, k, λ).
H i n t: Show that the matrix A^T is an incidence matrix of a (v, k, λ)-configuration, i.e. $A^T A = AA^T = (k - \lambda) I + \lambda J$. For this prove that A is nonsingular, $A^T J = = (k - \lambda + \lambda v) k^{-1} J = JA$ because $AJ = kJ$, hence $JAJ = (k - \lambda + \lambda v) k^{-1} vJ$. But also $JAJ = kvJ$, whence it follows that $k - \lambda = k^2 - \lambda v$, hence $JA = AJ = = kJ$.
Finally, $A^T A = A^{-1} (AA^T) A = (k - \lambda) I + \lambda A^{-1} JA = (k - \lambda) I + \lambda J$ [28].

8.　　Let $A_1 = \{a_{ij}^{(1)}\}_{i,j=1, ..., n}$ and $A_2 = \{a_{ij}^{(2)}\}_{i,j=1, ..., n}$ be two n by n Latin squares based on n elements labelled $1, 2, ..., n$ and let $n \geqslant 3$.

The Latin squares A_1 and A_2 are called orthogonal provided that the n^2 ordered pairs $(a_{ij}^{(1)}, a_{ij}^{(2)})$ $(i, j = 1, 2, ..., n)$ are distinct.

Let $n = p^\alpha$, where p is a prime and α is a positive integer. Prove that for $n \geqslant 3$ there exists a set of $n - 1$ orthogonal Latin squares of order n.
H i n t: Let the elements of the Galois field $GF(p^\alpha)$ be denoted by $a_0 = 0, a_1 = 1$, $a_2, ..., a_{n-1}$. Then the $n - 1$ matrices of order n, $A_k = \{a_k a_i + a_j\}_{i,j=0,1, ..., n-1}$ $(k = 1, 2, ..., n - 1)$ form a set of $n-1$ orthogonal Latin squares [28].

9.　　Let $N(i)$ denote the number of occurrences of i in an r by s Latin rectangle based on n elements labelled $1, 2, ..., n$.
If the Latin rectangle can be extended to a Latin square of order n, prove that

$$N(i) \geqslant r + s - n \quad (i = 1, 2, ..., n).$$

(Ryser, 1951)

This condition is also sufficient [29] (a generalization of Proposition 6).

10.　　Prove the following property: Let $\mathscr{F} = (M_i : i \in I)$ be a family of nonempty subsets of a set S. $M_i = M_j$ for $i \neq j$ is permitted.

If the cardinalities $|I| = f$ and $|S| = n$ are finite and $f > n(r - 1)$ for some integer r, then it is possible to find r nonempty disjoint subsets I_v $(v = 1, ..., r)$ of I such that

$$\bigcup_{i \in I_1} M_i = ... = \bigcup_{i \in I_r} M_i.$$

(Lindström, 1972)

H i n t: Apply induction on the number of elements in S and use Rado's theorem.

11. Let $A = (a_{ij})$ be a matrix of m rows and n columns whose entries are either zero or one with row i of sum r_i $(i = 1, 2, ..., m)$ and column j of sum s_j $(j = 1, 2, ..., n)$. The total sum σ of the elements of A is equal to $\sum_{i=1}^{m} r_i = \sum_{j=1}^{n} s_j$. Prove the following inequality:

$$\sum_{i=1}^{m} r_i^2 + \sum_{j=1}^{n} s_j^2 \leqslant \sigma \left(l + \frac{\sigma}{l} \right),$$

where $l = \max(m, n)$.

(Khintchine, 1933)

12. Let $A = \{a_{ij}\}_{i,j=1,...,n}$ be a $(0,1)$ − matrix such that $\sum_{i=1}^{n} a_{ij} = m$ $(j = 1, ..., n)$ and $\sum_{j=1}^{n} a_{ij} = m$ $(i = 1, ..., n)$ for $n \geqslant m$. Let $r \geqslant 1$ be the smallest number having the following property: For any such matrix A, there are r different numbers $i_1, i_2, ..., i_r \in \{1, 2, ..., n\}$ for which the submatrix $\{a_{ij}\}_{\substack{i=i_1, ..., i_r \\ j=1, ..., n}}$ has at least one one in each column. Prove that

$$\frac{n}{m} \leqslant r \leqslant \frac{mn}{2m-1}.$$

(Ghelfond, 1968)

H i n t: Deduce that $rm \geqslant n$ and $(n-r)m \geqslant r(m-1)$. The last inequality is derived from the minimality assumption, because for a row with index i_k $(1 \leqslant k \leqslant r)$ there is a column which contains exactly one one in this row in the submatrix.

13. Suppose that $n > 0$ and that the letters A_i denote finite sets. Let H_n be the set of all systems $(A_0, ..., A_{n-1})$ which satisfy the following $2^n - 1$ conditions: if $0 < k < n$ and $v_0 < ... < v_{k-1} < n$, then $|A_{v_0} \cup ... \cup A_{v_{k-1}}| \geqslant k$.

If $(A_0, ..., A_{n-1}) \in H_n$, $|A_0| \leqslant ... \leqslant |A_{n-1}|$, prove that the number of SDR's for $(A_0, ..., A_{n-1})$ is greater than or equal to $\prod(|A_k| - k)$, where the product is over all values of k which satisfy $0 \leqslant k < \min(n, |A_0|)$.

(Rado, 1967)

H i n t: If $n = 1$ the result is obvious. For $n \geqslant 2$ use induction with respect to n and distinguish between the cases: i) $|A_{v_0} \cup ... \cup A_{v_{k-1}}| > k$ for $0 < k < n$ and $0 < v_0 < ... < v_{k-1} < n$; ii) For some k, $v_0, ..., v_{k-1}$ such that $0 < k < n$ and $0 < v_0 < ... < v_{k-1} < n$ we have $|A_{v_0} \cup ... \cup A_{v_{k-1}}| = k$ [25].

14. Prove that the number of different SDR's for the collection of sets of integers $S_i = \{i, i+1, i+2\}$ $(i = 1, 2, ..., n)$ reduced modulo n, $n \geqslant 3$, is $L_n + 2$, where L_n denotes the n-th Lucas number, defined by $L_1 = 1$, $L_2 = 3$ and $L_n = L_{n-1} + L_{n-2}$ for $n = 3, 4, ...$

(B. Ross, 1972)

BIBLIOGRAPHY

1. Abramson, M., *A note on permanents*, Canad. Math. Bull., **14**, 1971, 1—4.
2. Berge, C., *Graphes et hypergraphes*, Dunod, Paris, 1970.
   English edition: *Graphs and hypergraphs*, North-Holland, Amsterdam, 1973.
3. Bregman, L. M., *Some properties of nonnegative matrices and of their permanents*, Dokl. Akad. Nauk SSSR, **211**, 1973, 27—30.
4. Brualdi, R. A., *A very general theorem on systems of distinct representatives*, Trans. Amer. Math. Soc., **140**, 1969, 149—160.
5. Eberlein, P. J., *Remarks on the van der Waerden conjecture* II, Linear Algebra and Appl., **2**, 1969, 311—320.
6. Erdös, P., Kaplansky, I., *The asymptotic number of Latin rectangles*, Amer. J. Math., **68**, 1946, 230—236.
7. Ghelfond, A., *Certain combinatorial properties of* (0, 1)-*matrices*, Mat. Sb., **75** (117), 1968, 3.
8. Hall, M., Jr., *Distinct representatives of subsets*, Bull. Amer. Math. Soc., **54**, 1948, 922—926.
9. Hall, P., *On representatives of subsets*, J. London Math. Soc., **10**, 1935, 26—30.
10. Hilton, A. J. W., *A simplification of Moore's proof of the existence of Steiner triple systems*, J. Combinatorial Theory (A), **13**, 1972, 422—425.
11. Kaufmann, A., *Introduction à la combinatorique en vue des applications*, Dunod, Paris, 1968.
12. Khintchine, A., *Über ein metrisches Problem der additieven Zahlentheorie*, Mat. Sb., **40**, 1933, 180—189.
13. König, D., *Theorie der endlichen und unendlichen Graphen*, Akademische Verlagsgesellschaft, Leipzig, 1936; reprinted Chelsea, New York, 1950.
14. Las Vergnas, M., *Sur les systèmes de représentants distincts d'une famille d'ensembles*, C. R. Acad. Sci. Paris, Ser. A, **270**, 1970, 501—503.
15. Lindström, B., *A theorem on families of sets*, J. Combinatorial Theory (A), **13**, 1972, 274—277.
16. Luxemburg, W. A. J., *On an inequality of A. Khintchine for zero-one matrices*, J. Combinatorial Theory (A), **12**, 1972, 289—296.
17. Marica, J., Schönheim, J., *Incomplete diagonals of Latin squares*, Canad. Math. Bull., **12** 1969, 235.
18. Mendelsohn, N. S., Dulmage, A. L., *Some generalizations of the problem of distinct representatives*, Canad. J. Math., **10**, 1958, 230—241.
19. Metropolis, N., Stein, M., Stein, P., *Permanents of cyclic* (0, 1) *matrices*, J. Combinatorial Theory, **7**, 1969, 291—321.
20. Mirsky, L., Perfect, H., *Systems of representatives*, J. Math. Anal. Appl., **15**, 1966, 520—568.
21. Mirsky, L., *Hall's criterion as a 'self-refining' result*, Monatsh. Math., **73**, 1969, 139—146.
22. Nijenhuis, A., Wilf, S. H., *On a conjecture in the theory of permanents*, Bull. Amer. Math. Soc., **76**, 1970, 738—739.
23. Perfect, H., *Independence spaces and combinatorial problems*, Proc. London Math. Soc., (3) **19**, 1969, 17—30.
24. Perfect, H., *Remark on a criterion for common transversals*, Glasgow Math. J., **10**, 1969, 66—67.
25. Rado, R., *On the number of systems of distinct representatives of sets*, J. London Math. Soc., **42**, 1967, 107—109.
26. Rado, R., *Note on the transfinite case of Hall's theorem on representatives*, J. London Math. Soc., **42**, 1967, 321—324.
27. Ross, B., *A Lucas number counting problem*, Fibonacci Quart., **10**, 1972, 325—328.
28. Ryser, H., *Combinatorial mathematics*, John Wiley, New York, 1963.
29. Ryser, H., *A combinatorial theorem with an application to Latin rectangles*, Proc. Amer. Math. Soc., **2**, 1951, 550—552.
30. Ryser, H., *An extension of a theorem of de Bruijn and Erdös on combinatorial designs*, J. Algebra, **10**, 1968, 246—261.
31. Tverberg, H., *A generalization of Rado's theorem*, J. London Math. Soc., **41**, 1966, 123—128.
32. Welsh, D. J. A., *Generalized versions of Hall's theorem*, J. Combinatorial Theory (B), **10**, 1971, 95—101.
33. Wesley, B. J., *Enumeration of Latin squares with applications to order* 8, J. Combinatorial Theory, **5**, 1968, 177—184.
34. Yamamoto, K., *On the asymptotic number of Latin rectangles*, Japan. J. Math., **21**, 1951, 113—119.

CHAPTER 11

Ramsey's Theorem

Ramsey's theorem may be considered a generalization of a simple principle: if we have a set containing sufficiently many elements and we partition it into not too many subsets, then at least one of these subsets must contain sufficiently many elements. This assertion is an important combinatorial theorem which draws its origin from various researches in mathematics and has been formulated by the English logician F. P. Ramsey. Let S be an n-element set and let $P_r(S)$ be the set of all subsets of S which contain r elements. Let

$$P_r(S) = A_1 \cup A_2 \cup \dots \cup A_t$$

be an ordered decomposition (in which the order of the sets $A_1, A_2, \dots, A_t$ is material) of the set $P_r(S)$ into pairwise disjoint subsets $A_1, \dots, A_t$; some of these subsets may be empty. Let $q_1, q_2, \dots, q_t$ and r be integers such that $1 \leqslant r \leqslant q_1, q_2, \dots, q_t$.

PROPOSITION 1 *(Ramsey). If the integers $q_1, q_2, \dots, q_t$ and r satisfy the inequalities $1 \leqslant r \leqslant q_1, q_2, \dots, q_t$, then there exists a least positive integer, denoted by $N(q_1, q_2, \dots, q_t, r)$ and called the Ramsey number with parameters $q_1, q_2, \dots, q_t, r$, such that the following property holds for every $n \geqslant N(q_1, q_2, \dots, q_t, r)$: if S is an n-element set and $P_r(S) = A_1 \cup A_2 \cup \dots \cup A_t$ is an ordered decomposition into t classes of the family $P_r(S)$ of the r-element subsets of S, then there exists an index i with $1 \leqslant i \leqslant t$, such that the set S contains a q_i-element subset having all its r-element subsets contained in A_i.*

Notice that for $r = 1$, $P_r(S) = S$ and $A_1 \cup A_2 \cup \dots \cup A_t$ is a decomposition of S. Let us prove that $N(q_1, q_2, \dots, q_t, 1) = q_1 + q_2 + \dots + q_t - t + 1$. If a set S contains n elements, where $n \geqslant q_1 + \dots + q_t - t + 1$, then for any decomposition $S = A_1 \cup \dots \cup A_t$ having t subsets, there is an index i, $1 \leqslant i \leqslant t$, such that the subset A_i contains a q_i-element subset of S, therefore $|A_i| \geqslant q_i$. For assuming that this is not true, it follows that $|A_i| \leqslant q_i - 1$, therefore $n = \sum_{i=1}^{t} |A_i| \leqslant q_1 + \dots + q_t - t$, which contradicts the hypothesis $n \geqslant q_1 + \dots + q_t - t + 1$. We also remark that $q_1 + \dots + q_t - t + 1$ represents the least value of n which satisfies this property, for if $n = q_1 + \dots + q_t - t$, we can choose the classes of the decomposition of S, so that $|A_i| = q_i - 1$ and there can be no index $1 \leqslant i \leqslant t$ for which $|A_i| \geqslant q_i$.

Another particular case is the following: suppose that $q_1 = q_2 = \dots = q_t = q \geqslant r \geqslant 1$. We prove that for every n-element set S with a sufficiently large n

and for every partition into t classes of the set of r-element subsets of S, there exists a q-element subset of S having all its r-element subsets in one of the classes of S. If there is an n with this property, it follows that there will also exist a least such integer, which is denoted by $N(q, q, ..., q, r)$. The existence of the numbers $N(q,q, ..., q, r)$ implies the existence of any number $N(q_1, ..., q_t, r)$, taking $q = \max(q_1, q_2, ..., q_t)$. For in every decomposition $P_r(S) = A_1 \cup A_2 \cup ... \cup A_t$ there exists a q-element subset of S having all r-element subsets in a class of the partition, say A_j, then by deleting $q - q_j$ elements from that subset we obtain a subset X of S with $|X| = q_j$ and $P_r(X) \subset A_j$.

For $t = 1$ the proposition is obvious and we get $N(q_1, r) = q_1$. Assuming that we have proved it for $t = 2$, we shall show that it follows for $t = 3$. It will also be valid for any t, the inference from t to $t + 1$ being demonstrated by induction in a similar manner. If $t = 3$, $P_r(S) = A_1 \cup A_2 \cup A_3$. Setting $q_2' = N(q_2, q_3, r)$, as soon as $n \geqslant N(q_1, q_2', r)$ the n-element set S contains either a subset U with $|U| = q_1$ and $P_r(U) \subset A_1$, or a subset T with $|T| = q_2'$ and $P_r(T) \subset A_2 \cup A_3$. If S contains the subset U, then the proposition will also hold for $t = 3$. Otherwise S contains a subset T having q_2' elements with all r-element subsets in $A_2 \cup A_3$. Letting A_2' denote the family of subsets of A_2 which contain only elements of T, and, similarly, A_3' the subsets of A_3 which contain only elements of T, then since $q_2' = N(q_2, q_3, r)$, it follows that there exists either a subset of $T \subset S$ having q_2 elements with all of its r-element subsets in A_2', and hence in A_2, or a q_3-element subset having all its r-element subsets in A_3', and hence in A_3, so the proposition holds for $t = 3$. Thus, if we prove the proposition is also true for $t = 2$, it will be valid, by induction, for any t.

Notice first that as there exists a bijection from the set of ordered decompositions of $P_r(S)$ into two classes onto itself, which can be defined by associating the decomposition $A_2 \cup A_1$ with the decomposition $A_1 \cup A_2$, we obtain $N(q_1, q_2, r) = N(q_2, q_1, r)$. We have already proved that $N(q_1, q_2, 1) = q_1 + q_2 - 1$. Let us now prove that $N(q_1, r, r) = q_1$.

If the number of elements of the set S is greater than or equal to q_1 and $P_r(S) = A_1 \cup A_2$, where $A_2 \neq \varnothing$, then the n-element set S will contain an r-element subset T, having all its r-element subsets (which reduce to T) in A_2. It is sufficient to choose a subset of S with r elements from A_2, which is a family of r-element subsets of S. If $A_2 = \varnothing$, then $P_r(S) = A_1$ and the set S, with $n \geqslant q_1$, will include a q_1-element subset having all its r-element subsets in $P_r(S)$. From this latter case we infer that the least value of n which satisfies the property is precisely q_1; hence $N(q_1, r, r) = q_1$.

According to the latter result, we can now assume that $1 < r < q_1, q_2$ when proving by induction the proposition in the case $t = 2$. We shall use induction on the three indices q_1, q_2, r: first on q_1 and q_2, then on r.

For instance, as $N(2, 2, 2)$ exists, we shall prove the existence of $N(3, 2, 2)$, $N(4, 2, 2), ..., N(q_1, 2, 2)$ which will imply the existence of $N(2, 3, 2)$, $N(3, 3, 2)$, $N(4, 3, 2), ..., N(q_1, 3, 2)$; $N(2, 4, 2)$, $N(3, 4, 2), ..., N(q_1, 4, 2)$ etc., which will imply the existence of the integers $N(q_1, q_2, 3)$ and so on.

According to the induction hypothesis, there exist $p_1 = N(q_1 - 1, q_2, r)$, $p_2 = N(q_1, q_2 - 1, r)$ and the integers $N(q_1', q_2', r - 1)$ for any q_1', q_2' with $1 \leqslant r - 1 \leqslant q_1', q_2'$, therefore in particular there exists $N(p_1, p_2, r - 1)$.

We shall establish the existence of $N(q_1, q_2, r)$ and we shall prove that

$$N(q_1, q_2, r) \leqslant N(p_1, p_2, r - 1) + 1.$$

Let $n \geqslant N(p_1, p_2, r - 1) + 1$ and let a be an element of the n-element set S. We put $T = S \setminus \{a\}$, hence $|T| = n - 1$. Starting from the decomposition $P_r(S) = A_1 \cup A_2$ of the family of r-element subsets of S we can define a partition $P_{r-1}(T) = B_1 \cup B_2$ of the family of $(r - 1)$-element subsets of T as follows. Let $R \subset T$ and let $|R| = r - 1$. If the set $R \cup \{a\} \in A_1$ we put R into B_1, and if $R \cup \{a\} \in A_2$ we put R into B_2; we shall do the same thing for all the $(r - 1)$-element subsets R of T. As T has $n - 1$ elements, it follows that $|T| \geqslant N(p_1, p_2, r - 1)$ hence T contains either a p_1-element subset having all its $(r - 1)$-element subsets in B_1, or a p_2-element subset having all its $(r - 1)$-element subsets in B_2.

Let us consider first the case when T contains a subset U with $|U| = p_1$ and $P_{r-1}(U) \subset B_1$. But we have already defined $p_1 = N(q_1 - 1, q_2, r)$ and as $P_r(S) = A_1 \cup A_2$ induces a decomposition $P_r(U) = C_1 \cup C_2$, where C_i consists of those r-element subsets of A_i which are contained in U for $i = 1, 2$, it follows that U contains either a subset having $q_1 - 1$ elements with all its r-element subsets in C_1, hence also in A_1, or a subset having q_2 elements with all its r-element subsets in C_2, hence also in A_2. In the latter case we have found a q_2-element subset V of U, and therefore of S, such that $P_r(V) \subset A_2$, which concludes the proof.

In the former case U contains a subset V with $|V| = q_1 - 1$, which satisfies $P_r(V) \subset A_1$. Set $W = V \cup \{a\} \subset S$ and $|W| = q_1$. If an r-element subset of W does not contain the element a, then it is an r-element subset of V and hence it is in A_1.

If an r-element subset of W does contain the element a, then it consists of a and an $(r - 1)$-element subset of V. But V, being a subset of U, has the property that $P_{r-1}(V) \subset B_1$. It follows from the definition of the partition $P_{r-1}(T) = B_1 \cup B_2$ that an r-element subset of W which contains the element a, is in the class A_1, hence W is a q_1-element subset of S with all r-element subsets in A_1. The case when T contains a p_2-element subset with all its $(r - 1)$-element subsets in B_2, is treated similarly. We have thus established the existence of a positive integer n such that the property stated in the proposition holds for $t = 2$; hence there also exists a least such number n, which is by definition the Ramsey number $N(q_1, q_2, r)$. The property stated in the proposition is obviously true for every $n \geqslant N(q_1, q_2, r)$, because from every partition $P_r(S) = A_1 \cup A_2$ we can obtain a partition $P_r(X) = Y_1 \cup Y_2$ where $|X| = N(q_1, q_2, r)$ and Y_i is obtained from A_i by eliminating those r-element subsets which contain elements from $S \setminus X$ for $i = 1, 2$.

The determination of the Ramsey numbers is a difficult combinatorial problem. For $t > 2$ it is known, for example, that $N(3, 3, 3, 2) = 17$.

A *good q-colouring* of the edges of a clique is a colouring of its edges with q colours such that the clique has no monochromatic triangles. Let n_q be the maximal number of vertices of a clique such that it has a good q-colouring of its edges.

Obviously $n_q = N(\underbrace{3, 3, ..., 3}_{q}, 2) - 1$.

PROPOSITION 2 *(Greenwood, Gleason). The Ramsey numbers $N(3, 3, ..., 3, 2)$*
satisfy

$$N(\underbrace{3, 3, ..., 3}_{q}, 2) \leqslant q! \sum_{k=0}^{q} \frac{1}{k!} + 1. \tag{11.1}$$

Let K_{n_q} be a clique with n_q vertices that has a good q-colouring of its edges.
For a vertex a of K_{n_q} let A_i denote the set of vertices of K_{n_q} joined to a by an edge
having the colour i.

The induced subgraph with vertex-set A_i is a clique that has a good $(q-1)$-
colouring of its edges, hence $|A_i| \leqslant n_{q-1}$. We deduce that $n_q - 1 = d(a) =$
$= \sum_{i=1}^{q} |A_i| \leqslant q\, n_{q-1}$, where $d(a)$ is the degree of a in K_{n_q}.

Because $n_1 = 2$, we may write:

$$n_q \leqslant q n_{q-1} + 1 \leqslant q((q-1)\, n_{q-2} + 1) + 1 \leqslant \cdots$$

$$\leqslant q! + q! + \frac{q!}{2!} + \cdots + \frac{q!}{q!} = \sum_{k=0}^{q} \frac{q!}{k!}.$$

This proves proposition 2. E. G. Whitehead, Jr. showed that the first value of q
having the property that the two members of (11.1) are different is $q = 4$ [24].
In the case $t = 2$ we have shown that $N(q_1, q_2, 1) = q_1 + q_2 - 1$ and
$N(q_1, r, r) = q_1$; the above established inequality

$$N(q_1, q_2, r) \leqslant N(N(q_1 - 1, q_2, r), N(q_1, q_2 - 1, r), r - 1) + 1 \tag{11.2}$$

will enable us to obtain an upper bound for the Ramsey numbers in the case $t = 2$.

PROPOSITION 3 *(Erdös, Szekeres). If $p, q \geqslant 2$, the Ramsey numbers satisfy*
the inequality

$$N(p, q, 2) \leqslant \binom{p + q - 2}{p - 1}. \tag{11.3}$$

We shall prove this inequality by induction on $p + q$. For $p + q = 4$ the
inequality (11.3) holds, for $N(2, q, 2) = q$ and $\binom{q}{1} = q$. Assuming (11.3) to hold
for all pairs (p', q') with $p', q' \geqslant 2$ and $p' + q' < p + q$, let us prove it for (p, q).
Using $N(p, q, 1) = p + q - 1$ and (11.2), we obtain

$$N(p, q, 2) \leqslant N(N(p-1, q, 2), N(p, q-1, 2), 1) + 1 = N(p - 1, q, 2) + N(p, q - 1, 2)$$

$$\leqslant \binom{p - 1 + q - 2}{p - 2} + \binom{p + q - 1 - 2}{p - 1} = \binom{p + q - 2}{p - 1},$$

by the inductive hypothesis and the recurrence formula for the calculation of
combinations.

The numbers $N(p, q, 2)$ have been determined for small values of the parameters p and q, for which they are very close to the binomial coefficients $\binom{p + q - 2}{p - 1}$. In Table 11.1, each cell contains the corresponding two numbers separated by a semicolon.

TABLE 11.1

$N(p, q, 2)$; $\binom{p + q - 2}{p - 1}$	$p = 2$	$p = 3$	$p = 4$	$p = 5$	$p = 6$	$p = 7$
$q = 2$	2; 2	3; 3	4; 4	5; 5	6; 6	7; 7
$q = 3$	3; 3	6; 6	9; 10	14; 15	18; 21	23; 28
$q = 4$	4; 4	9; 10	18; 20			
$q = 5$	5; 5	14; 15				
$q = 6$	6; 6	18; 21				
$q = 7$	7; 7	23; 28				

J. Yackel proved recently [26] the existence of a constant c such that

$$N(n + 1, n + 1, 2) \leqslant c \, \frac{\log \log n}{\log n} \binom{2n}{n}.$$

Application. Consider n points in arbitrary positions in a 3-dimensional space. Two points determine an edge of the complete graph having as vertices the n points. If we colour some of the edges of this graph in red and the remainder in blue, then the family of two-element subsets of the set of vertices can be partitioned into the set A_1 of red edges and the set A_2 of blue edges. If p and q are integers such that $2 \leqslant p, q$ and if $n \geqslant N(p, q, 2)$, the Ramsey theorem ensures the existence of p vertices which are joined by red edges only, or of q vertices which are joined by blue edges only; $N(p, q, 2)$ is the least integer with this property. These are the Ramsey numbers in graph theory. An equivalent definition is the following: the number $N(p, q, 2)$ is the least integer n such that every graph with n vertices includes either a complete subgraph with p vertices, or an independent set with q vertices. We have seen (Chap. 8) that there exists a bijection from the set of 2-colourings of the edges of the complete n-vertex graph onto the set of n-vertex graphs.

PROPOSITION 4 (*Erdös, Szekeres*). *For every integer $m \geqslant 3$ there exists a least positive integer N_m with the following property: for any integer $n \geqslant N_m$, if n points of the plane do not contain three collinear then m of them are the vertices of a convex polygon.*

We shall first prove that if five points of the plane do not contain three collinear points, then four of them are the vertices of a convex quadrilateral. The five points define $\binom{5}{2} = 10$ line segments which join the five points in all posibles ways and the perimeter of this configuration is a convex polygon. If the latter is a quadrilateral or a pentagon, the property is immediate, because in the case of a convex pentagon four out of its vertices form a convex quadrilateral. In the opposite case the convex polygon is a triangle, hence two out of the five points are inside the triangle.

The two interior points define a line and two of the three vertices of the triangle are situated on the same side of this line. In this case the two interior points together with the two vertices of the triangle situated on the same side of the line form a convex quadrilateral and the property is established.

Let us show further that if m points of the plane do not contain three collinear points and if every four-point subset defines a convex quadrilateral, then the m points are the vertices of a convex polygon.

The m points define $\dfrac{m(m-1)}{2}$ line segments that join them in all possible ways, the perimeter of this configuration being a convex polygon with q vertices. Let us prove that $m = q$. Let $V_1, V_2, ..., V_q$ be the vertices of the convex polygon in the order in which they are met when running around the perimeter in one of the two possible directions. If one of the chosen points is inside this convex polygon, it must belong to the interior of one of the triangles $V_1 V_2 V_3, V_1 V_3 V_4, ..., V_1 V_{q-1} V_q$, which contradicts the assertion that every four-point set defines a convex quadrilateral. Therefore $q = m$ and the m points are the vertices of a convex polygon.

Now proposition 4 follows from Ramsey's theorem. If $m = 3$, we get $N_3 = 3$ and the proposition is obvious, since three noncollinear points of the plane determine a triangle, which is a convex polygon. If $m \geqslant 4$, let $n \geqslant N(5, m, 4)$, and take n points in the plane. We form a partition of the family of four-point subsets of the set of the n points of the plane, namely the set of concave quadrilaterals and the set of convex quadrilaterals. In view of the Ramsey theorem, there exists either a pentagon for which all four-vertex subsets are concave quadrilaterals, or an m-vertex polygon for which all four-vertex subsets are convex quadrilaterals. But we have seen that the former alternative is impossible, because a pentagon contains at least one convex quadrilateral, while the latter alternative implies that the m-vertex polygon is convex, since all its four-vertex subsets determine convex quadrilaterals.

It follows from the above proof that $N_m \leqslant N(5, m, 4)$. It is known that $N_3 = 3 = 2 + 1$, $N_4 = 5 = 2^2 + 1$, $N_5 = 9 = 2^3 + 1$, thus leading to the conjecture $N_m = 2^{m-2} + 1$, which is neither proved nor disproved for the time being [19].

Problems

1.   Prove that $N(3, 3, 2) = 6$, i.e. any colouring of the edges of a complete hexagon K_6 with two colours contains at least one monochromatic triangle.

2.   Given the integers $l_i, k_i, i = 1, ..., n$, and r, which satisfy the properties $l_i \geqslant r \geqslant k_i > 0$, for $i = 1, ..., n$, prove that we may define an integer $N(l_1, k_1; l_2, k_2; ...; l_n, k_n; r) = m$ as the smallest integer with the following property: If S is a set containing m points and the r-subsets of S are partitioned arbitrarily into n classes, then for some i, $1 \leqslant i \leqslant n$, there exists an l_i-subset of S each of whose k_i-subsets lies in some r-subset of the ith class.

(Erdös, O'Neil, 1973)

H i n t: This assertion follows from the existence of the Ramsey number $N(l_1, l_2, ..., l_n; r)$.

3. Prove that the generalized Ramsey numbers have the properties

i) $N(r, k_1; l, k_2; r) = N(l, k_1; r, k_2; r) = l$;

ii) $N(l_1, k_1; l_2, k_2; r)$

$\leqslant N(N(l_1 - 1, k_1; l_2, k_2; r), k_1 - 1; N(l_1, k_1; l_2 - 1, k_2; r), k_2 - 1; r - 1) + 1$.

(Erdös, O'Neil, 1973)

H i n t: See the proof of Proposition 1.

4. Prove the following generalization of the pigeon-hole principle: If $k_1 + k_2 = r + 1$, then $N(l_1, k_1; l_2, k_2; r) = l_1 + l_2 - k_1 - k_2 + 1$. Further, if $k_1 + k_2 \leqslant r$, then $N(l_1, k_1; l_2, k_2; r) = \max(l_1, l_2)$.

(Erdös, O'Neil, 1973)

H i n t: Prove that $N(l_1, k_1; l_2, k_2; r) = l_1 + l_2 - k_1 - k_2 + 1$ (if $k_1 + k_2 = r + 1$) by double inequality.

5. Prove the following statement: If the points of the plane are arbitrarily partitioned into a finite number k of sets, then there exists at least one of these sets which contains the vertices of some equilateral triangle.

(Simmons, 1971)

H i n t: Apply repeatedly van der Waerden's theorem [22]: for any positive integers k, t, there exists a number denoted by $W(k, t)$, which is the smallest integer having the following property: If the set $\{1, 2, 3, ..., W(k,t)\}$ is arbitrarily partitioned into k sets, there exists a class of the partition which contains an arithmetical progression with $t + 1$ members. It is known that $W(1, t) = t + 1$, $W(k, 1) = k + 1$, $W(2, 2) = 9$, $W(2, 3) = 35$, $W(3,2) = 27$.

An upper bound for these numbers is obtained in [1].

6. Prove that $N(\underbrace{3, 3, ..., 3}_{k \text{ times}}; 2) \geqslant 2^k + 1$

for $k \geqslant 1$.

(Greenwood, Gleason, 1955)

7. Let $\overline{G}$ denote the complementary graph or complement of G (two vertices in $\overline{G}$ are adjacent if and only if they are not adjacent in G). The number of vertices of the graph G will be denoted by $v(G)$.

Let $g(k, l)$ denote the least integer for which $v(G) \geqslant g(k, l)$ implies that either G contains an elementary chain of length k, or $\overline{G}$ one of length l (the length of a chain is the number of edges in the chain).

Prove that for $k \geqslant l$ we have $g(k, l) = k + \left[\dfrac{l + 1}{2} \right]$.

(Gerencsér, Gyárfás, 1967)

H i n t: First prove by induction on k that $g(k, l) \leqslant k + \left[\dfrac{l+1}{2}\right]$ (for further details see [10]). For the reverse inequality consider a particular graph.

8.   Let $f_r(n)$ denote the greatest integer with the property that colouring arbitrarily the edges of a complete graph K_n with r colours, there always exists a one-coloured connected subgraph with at least $f_r(n)$ vertices.

Prove that $f_2(n) = n$ and $f_3(n) = \left[\dfrac{n+1}{2}\right]$.

(Gerencsér, Gyárfás, 1967)

9.   By definition the Ramsey number $r(F_1, F_2)$ for two graphs F_1, F_2, is the minimum p such that every 2-colouring of the edges of K_p contains a green F_1 or a red F_2. Prove that the Ramsey numbers for stars are given by the formula:

$$r(K_{1,m}, K_{1,n}) = \begin{cases} m + n & \text{if } m \text{ or } n \text{ is odd,} \\ m + n - 1 & \text{if } m \text{ and } n \text{ are both even.} \end{cases}$$

(Harary, 1972)

10.   We may extend Ramsey theory to digraphs (directed graphs), following Harary and Hell, by letting $\vec{K}_p$ denote the complete digraph with a symmetric pair of arcs joining each pair of its p vertices.

Then $r(D_1, D_2)$ is defined as the smallest p such that any 2-colouring of the arcs of $\vec{K}_p$ has a red D_1 or a green D_2 where D_1, D_2 are directed graphs.

Prove that the Ramsey number $r(D_1, D_2)$ exists if and only if at least one of D_1 or D_2 is acyclic.

(Harary, Hell, 1972)

H i n t: To prove necessity colour red the arcs of the transitive tournament T_p (an acyclic digraph with vertex set $\{1, 2, ..., p\}$ and the arcs (i, j) where $i > j$), and colour green the remaining arcs, which also form a transitive tournament $\bar{T}_p$.

11.   Show that the Ramsey numbers for transitive tournaments and complete graphs are related by the equality $r(T_m, T_n) = r(K_m, K_n)$.

(Harary, 1972)

H i n t: Use the property that $\vec{K}_p$ is a superposition of two transitive tournaments, T_p and $\bar{T}_p$.

12.   Deduce that the Ramsey number $r(K_m, K_{1,n})$ is given by the formula

$$r(K_m, K_{1,n}) = (m - 1)n + 1.$$

(Schwenk, 1972)

H i n t: To get the lower bound colour red the edges of the complete $(m - 1)$-partite graph $K_{\underbrace{n, n, ..., n}_{m-1}}$ and colour green the remaining edges of $K_{(m-1)n+1}$.

The upper bound can be obtained as a corollary of Turán's theorem.

13.   Prove that every 2-colouring of K_6 has at least two monochromatic quadrilaterals (C_4).

14.   If C_n denotes a cycle with n vertices, the cyclic Ramsey numbers $r(C_m, C_n)$ are defined for $m \leqslant n$.

Show that $r(C_3, C_3) = r(C_4, C_4) = 6$ and $r(C_4, C_5) = 7$. (R. Faudree, R. Schelp and V. Rosta [18] showed that excepting $r(C_3, C_3)$ and $r(C_4, C_4)$, the numbers $r(C_m, C_n)$ are equal to: i) $2n - 1$ for $3 \leqslant m \leqslant n$, for m odd; ii) $n + \dfrac{m}{2} - 1$ for $4 \leqslant m \leqslant n$, m and n even; iii) $\max\left(n + \dfrac{m}{2} - 1, 2n - 1\right)$ for $4 \leqslant m < n$, m even, n odd).

15.   Show that the generalized Ramsey numbers $r(F_1, F_2)$, where F_1, F_2 are two graphs with at most 4 vertices, none of which is isolated, are given by the table:

TABLE 11.2

	K_2	P_3	$2K_2$	K_3	P_4	$K_{1,3}$	C_4	$K_{1,3}+x$	K_4-x	K_4
K_2	2	3	4	3	4	4	4	4	4	4
P_3		3	4	5	4	5	4	5	5	7
$2K_2$			5	5	5	5	5	5	5	6
K_3				6	7	7	7	7	7	9
P_4					5	5	5	7	7	10
$K_{1,3}$						6	6	7	7	10
C_4							6	7	7	10
$K_{1,3}+x$								7	7	10
K_4-x									10	11
K_4										18

These graphs are illustrated in Fig. 11.1.

(Chvátal, Harary, 1972)

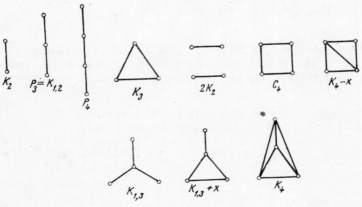

K_2 $P_3 = K_{1,2}$ P_4 K_3 $2K_2$ C_4 K_4-x

$K_{1,3}$ $K_{1,3}+x$ K_4

Fig. 11.1

16.   Let F be a graph having no isolated vertex, G a given graph, and c a positive integer. We denote by $R(G, F; c)$ the greatest integer n with the property that, in every c-colouring of the edges of G, there are at least n monochromatic occurrences of F.

The cube Q_n is defined as usual as the graph with 2^n vertices which can be taken as all binary n-sequences, with two points adjacent whenever their sequences differ in just one place.

Demonstrate that the numbers $R(Q_n, Q_m; c)$ for $n \geqslant m$ are given by the formula:

$$R(Q_n, Q_m; c) = \begin{cases} \binom{n}{m} 2^{n-m} & \text{if } \min{(m, c)} = 1, \\ \\ 0 & \text{otherwise.} \end{cases}$$

(Harary, 1972)

17.   For any two trees T_1 and T_2 such that T_1 is not a star $(K_{1, p})$, prove that $R(T_2, T_1; c) = 0$ whenever $c \geqslant 2$.

(Harary, 1972)

18.   Denote by $P(k, l)$ the smallest integer having the property that any planar graph with $P(k, l)$ or more vertices contains as a subgraph the complete graph with k vertices or a set of l independent vertices.

The existence of $P(k, l)$ follows from the inequality $P(k, l) \leqslant N(k, l, 2)$, where $N(k, l, 2)$ denotes the ordinary Ramsey number. Prove that $P(2, l) = l$, $P(4, 2) = 4$, $P(3, l) = 3(l - 1)$ for $l \geqslant 2$, $P(k, l) = P(5, l)$ for $k \geqslant 5$.

(Walker, 1969)

H i n t: No planar graph contains as a subgraph K_5 or $K_{3,3}$ or these graphs with some vertices inserted on their edges.

19.   Show that $r(P_m, K_n) = (m - 1)(n - 1) + 1$.

(T. D. Parsons, 1973)

H i n t: Prove that $r(P_m, K_n) > (m - 1)(n - 1)$ by considering $n - 1$ disjoint copies of K_{m-1}. The reverse inequality is obtained from Erdös' result [8] which asserts that if a graph G has $(m - 1)(n - 1) + 1$ vertices, then G contains P_m or $\overline{G}$ contains K_n.

20.   Prove the following generalization of proposition 3:

For $k_1, k_2, ..., k_r \geqslant 1$ the following inequality holds:

$$N(k_1 + 1, k_2 + 1, ..., k_r + 1, 2) \leqslant \frac{(k_1 + k_2 + ... + k_r)!}{k_1! \, k_2! \, ... \, k_r!}.$$

(Greenwood, Gleason, 1955)

H i n t: Prove that $N(k_1, ..., k_r, 2) \leqslant \sum_i N(k_1, ..., k_i - 1, ..., k_r, 2)$ and use (2.9).

BIBLIOGRAPHY

1. Abott, H. L., Liu, A. C., *On partitioning integers into progression free sets*, J. Combinatorial Theory (A), **13**, 1972, 432—436.
2. Berge, C., *Graphes et hypergraphes*, Dunod , Paris, 1970.
   English edition: *Graphs and hypergraphs*, North-Holland , Amsterdam, 1973.
3. Bondy, J. A., Erdös, P., *Ramsey numbers for cycles in graphs*, J. Combinatorial Theory (B) **14**, 1973, 46—54.
4. Chvátal, V., Harary, F., *Generalized Ramsey theory for graphs* I: *Diagonal numbers*, Period. Math. Hungar., **3**, 1973, 115—124.
6. Chvátal, V., Harary, F., *Generalized Ramsey theory for graphs* II: *Small diagonal numbers*, Proc. Amer. Math. Soc., **32**, 1972, 389—394.
6. Chvátal, V., Harary, F., *Generalized Ramsey theory for graphs* III: *Small off-diagonal numbers*, Pacific J. Math., **41**, 1972, 335—345.
7. Erdös, P., Szekeres, G., *A combinatorial problem in geometry*, Compositio Math., **2**, 1935, 463—470.
8. Erdös, P., *Some remarks on the theory of graphs*, Bull. Amer. Math. Soc., **53**, 1947, 292—294.
9. Erdös, P., O'Neil, P. E., *On a generalization of Ramsey numbers*, Discrete Math., **4**, 1973, 29—35.
10. Gerencsér, L., Gyárfás, A., *On Ramsey-type problems*, Ann. Univ. Sci. Budapest Eötvös Sect. Mat., **10**, 1967, 167—170.
11. Graham, R. L., Rothschild, B. L., *Ramsey's theorem for n-dimensional arrays*, Bull. Amer. Math. Soc., **75**, 1969, 418—422.
12. Graham, R. L., Leeb, K., Rothschild, B. L., *Ramsey's theorem for a class of categories*, Advances in Math., **8**, 1972, 417—433.
13. Graver, J. E., Yackel, J., *Some graph theoretic results associated with Ramsey's theorem*, J. Combinatorial Theory, **4**, 1968, 125—175.
14. Greenwood, R. E., Gleason, A. M., *Combinatorial relations and chromatic graphs*, Canad. J. Math., **7**, 1955, 9—20.
15. Harary, F., *Recent results on generalized Ramsey theory for graphs*, Graph Theory and Applications (ed: Y. Alavi, D. R. Lick and A. T. White), Lecture Notes in Math. N° 303, Springer, Berlin, 1972, 125—138.
16. Motzkin, T., *Shadows of finite sets*, J. Combinatorial Theory, **4**, 1968, 40—48.
17. Parsons, T. D., *The Ramsey numbers $r(P_m, K_n)$*, Discrete Math., **6**, 1973, 159—162.
18. Rosta, V., *On a Ramsey-type problem of J . A. Bondy and P. Erdös*, I; II, J. Combinatorial Theory (B), **15**, 1973, 94—104; 105—120.
19. Ryser, H., *Combinatorial mathematics*, Wiley, New York, 1963.
20. Simmons, G. J., *Combinatorial properties of plane partitions*, Studia Sci. Math. Hungar., **6**, 1971, 335—339.
21. Sobczyk, A., *Graph-colouring and combinatorial numbers*, Canad.J. Math., **20**, 1968, 520—534.
22. van der Waerden, B. L., *Beweis einer baudetschen Vermutung*, Nieuw Arch. Wisk., **15**, 1927, 212—216.
23. Walker, K., *The analogue of Ramsey numbers for planar graphs*, Bull. London Math. Soc., **1**, 1969, 187—190.
24. Whitehead, E. G., Jr., *The Ramsey number N(3, 3, 3, 3; 2)*, Discrete Math., **4**, 1973, 389—396.
25. Williamson, J. E., *A Ramsey-type problem for paths in digraphs*, Math. Ann., **203**, 1973, 117—118.
26. Yackel, J., *Inequalities and asymptotic bounds for Ramsey numbers*, J. Combinatorial Theory (B), **13**, 1972, 56—68.

CHAPTER 12

Minimum Distances and Paths in Graphs

The problem of smallest distances and optimal paths in graphs has been much studied in the literature and has led to algorithms both efficient and elegant from the mathematical point of view. Among the recent surveys on this subject we quote [9], [29]. On the other hand, as similar operations occur when solving some related problems (for example, in the determination of the transitive closure and of the algebraic closure of a graph [28], the determination of the maximum transport capacity between two nodes, of the total conductibility of a multipole [19] or of the set of paths joining two nodes of a multigraph), certain algebraic structures which enable a unified presentation of the algorithms for the above problems have been suggested by Gr. C. Moisil [20], M. Yoeli [41], G. Povarov [25], L. Nolin [21], C. Benzaken [3], V. Peteanu [23].

We shall present below the problem of determining minimum distances and paths in graphs, as well as two matrix algorithms for the determination of least distances: the algorithm of B. Roy for the determination of the algebraic closure of a graph, transposed for minimum distances, and an improvement of the Bellman-Kalaba algorithm.

Consider a loop-free directed graph $G = (X, \Gamma)$ with p nodes; so $x \cap \Gamma x = \varnothing$ for every $x \in X$. Let U be the set of arcs of the graph, and $l : U \to R_+$ a mapping which associates with every arc $u \in U$ a nonnegative real number $l(u) \geqslant 0$, called the *length (cost, weight)*, of the arc u. The length of a path $\mu = (x_i, x_{k_1}, x_{k_2}, \ldots \ldots, x_{k_s}, x_j)$ from node x_i to node x_j is by definition the sum of the lengths of the arcs of the path, i.e.

$$l(\mu) = l(x_i, x_{k_1}) + l(x_{k_1}, x_{k_2}) + \ldots + l(x_{k_s}, x_j). \tag{12.1}$$

If there exists at least one path joining the nodes x_i and x_j, the smallest distance between these nodes will be equal, by definition, to $\min\limits_{\mu} \sum\limits_{u \in \mu} l(u)$, the minimum being taken with respect to all paths μ going from x_i to x_j; this smallest distance will be denoted by $\hat{a}_{ij}$, therefore

$$\hat{a}_{ij} = \min_{\mu = (x_i, \ldots, x_j)} \sum_{u \in \mu} l(u). \tag{12.2}$$

We shall define $\hat{a}_{ii} = 0$ for every i, while in the case where there is no path in the graph G from x_i to x_j, we shall say that $\hat{a}_{ij} = \infty$. The matrix of the least distances will be denoted by $\hat{A} = \{\hat{a}_{ij}\}_{i,j=1,\ldots,p}$. The problem is to determine the least distances and paths between two given nodes of the graph, between a fixed node and all the other ones or between any two nodes of the graph.

In order to present the matrix algorithms for determining the shortest distances, we shall define a multiplication of distance matrices as follows:

If $A = \{a_{ij}\}_{i,j=1,\ldots,p}$ and $B = \{b_{ij}\}_{i,j=1,\ldots,p}$

are two square matrices with $a_{ij}, b_{ij} \in \{x|\, x \geqslant 0\} \cup \{\infty\}$, their product will be defined by

$$\{A \times B\}_{ij} = \min_{k=1,\ldots,p} (a_{ik} + b_{kj}) \tag{12.3}$$

for any $i, j = 1, \ldots, p$ where by definition the symbol ∞ possesses the property $\infty + a = a + \infty = \infty$ and $\infty > a$ for any real number a. This operation is associative, i.e. $A \times (B \times C) = (A \times B) \times C$, because $\{A \times (B \times C)\}_{ij} = \min_{k_1}(a_{ik_1} + \min_{k_2}(b_{k_1 k_2} + c_{k_2 j})) = \min_{k_1} \min_{k_2}(a_{ik_1} + b_{k_1 k_2} + c_{k_2 j}) = \min_{k_2}(\min_{k_1}(a_{ik_1} + b_{k_1 k_2}) + c_{k_2 j}) = \{(A \times B) \times C\}_{ij}$. The order relation between distance matrices will be defined by: $A \leqslant B$ if $a_{ij} \leqslant b_{ij}$ for all $i, j = 1, \ldots, p$. If $A \leqslant B$ and $C \leqslant D$, then $a_{ik} + c_{kj} \leqslant b_{ik} + d_{kj}$, therefore

$$\min_k (a_{ik} + c_{kj}) \leqslant \min_k (b_{ik} + d_{kj})$$

for any $i, j = 1, \ldots, p$ and also $A \times C \leqslant B \times D$, i.e. the multiplication operation of distance matrices is isotone with respect to the order relation.

Given a graph $G = (X, U)$ with the distance function $l : U \to R_+$, the matrix $A = \{a_{ij}\}_{i,j=1,\ldots,p}$ of direct distances between the nodes of the graph will be defined as follows:

$$a_{ii} = 0, \quad a_{ij} = l(x_i, x_j) \text{ if } (x_i, x_j) \in U$$

and

$$a_{ij} = \infty \text{ if } (x_i, x_j) \notin U$$

for $i, j = 1, \ldots, p$.

We shall study the powers of the matrix of direct distances A with respect to the multiplication of matrices defined by (12.3). We obtain

$$\{A^2\}_{ii} = 0$$

and

$$\{A^2\}_{ij} = \min_k (a_{ik} + a_{kj}) = \min (a_{ij}, \min_{k \neq i, j} (a_{ik} + a_{kj}))$$

since $a_{ii} = a_{jj} = 0$.

We shall show by induction that

$$\{A^r\}_{ii} = 0$$

and

$$\{A^r\}_{ij} = \min\left(\{A^{r-1}\}_{ij}, \min_{(k_1, \ldots, k_{r-1})}(a_{ik_1} + a_{k_1 k_2} + \ldots + a_{k_{r-1} j})\right),$$

where the numbers $k_1, \ldots, k_{r-1} \in \{1, 2, \ldots, p\}\setminus\{i, j\}$ and are pairwise distinct. For $r = 2$ this reduces to the expression for the elements of the matrix A^2. If the property is true for r, then

$$\{A^{r+1}\}_{ii} = \{A^r \times A\}_{ii} = \min\left(0, \min_{k \neq i}(\{A^r\}_{ik} + a_{ki})\right) = 0$$

and

$$\{A^{r+1}\}_{ij} = \{A^r \times A\}_{ij} = \min_k(\{A^r\}_{ik} + a_{kj}) =$$

$$= \min\left(\{A^r\}_{ij}, a_{ij}, \min_{k \neq i, j}(\{A^r\}_{ik} + a_{kj})\right) = \min\left(\{A^r\}_{ij}, \min_{k \neq i, j}(\{A^r\}_{ik} + a_{kj})\right),$$

since $\{A^r\}_{ij} \leqslant a_{ij}$. According to the induction hypothesis, the term $\{A^r\}_{ik}$ represents the minimum of the sums $a_{ik_1} + a_{k_1 k_2} + \ldots + a_{k_q k}$ where $q \leqslant r - 1$, and the indices $k_1, k_2, \ldots, k_q$ are pairwise distinct and different from i and k. If in the expression $\min_{k \neq i, j}(a_{ik_1} + a_{k_1 k_2} + \ldots + a_{k_q k} + a_{kj})$, there are two equal indices, i.e. there is a $k_s = j$, then

$$a_{ik_1} + a_{k_1 k_2} + \ldots + a_{k_{s-1} j} \leqslant a_{ik_1} + a_{k_1 k_2} + \ldots + a_{kj},$$

since the left hand term appears also in the expression of $\{A^r\}_{ij}$, so that it is not necessary to take it into consideration in the expression $\min_{k \neq i, j}(\{A^r\}_{ik} + a_{kj})$. A similar result is obtained when the number of terms $q < r - 1$, these terms being absorbed by $\{A^r\}_{ij}$. Hence the only terms to be taken into account in the expression $\min_{k \neq i, j}(\{A^r\}_{ik} + a_{kj})$, are the terms $a_{ik_1} + a_{k_1 k_2} + \ldots + a_{k_{r-1} k} + a_{kj}$, where the indices $k_1, k_2, \ldots, k_{r-1}, k$ are pairwise distinct and different from i and j. Therefore we obtain

$$\{A^{r+1}\}_{ij} = \min\left(\{A^r\}_{ij}, \min_{(k_1, \ldots, k_r)}(a_{ik_1} + a_{k_1 k_2} + \ldots + a_{k_r j})\right),$$

where the numbers $k_1, \ldots, k_r \in \{1, 2, \ldots, p\}\setminus\{i, j\}$ are pairwise distinct. Since the greatest number of pairwise distinct elements from the set $\{1, 2, \ldots, p\}\setminus\{i, j\}$ is equal to $p - 2$, it follows that the powers of the matrix of direct distances A become stationary beginning with a power at most equal to $p - 1$, and therefore there exists an $r \leqslant p - 1$ such that $A^r = A^{r+1} = A^{r+2} = \ldots$. Since $\{A^2\}_{ij} \leqslant a_{ij}$, we shall have $A^2 \leqslant A$, and, by multiplying both members of this inequality by A and taking into account the property of isotony, we obtain $A^3 \leqslant A^2$, etc., therefore the above equations can be completed as follows

$$A \geqslant A^2 \geqslant A^3 \geqslant \ldots \geqslant A^r = A^{r+1} = \ldots \qquad (12.4)$$

where A is a matrix of distances with $a_{ii} = 0$ for $i = 1, \ldots, p$ and $r \leqslant p - 1$.

PROPOSITION 1. *If the matrix of direct distances between the vertices of a graph having p vertices is A, the matrix of smallest distances will be A^{p-1}.*

If we adopt the notation $a_{ij} = \infty$ to indicate that there is no arc from x_i to x_j, we see that

$$\min_{\mu=(x_i,\ldots,x_j)} \sum_{u \in \mu} l(u) = \min\left(a_{ij}, \min_{r=1,\ldots,p-2} \min_{(k_1,\ldots,k_r)} (a_{ik_1} + a_{k_1k_2} + \ldots + a_{k_rj})\right) = \{A^{p-1}\}_{ij},$$

where the numbers $k_1, \ldots, k_r \in \{1, \ldots, p\} \setminus \{i, j\}$ are pairwise distinct.

We remark that $a_{ik_1} + \ldots + a_{k_rj}$ represents the length of a path from x_i to x_j, that has r intermediate nodes, namely $x_{k_1}, x_{k_2}, \ldots, x_{k_r}$.

As concerns the number of additions and comparisons, an effective method for computing the matrix A^{p-1} is offered by the algorithm of B. Roy [27] and S. Warshall [39], and extended to more general situations by several authors [13], [21], [26], [36].

The elementary operators occurring in the Roy algorithm are matrix operators defined as follows:

$T_{ij}^{(k)}(A) = B$ if matrices A and B have p rows and p columns, where $b_{ij} = \min(a_{ij}, a_{ik} + a_{kj})$ and $b_{mn} = a_{mn}$ for any $(m, n) \neq (i, j)$. It follows from this definition that $A \geqslant T_{ij}^{(k)}(A) \geqslant A^2$, hence, using isotony, we deduce that

$$A^{p-1} \geqslant (T_{ij}^{(k)}(A))^{p-1} \geqslant (A^2)^{p-1} = A^{p-1},$$

therefore

$$(T_{ij}^{(k)}(A))^{p-1} = A^{p-1}.$$

Since $a_{ii} = 0$ for any $i = 1, \ldots, p$, the operators $T_{ij}^{(k)}$ with $i = j$ or $i = k$ or $j = k$ reduce to the identity operator. It also follows from the way of defining the operators $T_{ij}^{(k)}$, that

$$T_{i_1j_1}^{(k_1)} T_{i_2j_2}^{(k_2)}(A) = T_{i_2j_2}^{(k_2)} T_{i_1j_1}^{(k_1)}(A)$$

and

$$T_{ij}^{(k)} T_{ij}^{(k)}(A) = T_{ij}^{(k)}(A) \qquad (12.5)$$

for any matrix A with p rows and p columns. We shall also introduce the operators

$$V_k = \prod_{\substack{i, j=1 \\ i \neq j, \, i \neq k, \, j \neq k}}^{p} T_{ij}^{(k)} \quad \text{and} \quad V = \prod_{k=1}^{p} V_k. \qquad (12.6)$$

The order of the operators V_k is immaterial in the definition of V, because

$$V_{k_1} V_{k_2}(A) = V_{k_2} V_{k_1}(A)$$

for every k_1, k_2. Indeed,

$$\{V_{k_1} V_{k_2}(A)\}_{ij} = \min (a_{ij}, a_{ik_2} + a_{k_2 j}, \min (a_{ik_1}, a_{ik_2} + a_{k_2 k_1}) + \min (a_{k_1 j}, a_{k_1 k_2} + a_{k_2 j}))$$

$$= \min (a_{ij}, a_{ik_1} + a_{k_1 j}, a_{ik_2} + a_{k_2 j}, a_{ik_1} + a_{k_1 k_2} + a_{k_2 j}, a_{ik_2} + a_{k_2 k_1} + a_{k_1 j})$$

$$= \min (a_{ij}, a_{ik_1} + a_{k_1 j}, \min (a_{ik_2}, a_{ik_1} + a_{k_1 k_2}) + \min (a_{k_2 j}, a_{k_2 k_1} + a_{k_1 j}))$$

$$= \{V_{k_2} V_{k_1}(A)\}_{ij}.$$

Taking into account the relation $(T_{ij}^{(k)}(A))^{p-1} = A^{p-1}$, we obtain, by iteration, $(V_k(A))^{p-1} = A^{p-1}$ and $(V(A))^{p-1} = A^{p-1}$.

PROPOSITION 2 *(Roy, Floyd). For every matrix of distances A of order p with $a_{ii} = 0$, we have $V(A) = A^{p-1}$.*

Taking into account the commutativity of the product of the operators V_k and the idempotency of these operators, induced by the commutativity and the idempotency of the operators $T_{ij}^{(k)}$, given by (12.5), we obtain $V(A) = V^2(A)$.

Applying the operator V to both members, we obtain

$$V(A) = V^2(A) = V^3(A) = \dots$$

We prove by induction upon s, that

$$\{V_{k_s} V_{k_{s-1}} \dots V_{k_1}(A)\}_{ij} \leqslant \min (a_{ij}, \min_{k \in \{k_1, \dots, k_s\} \setminus \{i, j\}} (a_{ik} + a_{kj})).$$

For $s = 1$, the equality is deduced from the definition of the operators V_k. If the inequality is true for $s - 1$, then with the notation

$$V_{k_{s-1}} \dots V_{k_1} = W,$$

we obtain

$$\{V_{k_s} W(A)\}_{ij} = \min (\{W(A)\}_{ij}, \{W(A)\}_{ik_s} + \{W(A)\}_{k_s j})$$

$$\leqslant \min (a_{ij}, \min_{k \in \{k_1, \dots, k_s\} \setminus \{i, j\}} (a_{ik} + a_{kj}))$$

if we take into account the induction hypothesis and the inequality

$$\{W(A)\}_{ik_s} + \{W(A)\}_{k_s j} \leqslant a_{ik_s} + a_{k_s j}.$$

If $s = p$, then $\{k_1, \dots, k_s\} = \{1, 2, \dots, p\}$ assuming that the numbers $k_1, \dots, k_s$ are pairwise different, hence $V(A) \leqslant A^2$, from which, according to isotony, we deduce that $V^2(A) \leqslant A^4, ..., V^n(A) \leqslant A^{2^n}$ for any $n \in N$. If $2^n \geqslant p - 1$, then $A^{2^n} = A^{p-1}$ and therefore $V^n(A) \leqslant A^{p-1}$. The contrary inequality also holds:

$$V^n(A) \geqslant (V^n(A))^{p-1} = (V(A))^{p-1} = A^{p-1},$$

therefore

$$V(A) = V^2(A) = ... = V^n(A) = A^{p-1}.$$

The number of computations required by the Roy algorithm is equal to $p(p - 1)(p - 2)$ additions and as many comparisons and the programming of the algorithm for a computer is simple.

In some cases, a single row of the matrix A^{p-1} interests us, namely the smallest distances from a fixed origin to all the other nodes of the graph and, to this end, we shall define an operator that transforms a single row of the matrix A.

We introduce the operator $T_{ij} = \prod\limits_{\substack{k=1 \\ k \neq i, j}}^{n} T_{ij}^{(k)}$. This operator acts upon the

matrices with p rows and p columns as follows:

$$T_{ij}(A) = B \text{ when } b_{ij} = \min(a_{ij}, \min_{k \neq i, j}(a_{ik} + a_{kj})) \text{ and } b_{mn} = a_{mn} \text{ for any}$$

$$(m, n) \neq (i, j).$$

Now set $U_1 = T_{1p}T_{1p-1} ... T_{12}$. The operator U_1 has the property that $(U_1^n)^{p-1} = A^{p-1}$ for any $n \in N$, because its component operators $T_{ij}^{(k)}$ have this property.

PROPOSITION 3. *The inequality $\{U^n(A)\}_{1i} \leqslant \{A^{n+1}\}_{1i}$ holds for any $i = 1, ..., p$ and $n \in N$.*

The proof is by induction on n. For $n = 1$, we obtain

$$\{U_1(A)\}_{12} = \min(a_{12}, \min_{k=3,...,p}(a_{1k} + a_{k2})) = \{A^2\}_{12}.$$

If it is already proved that

$$\{U_1(A)\}_{13} \leqslant \{A^2\}_{13}, ..., \{U_1(A)\}_{1i-1} \leqslant \{A^2\}_{1i-1},$$

it follows that

$$\{U_1(A)\}_{1i} = \min(a_{1i}, \min_{k=2,...,i-1}(\{U_1(A)\}_{1k} + a_{ki}), \min_{k=i+1,...,p}(a_{1k} + a_{ki}))$$

$$\leqslant \min(a_{1i}, \min_{\substack{k=2,...,p \\ k \neq i}}(a_{1k} + a_{ki})) = \{A^2\}_{1i} \text{ for any } 3 \leqslant i \leqslant p, \text{ since}$$

$\{U_1(A)\}_{1k} \leqslant \{A^2\}_{1k} \leqslant a_{1k}$ for $k = 2, ..., i - 1$. It follows that $\{U_1(A)\}_{1i} \leqslant \{A^2\}_{1i}$ for $i = 1, ..., p$, because $\{U_1^n(A)\}_{11} = 0$ for any $n \in N$.

Assuming that $\{U_1^n(A)\}_{1i} \leqslant \{A^{n+1}\}_{1i}$ for $i = 1, ..., p$, it follows that

$$\{U_1^{n+1}(A)\}_{12} = \min(\{U_1^n(A)\}_{12}, \min_{k=3,...,p}(\{U_1^n(A)\}_{1k} + a_{k2}))$$

$$\leqslant \min(\{A^{n+1}\}_{12}, \min_{k=3,...,p}(\{A^{n+1}\}_{1k} + a_{k2})) = \{A^{n+2}\}_{12}$$

since $\{A^{n+1}\}_{12} \leqslant a_{12}$ and therefore $\min(\{A^{n+1}\}_{12}, a_{12}) = \{A^{n+1}\}_{12}$. If it is already proved that

$$\{U_1^{n+1}(A)\}_{13} \leqslant \{A^{n+2}\}_{13}, \ldots, \{U_1^{n+1}(A)\}_{1i-1} \leqslant \{A^{n+2}\}_{1i-1},$$

it follows that

$$\{U_1^{n+1}(A)\}_{1i} = \min(\{U_1^n(A)\}_{1i}, \min_{k=2,\ldots,i-1}(\{U_1^{n+1}(A)\}_{1k} + a_{ki}, \min_{k=i+1,\ldots,p}(\{U_1^n(A)\}_{1k} + a_{ki}))$$

$$\leqslant \min(\{A^{n+1}\}_{1i}, \min_{\substack{k=2,\ldots,p \\ k \neq i}}(\{A^{n+1}\}_{1k} + a_{ki})) = \{A^{n+2}\}_{1i} \text{ for } 3 \leqslant i \leqslant p.$$

Indeed,

$$\{U_1^{n+1}(A)\}_{1k} \leqslant \{A^{n+2}\}_{1k} \leqslant \{A^{n+1}\}_{1k}$$

for $k = 2, \ldots, i - 1$ and $A^{n+1} \leqslant A$, therefore

$$\min(\{A^{n+1}\}_{1i}, a_{1i}) = \{A^{n+1}\}_{1i}.$$

Therefore $\{U_1^{n+1}(A)\}_{1i} \leqslant \{A^{n+2}\}_{1i}$ for any $i = 1, \ldots, p$.

If r is the smallest index having the property $A^r = A^{p-1}$, we get

$$\{U_1^{r-1}(A)\}_{1i} \leqslant \{A^r\}_{1i} = \{A^{p-1}\}_{1i};$$

but the contrary inequality

$$\{U_1^{r-1}(A)\}_{1i} \geqslant \{(U^{r-1}(A))^{p-1}\}_{1i} = \{A^{p-1}\}_{1i},$$

also holds, therefore $\{U_1^{r-1}(A)\}_{1i} = \{A^{p-1}\}_{1i}$ for any $i = 1, \ldots, p$. At the moment when $U_1^n(A) = U_1^{n+1}(A)$, we obtain

$$\{U_1^n(A)\}_{1i} = \{U_1^{n+1}(A)\}_{1i} = \ldots = \{U_1^{p-2}(A)\}_{1i} = \{A^{p-1}\}_{1i}$$

and therefore $\{U_1^n(A)\}_{1i} = \{A^{p-1}\}_{1i}$ for any $i = 1, \ldots, p$. The elements of the first row of the matrix $U_1^{p-2}(A)$ are equal to the elements of the first row of the matrix A^{p-1}, and the elements of the other rows are equal to the corresponding elements of the matrix A. It is proved in [35] that almost all the elements of the first row of the matrix A^{p-1} are obtained after at most $p - 3$ steps, namely $\{U_1^{p-3}(A)\}_{1i} = \{A^{p-1}\}_{1i}$ for any $4 \leqslant i \leqslant p$, the computing process being finished as soon as $U_1^n(A) = U_1^{n+1}(A)$. Therefore, to obtain the elements $\hat{a}_{12}$ and $\hat{a}_{13}$ we also apply the operators T_{12} and T_{13} and $\{T_{13}T_{12}U_1^{p-3}(A)\}_{1i} = \{A^{p-1}\}_{1i}$ for any $i = 1, \ldots, p$, an upper bound for the number of additions as well as for the number of comparisons required by this algorithm being equal to $(p - 2)(p^2 - 4p + 5)$. It is obvious that any shortest path consists of subpaths of shortest length, therefore the smallest distances $\hat{a}_{1i}$ between node x_1 and the other nodes of the graph satisfy the functional equation

$$\hat{a}_{1i} = \min_{\substack{j=1,\ldots,p \\ j \neq i}} (\hat{a}_{1j} + a_{ji}) \ (i = 2, \ldots, p), \tag{12.7}$$

with the definition $\hat{a}_{11} = 0$. This functional equation is solved iteratively, starting from certain initial values $a_{1i}^{(0)}$ and then applying the recurrence relations:

$$a_{1i}^{(k)} = \min_{\substack{j=1,\ldots,p \\ j \neq i}} (a_{1j}^{(k-1)} + a_{ji}) \quad (i = 2, \ldots, p), \tag{12.8}$$

if we adopt the definition $a_{11}^{(0)} = a_{11}^{(1)} = \ldots = 0$. The computing process is concluded when $a_{1i}^{(k)} = a_{1i}^{(k-1)}$ for any $i = 2, \ldots, p$ and this situation will occur after a finite number of steps, whatever the initial values $a_{1i}^{(0)}$ may be [2].

This way of solving the problem is called the *Bellman-Kalaba algorithm*.

We usually choose the initial values equal to the direct distances between the node x_1 and the other nodes of the network, i.e. $a_{1i}^{(0)} = a_{1i}$ for $i = 2, \ldots, p$. In this case, we remark that

$$\{A^k\}_{1i} = \min_{j=1,\ldots,p} (\{A^{k-1}\}_{1j} + a_{ji}) = \min_{\substack{j=1,\ldots,p \\ j \neq i}} (\{A^{k-1}\}_{1j} + a_{ji}),$$

because $\{A^{k-1}\}_{1i}$ represents the minimum of certain sums that also appear in the expression $\min\limits_{\substack{j=1,\ldots,p \\ j \neq i}} (\{A^{k-1}\}_{1j} + a_{ji})$, so that the Bellman-Kalaba algorithm reduces to the computation of the first row of the matrix A^k and involves at most $p - 2$ steps if the chosen initial values are $a_{1i}^{(0)} = a_{1i}$.

According to proposition 3, the algorithm, consisting of the computation of the sequence $\{U_1^n(A)\}_{n \in N}$ yields the result faster than the Bellman-Kalaba algorithm, provided the initial values $a_{1i}^{(0)}$ are chosen equal to the direct distances a_{1i}; moreover, this algorithm requires a smaller space in the computer memory, because the first row of the matrix A^k no longer has to be separately stored, since the intermediate results are directly inscribed in the matrix.

If the graph G does not contain circuits of length zero, any path of minimum length between two nodes is elementary, i.e. it does not pass twice through the same vertex. To determine the shortest elementary paths when one knows the matrix A^{p-1} of least distances or only its first row, one can use a simple algorithm as follows:

If $\hat{a}_{1i} \neq \infty$, then paths do exist, therefore shortest paths between x_1 and x_i also exist. If $\hat{a}_{1i} = a_{1i}$, then the arc (x_1, x_i) is a path of minimum length between x_1 and x_i. If there is an index $k \neq 1, i$ satisfying the equation

$$\hat{a}_{1i} = \hat{a}_{1k} + a_{ki}, \tag{12.9}$$

a path of smallest length from x_1 to x_i will be (μ_{1k}, x_i) provided that $\mu_{1k} = (x_1, \ldots, x_k)$, is a shortest path from x_1 to x_k. When there is no $k \neq 1, i$ satisfying

(12.9), then the only path of smallest length from x_1 to x_i is arc (x_1, x_i). By repeating the process indicated for the path μ_{1k_0}, where k_0 is a solution of equation (12.9), i.e. by solving again equation (12.9) where k_0 is substituted for i, etc., we shall find by recurrence all elementary shortest paths from x_1 to x_i.

The method of forming the sequence $\{U_1^n(A)\}_{n \in N}$, where A is the matrix of direct distances between the nodes of the graph, the computation process being finished when $U_1^n(A) = U_1^{n+1}(A)$ or when $n = p - 2$, has been programmed in FORTRAN IV for the IBM 360/30 computer of the Computing Centre of the Bucharest University, this program being described in [37]. The matrix of direct distances is $MDIS(N, N)$, where N denotes the number of nodes of the graph, while M is the node for which a shortest path from 1 to M is determined. The example refers to an oriented graph with 20 vertices and 47 arcs and was solved in two minutes, including the program compilation (the computation times of all the programs include the compilation time of the source program). In this variant, the symbol ∞ was represented in the computer by the integer -1. In another variant of the program, the same example is solved, but the organization of the distance matrix in the computer memory is different. In the case of the former program, a space was allocated in the memory for a matrix $MDIS$ (20, 20), where the figures within parentheses represent the greatest values of the row and column indices, their lowest values being equal to 1, therefore of the 400 cells, only 47 cells actually represent distances, the remainder being occupied by -1, the code for ∞ for the computer.

The latter variant makes use of a single vector $MAT(141)$ with 141 components, three cells containing the numbers i, j and a_{ij} being reserved for each distance $a_{ij} < \infty$, in increasing order of the indices i and j. The existing arcs between the nodes are found by the subroutine $ADRES$. The former variant contains 59 instructions and the latter 100 instructions. The example treated is represented in Fig. 12.1.

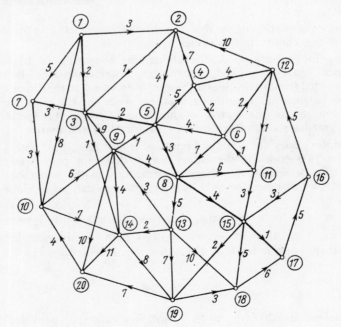

Fig. 12.1

The results supplied by the computer are now given:

THE MINIMAL DISTANCES BETWEEN NODE 1 AND THE OTHER NODES ARE:

(1, 2)	3
(1, 3)	2
(1, 4)	9
(1, 5)	4
(1, 6)	11
(1, 7)	5
(1, 8)	7
(1, 9)	5
(1, 10)	8
(1, 11)	13
(1, 12)	13
(1, 13)	12
(1, 14)	3
(1, 15)	11
(1, 16)	17
(1, 17)	12
(1, 18)	14
(1, 19)	11
(1, 20)	14

A PATH OF MINIMAL LENGTH FROM 1 TO 17 IS THE FOLLOWING:

17
15
8
5
3
1

Problems

1. Let $A = \{a_{ij}\}_{i,j=1,\ldots,n}$ be the distance matrix of a graph G and $\hat{A} = \{\hat{a}_{ij}\}_{i,j=1,\ldots,n}$ be the matrix of shortest distances for G.

If $B = \{b_{ij}\}_{i,j=1,\ldots,n-1}$ is defined such that $b_{ij} = \min(a_{ij}, a_{in} + a_{nj})$ for $i, j = 1, \ldots, n-1$, prove that $\hat{b}_{ij} = \hat{a}_{ij}$ for $i, j = 1, \ldots, n-1$.

From this property deduce an algorithm for obtaining the shortest distance between two given vertices of a graph using $\dfrac{n(n-1)(n-2)}{3}$ additions and the same number of comparisons.

2. For a nonoriented connected graph $G = (X, U)$ we define the distance between two vertices x and y, denoted by $d(x, y)$, as the smallest number of edges occuring in a chain which joins x to y.

A centre is a vertex x for which the value $d(x) = \max_{y \in X} d(x, y)$ is minimum.

Prove that a tree has a unique centre or two centres joined by an edge.

3.   For a finite oriented graph G with n vertices we define its adjacency matrix $A = \{a_{ij}\}_{i,j=1,\ldots,n}$ as follows: $a_{ij} = 1$ if (i, j) is an arc of G and $a_{ij} = 0$ otherwise.

Prove that the matrix $A^k = \{a_{ij}^{(k)}\}_{i,j=1,\ldots,n}$ (AB denotes the usual matrix multiplication) has the following property: $a_{ij}^{(k)}$ equals the number of paths from i to j having exactly k arcs.

H i n t: Use induction upon k.

4.   For a directed weighted graph $G = (X, U)$ which has a root $a \in X$, i.e. a vertex such that for any $x \in X$ there is a path going from a to x, prove that there exists an arborescence (X, V) with root a such that any path joining a to x in this arborescence is a shortest path of G.

H i n t: This arborescence may be defined inductively using the shortest paths starting from a.

5.   Prove the validity of the following shortest-path algorithm due to G. Dantzig:

For a directed weighted graph $G = (X, U)$ having the vertices $x_0, x_1, \ldots, x_n$, let $t(x)$ be the shortest length of the paths going from x_0 to x. This algorithm has the following steps:

1) The mapping t is defined on the set $A_0 = \{x_0\}$ by $t(x_0) = 0$.

2) If the mapping t is defined on the set $A_k = \{x_0, x_1, \ldots, x_k\}$, then we proceed as follows: For any vertex $x_j \in A_k$ we choose a vertex $y_j \notin A_k$ such that $(x_j, y_j) \in U$ and the length $l(x_j, y_j)$ is minimal. The vertex $x_q \in A_k$ is defined such that

$$t(x_q) + l(x_q, y_q) = \min_j (t(x_j) + l(x_j, y_j)).$$

Put $A_{k+1} = A_k \cup \{y_q\}$ and $t(y_q) = t(x_q) + l(x_q, y_q)$. When $k = n$, the mapping t is defined on $X = \{x_0, x_1, \ldots, x_n\}$ such that $t(x)$ is the shortest length of the paths going from x_0 to x.

(Dantzig, 1960)

H i n t: Prove that $t(y_q)$ is the shortest distance between x_0 and y_q and that the shortest path going from x_0 to y_q has the form $(x_0, \ldots, x_q, y_q)$.

6.   Let us consider an operator $U = \prod\limits_{\substack{i,j=1 \\ i \neq j}}^{p} U_{ij}$ where the U_{ij} are defined

on the set of distance matrices A of order p by $U_{ij}(A) = B$ where $b_{ij} = \min \{a_{ij}, \min\limits_{k=1,\ldots,\min(i,j)-1} (a_{ik} + a_{kj})\}$ and $b_{mn} = a_{mn}$ for $(m, n) \neq (i, j)$. We also

define an operator $V = \prod\limits_{\substack{i,j=1 \\ i \neq j}}^{n} V_{ij}$ where the V_{ij} are defined by $V_{ij}(A) = B$, where

$b_{ij} = \min \{a_{ij}, \min\limits_{\substack{k=\min(i,j)+1,\ldots,p \\ k \neq i,j}} (a_{ik} + a_{kj})\}$ and $b_{mn} = a_{mn}$ for $(m, n) \neq (i, j)$.

We define a partial order relation between the pairs (i, j) and (k, l) with $i \neq j$ and $k \neq l$ and we write $(i, j) \subset_T (k, l)$ if

$$T = T_{i_1 j_1} \dots T_{ij} \dots T_{kl} \dots .$$

Prove that if U and V satisfy the conditions:

i) $(i, j) \subset_U (k, j)$, $(j, i) \subset_U (j, k)$ and

ii) $(k, j) \subset_V (i, j)$, $(j, k) \subset_V (j, i)$

for any $i \neq j$, $k \neq j$ and $k < i$, then $VU(A) = A^{p-1}$.

For example $U = U_{n,n-1} U_{n,n-2} \dots U_{n,1} U_{n-1,n} U_{n-1,n-2} \dots U_{1,n} \dots U_{1,2}$

and $V = V_{1,2} V_{1,3} \dots V_{1,n} V_{2,1} V_{2,3} \dots V_{n,1} V_{n,2} \dots V_{n,n-1}.$

Prove that the total number of operations for this algorithm equals that required by Roy-Floyd's algorithm.

(Tomescu, 1971)

This algorithm generalizes those proposed by Farbey, Land, Murchland [12], Hu [15] and Bilde, Krarup [4].

H i n t : Prove that $\{VU(A)\}_{ij} \leqslant \min(\{U(A)\}_{ik_1} + \{U(A)\}_{k_1 k_2} + \dots + \{U(A)\}_{k_r j}) \leqslant$ $\leqslant \min(a_{ik_1} + a_{k_1 k_2} + \dots + a_{k_r j})$ for any $k_1, \dots, k_r \in \{1, 2, \dots, n\}$ and $k_p \neq i, j$; $k_p \neq k_q$ for $p \neq q$.

7.   Prove that the arc (k, l) belongs to a shortest path going from i to j for a weighted graph G whose distance matrix is A if and only if

$$\hat{a}_{ij} = \hat{a}_{ik} + a_{kl} + \hat{a}_{li}$$

where $\hat{A}$ is the matrix of shortest distances.

8.   Let $G = (X, \Gamma)$ be a finite graph, with $X = \{x_1, x_2, \dots, x_n\}$. Let $X_m = \{x_1, x_2, \dots, x_m\}$, $m \leqslant n$ and let G^m be the subgraph of G built on X_m. Let a_{ij} be the length of the arc (x_i, x_j) if $(x_i, x_j) \in \Gamma$ and $a_{ij} = \infty$ otherwise. Let $\{d_{ij}^m\}$ be the matrix of shortest distances in G^m. Prove the validity of Dantzig's inductive algorithm to find all shortest distances in G [8]: Set $d_{11}^1 = 0$ and then for $m = 2, 3, \dots, n$, execute steps $1 - 4$:

S t e p 1. Compute $d_{im}^m = \min_{1 \leqslant k < m} (d_{ik}^{m-1} + a_{km})$, $i = 1, 2, \dots, m - 1$.

(Since a_{ii} is assumed to be 0, so is d_{ii}^{m-1} if G^{m-1} has no negative circuit, which is assumed by step 3).

S t e p 2. Compute $d_{mj}^m = \min_{1 \leqslant k < m} (a_{mk} + d_{kj}^{m-1})$, $j = 1, 2, \dots, m - 1$.

S t e p 3. Ensure that $d_{im}^m + d_{mi}^m \geqslant 0$ for each $i = 1, 2, \dots, m - 1$ (if not $(m, \dots, i, \dots, m)$ is a negative circuit: stop) and let $d_{mm}^m = 0$.

S t e p 4. For each i, $1 \leqslant i < m$, and for each j, $1 \leqslant j < m$, compute $d_{ij}^m = \min(d_{ij}^{m-1}; d_{im}^m + d_{mj}^m)$.

9.   For a weighted connected graph G a shortest spanning tree is a spanning tree for which the sum of the lengths of its edges is minimal. Prove the validity of the following algorithm for obtaining a shortest spanning tree of G:

i) We choose an edge u_1 of minimal length;

ii) If the edges $u_1, u_2, ..., u_k$ are chosen, we select a new edge u_{k+1} having the following property: it has a minimal length in the set of edges u_l ($l \geqslant k + 1$) with the property that the graph spanned by the set of edges $\{u_1, u_2, ..., u_k, u_l\}$ contains no cycle. If for any selection of an edge u_l with $l \geqslant k + 1$, the graph spanned by $\{u_1, ..., u_k, u_l\}$ contains a cycle, then the procedure stops.

(Kruskal, 1956)

10.   Consider a non-empty set S with two binary composition laws denoted by $\circ$, $\wedge$ and called multiplication and addition respectively, which satisfy the following axioms:

1) $a \circ (b \circ c) = (a \circ b) \circ c$

2) $a \circ e = e \circ a = a$

3) $a \wedge (b \wedge c) = (a \wedge b) \wedge c$

4) $a \wedge b = b \wedge a$

5) $a \circ (b \wedge c) = (a \circ b) \wedge (a \circ c)$

6) $(b \wedge c) \circ a = (b \circ a) \wedge (c \circ a)$

7) $a \wedge (a \circ b) = a \wedge (b \circ a) = a$

for any $a, b, c \in S$, where e is the unity element for the multiplication.

S is called an S-semiring.

Prove that S is a partially ordered set, having a partial order relation defined by $a < b$ if $a \wedge b = a$, which is isotonic with respect to addition and multiplication, i.e. from $a < c$ and $b < d$ we deduce that $a \wedge b < c \wedge d$ and $a \circ b < < c \circ d$.

(Moisil, 1960; Yoeli, 1961)

11.   For two square matrices $A = \{a_{ij}\}_{i,j=1,...,p}$ and $B = \{b_{ij}\}_{i,j=1,...,p}$ whose elements belong to an S-semiring S, let us define the following operations:

$$A \wedge B = \{a_{ij} \wedge b_{ij}\}_{i,j=1,...,p}; \; A \times B = \{ \bigwedge_{k=1,...,p} (a_{ik} \circ b_{kj})\}_{i,j=1,...,p}.$$

Prove that these operations are associative and there exists an index $r \leqslant p$ such that

$$A > A \wedge A^2 > A \wedge A^2 \wedge A^3 > ... > \bigwedge_{k=1}^{r} A^k = \bigwedge_{k=1}^{r+1} A^k = ...$$

where by $A < B$ we mean $a_{ij} < b_{ij}$ for all $i, j = 1, ..., p$. If $a_{ii} = e$ for $i = 1, ..., p$ then we have $A > A^2 > A^3 > ... > A^r = A^{r+1} = ...$ and $r \leqslant p - 1$.

(Moisil, 1960; Yoeli, 1961; Nolin, 1964; Benzaken, 1966)

H i n t: Use induction on k and the absorption property of S.

12. Prove that Proposition 2 and Proposition 3 remain valid by considering the matrix $A \wedge A^2 \wedge \ldots \wedge A^p$ instead of A^{p-1}, where A has its elements belonging to an S-semiring S having $(\circ, \wedge, e, <)$ instead of R_+ with $(+, \min, 0, \geqslant)$.

(Robert, Ferland, Tomescu, 1968 and Nolin, 1964)

BIBLIOGRAPHY

1. Bellman, R., *On a routing problem*, Quart. Appl. Math., **16**, 1958, 87−90.
2. Bentley, D., Cooke, K., *Convergence of successive approximations in the shortest route problem*, J. Math. Anal. Appl., **10**, 1965, 269−274.
3. Benzaken, C., *Pseudo-treillis distributifs et applications*, I; II; III, Bul. Inst. Politehn. Iaşi, serie nouă, **12**, Fasc. 3−4, 1966, 13; **13**, Fasc. 3−4, 1967, 11−15; **14**, Fasc. 3−4, 1968, 25.
4. Bilde, O., Krarup, J., *A modified cascade algorithm for shortest paths*, Metra, **8**, 1969, 231−241.
5. Carré, B. A., *An algebra for network routing problems*, J. Inst. Math. Appl., **7**, 1971, 273−294.
6. Dantzig, G., *On the shortest route through a network*, Management Sci., **6**, 1960, 187.
7. Dantzig, G., *All shortest routes in a graph*, Theory of Graphs (International Symposium), International Computer Centre, Rome, Gordon and Breach, New York and Dunod, Paris, 1967, 91−92.
8. Dantzig, G., *All shortest routes from a fixed origin in a graph*, Theory of Graphs (International Symposium), International Computer Centre, Rome, Gordon and Breach, New York and Dunod, Paris, 1967, 85−90.
9. Domschke, W., *Kürzeste Wege in Graphen: Algorithmen, Verfahrensvergleiche*, Math. Systems in Economics, N° 2, Anton Hain, Meisenheim, 1972.
10. Dragomirescu, M., *L'algorithme de min-addition et les chemins critiques dans un graphe*, Rev. Roumaine Math. Pures Appl., **12**, 1967, 1045−1051.
11. Dreyfus, S., *An appraisal of some shortest-path algorithms*, Operations Res., **17**, 1969, 395−412.
12. Farbey, B., Land, A., Murchland, J., *The cascade algorithm for finding all shortest distances in a directed graph*, Management Sci., **14**, 1967, 19−28.
13. Floyd, R., *Algorithm* 97: *Shortest path*, Comm. ACM, **5**, 1962, 345.
14. Hasse, M, *Über die Behandlung graphentheoretischer Probleme unter Verwendung der Matrizenrechnung*, Wiss. Z. Techn. Univ. Dresden, **10**, 1961, 1313−1316.
15. Hu, T. C., *Revised matrix algorithms for shortest paths*, SIAM J. Appl. Math., **15**, 1967, 207−218.
16. Hu, T. C., *A decomposition algorithm for shortest paths in a network*, Operations Res., **16**, 1968, 91.
17. Joksch, H., *The shortest route problem with constraints*, J. Math. Anal. Appl., **14**, 1966, 191−197.
18. Kruskal, J. B., *On the shortest spanning subtree of a graph*, Proc. Amer. Math. Soc., **71**, 1956, 48−50.
19. Lunc, A. G., *Algebraic methods for the analysis and synthesis of contact circuits* (in Russian), Izv. Akad. Nauk SSSR Ser. Mat., **16**, 1952, 405−426.
20. Moisil, G. C., *On certain representations of graphs which occur in transportation problems* (in Romanian), Comunicările Academiei R.P.R., **10**, 1960, 647−652.
21. Nolin, L., *Traitement des données groupées*, Publication de l'Institut Blaise-Pascal, Paris, 1964.
22. Pandit, S., *The shortest route problem − an addendum*, Operations Res., **9**, 1961, 129.
23. Peteanu, V., *Optimal paths in networks and generalizations*, I; II, Mathematica (Cluj), **11** (34), 1969, 311−327; **12** (35), 1970, 159−186.
24. Pollack, M., Wiebenson, W., *Solutions of the shortest-route problem − a review*, Operations Res., 1960, 224−230.
25. Povarov, G. N., *Fundamentals of the theory of cumulative networks* (in Russian), Bul. Inst. Politehn. Iaşi, serie nouă, **6** (10), Fasc. 1−2, 1960, 29−36.
26. Robert, P., Ferland, J., *Généralisation de l'algorithme de Warshall*, Revue Française Informat. Recherche Opérationelle, N° 7, 1968, 71−85.
27. Roy, B., *Transitivité et connexité*, C. R. Acad. Sci. Paris, **249**, 1959, 216−218.

28. Roy, B., *Cheminement et connexité dans les graphes*: *Application aux problèmes d'ordonnancement*, Metra, Série spéciale, N° 1, 1962.
29. Roy, B., *Algèbre moderne et théorie des graphes (orientées vers les sciences économiques et sociales)*, Vol. II, Dunod, Paris, 1970.
30. Saksena, J., Santosh, K., *The routing problem with "K" specified nodes*, Operations Res., **14**, 1966, 909.
31. Thorelli, L. E., *An algorithm for computing all paths in a graph*, BIT, **6**, 1966, 347—349.
32. Tomescu, I., *Method for the determination of the total conductibilities of a multipole* (in Romanian), Stud. Cerc. Mat., **17**, 1965, 1109—1115.
33. Tomescu, I., *Sur les méthodes matricielles dans la théorie des réseaux*, C. R. Acad. Sci. Paris Sér. A., **263**, 1966, 826—829.
34. Tomescu, I., *Méthode pour la détermination de la fermeture transitive d'un graphe fini*, Revue Française Informat. Recherche Opérationnelle, N° 4, 1967, 33—37.
35. Tomescu, I., *Un algorithme pour la détermination des plus petites distances entre les sommets d'un réseau*, Revue Française Informat. Recherche Opérationnelle, N° 5, 1967, 133—139.
36. Tomescu, I., *Sur l'algorithme matriciel de B. Roy*, Revue Française Informat. Recherche Opérationnelle, N° 7, 1968, 87—91.
37. Tomescu, I., *Méthodes combinatoires dans la théorie des automates finis*, Logique, Automatique, Informatique. Ed. Acad. RSR, Bucharest, 1971, 269—423.
38. Tomescu, I., *Ordered algebraic structures in graph theory* (in Romanian), Stud. Cerc. Mat., **24**, 1972, 469—476.
39. Warshall, S., *A theorem on Boolean matrices*, J. Assoc. Comput. Mach., **9**, 1962, 11—13.
40. Yen, Y. J., *Finding the lengths of all shortest paths in n-node nonnegative-distance complete networks using $\frac{1}{2}$ n^3 additions and n^3 comparisons*, J. Assoc. Comput. Mach., **19**, 1972, 423—424.
41. Yoeli, M., *A note on a generalization of Boolean matrix theory*, Amer. Math. Monthly, **68**, 1961, 552—557.

CHAPTER 13

Determination of the Maximal Independent Sets and of the Chromatic Number of a Graph

13.1. THE BEDNAREK — TAULBEE ALGORITHM

In this chapter we shall work with undirected graphs $G = (X, U)$, where X is the set of vertices and U is the set of edges, with no edges of the form $[x, x]$.

An *independent set* $I \subset X$ is a set of vertices which are not joined by any edge, that is, the subgraph determined by the set I contains only isolated vertices. An independent set is *maximal* if it is maximal with respect to set inclusion, i.e. it is not included in any other independent set.

The equivalent concepts for the complementary graph $\bar{G} = (X, \bar{U})$, where $\bar{U}$ is the complement of the set U with respect to the set of edges of the complete graph on the vertex-set X, are those of complete subgraph and maximal complete subgraph, respectively. A complete subgraph of the graph G is defined as a set $C \subset X$ of vertices that are pairwise joined by edges of the graph G in all possible ways. For this reason every algorithm for the determination of maximal independent sets is also an algorithm for the determination of maximal complete subgraphs of the graph G, that is, of maximal independent sets of the complementary graph $\bar{G}$. Clearly the adjacency matrix of the graph $\bar{G}$ is obtained from the adjacency matrix of the graph G by replacing each element a_{ij} by $\bar{a}_{ij} = 1 - a_{ij}$, where $a_{ij} = 1$ or 0 according as there does or does not exist an edge that joins vertices x_i and x_j, for $i, j = 1, ..., |X|$.

Several algorithms, either algebraic or combinatorial, have been given for the determination of maximal independent sets. We quote: a method based on the calculus of Boolean expressions, suggested by Maghout [26] and Weissman [56], the algorithm of Malgrange [28] which constructs all square submatrices which contain only the element zero and are principal submatrices of the adjacency matrix of the graph, Rudeanu's method [40] which utilizes the Boolean equations that characterize the maximal independent sets of the graph and the Bednarek-Taulbee algorithm [3]. We also mention the combinatorial methods of Paull and Unger [34] and Ioanin's method [8], which utilizes a technique similar to the McCluskey algorithm for the determination of prime implicants of a Boolean function. These methods are used for determining the maximal compatible sets, i.e. the maximal independent sets of the graph of pairs of incompatible states, and they occur in the minimization problem for incompletely specified automata.

In the sequel we present the algorithm suggested by Bednarek and Taulbee which constructs recursively the maximal independent sets and a computer program is relatively easy to write, because of the nature of the operations involved in this algorithm. Given a graph $G = (X, U)$ with n vertices $x_1, x_2, ..., x_n$, the algorithm runs as follows:

For every k with $1 \leqslant k \leqslant n$, set $X_k = \{x_1, x_2, ..., x_k\}$ and let L_k be the family of maximal independent sets of the subgraph of G defined by the vertex-set X_k. Denote by Y_k the set of vertices of X_k which are not joined by an edge to x_k, i.e.

$$Y_k = \{y \mid y \in X_k, [x_k, y] \notin U\}.$$

S t e p 1. Take $Y_1 = \{x_1\}$, $L_1 = \{x_1\}$ and $k = 1$.

S t e p 2. Determine the family $I_k = \{S \mid S = M \cap Y_{k+1}, M \in L_k\}$.

S t e p 3. Determine I'_k = the family of subsets of I_k maximal with respect to set inclusion.

S t e p 4. Define the family of sets L^*_{k+1} as follows:

For every $M \in L_k$:

— if $Y_{k+1} \supset M$, then $M \cup \{x_{k+1}\} \in L^*_{k+1}$;

— if $Y_{k+1} \not\supset M$, then $M \in L^*_{k+1}$ and $\{x_{k+1}\} \cup (M \cap Y_{k+1}) \in L^*_{k+1}$ if and only if $M \cap Y_{k+1} \in I'_k$.

The family L^*_{k+1} consists solely of the above mentioned sets.

S t e p 5. Determine L_{k+1} as the family of subsets of L^*_{k+1} maximal with respect to inclusion.

S t e p 6. Repeat steps 2, 3, 4 and 5 for $k = 2, 3, ..., n - 1$.

At the end of the algorithm one obtains the family L_n, that is, the family of maximal independent sets of the graph G. We shall present below an application of this algorithm to the determination of the chromatic number of a seven-vertex graph. The justification of the algorithm results from the construction of the family L_{k+1} of maximal independent sets. For, if the new vertex x_{k+1} is not joined by any edge to the vertices of the maximal independent set M of the subgraph with the vertex-set $\{x_1, ..., x_k\}$, then $M \cup \{x_{k+1}\} \in L^*_{k+1}$. Let us prove that L^*_{k+1} is a family of independent sets which contains all maximal independent sets of the subgraph with vertex-set $\{x_1, ..., x_{k+1}\}$. If x_{k+1} is joined by edges to some of the vertices in M, then the set M is added to L^*_{k+1} and for each $M \in L_k$ with this property one determines the set of vertices that are not joined by edges to x_{k+1}. From these sets we keep the maximal ones with respect to set inclusion, to which we add the new vertex x_{k+1}; the independent sets obtained in this way are added to L^*_{k+1}. Moreover, the family L^*_{k+1} will contain all maximal independent sets of the subgraph with vertex-set

$\{x_1, ..., x_{k+1}\}$. To prove this, let Z be such a maximal independent set. If $x_{k+1} \notin Z$, then Z is a maximal independent set of the subgraph having the vertex-set $X_k = \{x_1, ..., x_k\}$, whereas in the opposite case $Z \setminus \{x_{k+1}\}$ is an independent set of the subgraph with vertex-set X_k, hence it is either maximal or included in a maximal independent set from the family L_k. Taking into account the previous construction, it follows that $Z \in L^*_{k+1}$.

The program based on this algorithm and written in FORTRAN IV [48] has 155 instructions and prints both the maximal independent sets of the graph (each of them on a separate line) and the independence number $s(G)$, which is the maximum number of elements in an independent set (which is necessarily maximal).

The following upper bound for the independence number has been used:

$$\max_{1 \leqslant i \leqslant n} (n - 1 - d(x_i)) + 1 = n - \min_{1 \leqslant i \leqslant n} d(x_i),$$

this following from the fact that for a complete k-vertex subgraph, all vertices have degree $\geqslant k - 1$.

The difficulty in applying this algorithm consists of the fact that the number of maximal independent sets increases rapidly as the number of vertices of the graph G becomes greater.

Moon and Moser proved [30] that the maximum number of maximal independent sets of a graph with n vertices is equal to $3^{n/3}$ for $n = 3k$, to $4.3^{(n-1)/3-1}$ for $n = 3k + 1$ and to $2 \cdot 3^{(n-2)/3}$ for $n = 3k + 2$.

The example treated is a 20-vertex loop-free graph, the complement of which is drawn in Fig. 13.1. The 18 maximal independent sets and the independence number, equal to 5, have been determined with the aid of an IBM 360/30. The results printed by the computer are the following:

THE MAXIMAL INDEPENDENT SETS OF THE GRAPH ARE THE FOLLOWING:

1	18			
2	13			
3	4	8		
5	14	19		
6	7	15		
4	8	9		
7	9	10	11	
8	9	10		
10	12	17		
3	13			
6	14	15	19	
7	11	15		
12	16	17	20	
13	16	17	18	20
14	18			
11	15	19		
12	19			
8	9	20		

THE INDEPENDENCE NUMBER OF THE GRAPH IS EQUAL TO 5

This program can also be used for the determination of maximal complete subgraphs of a graph G, by constructing the maximal independent sets of the complementary graph $\bar{G}$.

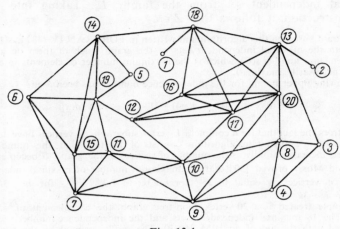

Fig. 13.1

13.2. Algorithms for the Determination of a Minimum Covering

We present below several combinatorial algorithms for finding a minimum covering of a set by subsets from a given family, with applications to the determination of the matchings of a bipartite graph and to the determination of the chromatic number.

Let $X = \{x_1, \ldots, x_n\}$ be a finite set with n elements and $E = (E_i)_{1 \leqslant i \leqslant m}$ a family of subsets of X such that $\bigcup_{i=1}^{m} E_i = X$.

We mention that the pair (X, E) is called a *hypergraph*. We shall assume that the sets E_i are incomparable with respect to set inclusion, i.e. the relation $E_i \not\subset E_j$ holds for every $i \neq j$.

A *minimum covering* of the set X by elements of the family E is a subfamily of E, say $(E_{i_k})_{1 \leqslant k \leqslant r}$ with $E_{i_k} \in E$, which consists of a minimal number of sets E_{i_k} that cover X, i.e. $\bigcup_{k=1}^{r} E_{i_k} = X$.

This problem can be formulated as a pseudo-Boolean linear programming problem [20].

An algebraic method for finding all minimum coverings is Petrick's algorithm [35], also suggested by Weissman [56] and Maghout [27].

Let: $E_{j_1}, E_{j_2}, ..., E_{j_s}$ be the sets of E that contain the element x_1; $E_{k_1}, E_{k_2}, ..., E_{k_t}$ the sets of E that contain x_2; ... ; $E_{v_1}, E_{v_2}, ..., E_{v_z}$ the sets of E that contain x_n. In this case we form the product

$$(a_{j_1} \cup a_{j_2} \cup ... \cup a_{j_s})(a_{k_1} \cup a_{k_2} \cup ... \cup a_{k_t}) ... (a_{v_1} \cup a_{v_2} \cup ... \cup a_{v_z}), \qquad (13.1)$$

where the operations $\cup$ and $\cdot$ are the disjunction and conjunction, respectively, of a distributive lattice, while $a_{j_1}, ..., a_{v_z}$ are two-valued variables in this distributive lattice, having the following significance: the set $E_k \in E$ belongs or does not belong to the subfamily of E associated with the corresponding system of values of $a_{j_1}, ..., a_{v_z}$, according as $a_k = 1$ or $a_k = 0$.

It follows from the construction of the product (13.1) that for every system of 0-1 values given to the variables $a_{j_1}, ..., a_{v_z}$, the expression (13.1) has the value 1 if and only if the family of sets from E associated with the variables $a_{j_1}, ..., a_{v_z}$ by the above rule, is a covering of the set X. For, if (13.1) has the value 1, this means that each factor of the product has the value 1, hence each of the elements $x_1, ..., x_n$ is covered by one or more sets from the family E. By using the properties of distributivity, commutativity, idempotency and absorption of the two operations in a distributive lattice, we can write the product (13.1) as a disjunction of products for which we have performed all possible absorptions ($a \cup b = b \cup a, ab = ba, a(b \cup c) = = ab \cup ac, a \cup a = a, aa = a, a \cup ab = a, a(a \cup b) = a$). Each product $a_{l_1} a_{l_2} ... a_{l_p}$ will correspond to an irreducible covering $(E_{l_i})_{1 \leqslant i \leqslant p}$ that is, to a covering with the property that $\bigcup_{\substack{i=1 \\ i \neq k}}^{p} E_{l_i} \neq X$ for each $k = 1, 2, ..., p$, because of the absorption properties. We thus obtain all irreducible coverings of the set X, hence in particular all minimal coverings, since every minimal covering is irreducible; the minimal coverings correspond to the products with a minimum number of factors.

A similar result can be obtained by using a combinatorial algorithm which consists of representing all possible coverings by paths joining the root with the terminal nodes of a certain rooted directed tree, denoted by A_E.

This rooted tree (arborescence) is constructed as follows:

Let $x_i \in X$ be an element which belongs to the following sets from the family E: $E_{i_1}, E_{i_2}, ..., E_{i_k}$. The rooted tree $A_E = (\tilde{X}, \Lambda)$ has a root denoted by x_i, the mapping Λ being defined by

$$\Lambda(x_i) = \{x_{i,r} \mid r = 1, 2, ..., k\}.$$

Associate every arc of the form $(x_i, x_{i,r})$ with the set $E_{i_r} \in E$. Choose an element $y_{i,1} \in X \setminus E_{i_1}$ which belongs to the following sets from the family E: $E_{j_1}, E_{j_2}, ..., E_{j1}$ and define $\Lambda(x_{i,1}) = \{x_{i,1,r} \mid r = 1, 2, ..., l\}$ by associating each arc $(x_{i,1}, x_{i,1,r})$ with the set $E_{j_r} \in E$; continue this procedure for the other vertices $x_{i,2}, ..., x_{i,k}$, etc.

Consider the path $(x_i, x_{i,n_1}, ..., x_{i,n_1,...,n_s})$ from the root x_i to the node $x_{i,n_1,...,n_s}$ and denote by $E_{i,n_1,...,n_s}$ the union of the sets associated with the arcs $(x_i, x_{i,n_1}), ...$ ..., $(x_{i,n_1,...,n_{s-1}}, x_{i,n_1,...,n_s})$. If $X = E_{i,n_1,...,n_s}$, define $\Lambda(x_{i,n_1,...,n_s}) = \varnothing$; in this case $n_s = 1$ and the vertex $x_{i, n_1, \ldots, n_s}$ will be called a terminal vertex of the rooted tree. In the opposite case consider an arbitrary element $y_{i, n_1, \ldots, n_s} \in X \setminus E_{i, n_1, \ldots, n_s}$ which belongs to the following sets from the family E: $E_{h_1}, E_{h_2}, \ldots, E_{h_q}$, then define $\Lambda(x_{i, n_1, \ldots, n_s}) = \{x_{i, n_1, \ldots, n_s, r} \mid r = 1, 2, \ldots, q\}$ and associate each arc of the form $(x_{i, n_1, \ldots, n_s}, x_{i, n_1, \ldots, n_s, r})$ with the set $E_{h_r} \in E$.

The construction of the rooted tree A_E is finished when all terminal vertices have been obtained. It follows from this construction that every path $(x_i, x_{i, n_1}, \ldots, x_{i, n_1, \ldots, n_s})$ of length s that joins the root x_i to the terminal vertex $x_{i, n_1, \ldots, n_s}$ $(n_s = 1)$ corresponds to a covering of the set X consisting of the s sets of the family E associated with the arcs of this path.

PROPOSITION 1. *The set of coverings associated with the paths that join the root x_i with the terminal vertices of the rooted tree A_E, includes the set of irreducible coverings of the set X with sets of the family E.*

Let $(E_{t_r})_{1 \leqslant r \leqslant p}$ be an irreducible covering of the set X with sets from E. The element x_i belongs to a set E_{t_q}, but, by renumbering the sets of the covering, we may assume that $x_i \in E_{t_1}$, hence there exists an index m_1 with $1 \leqslant m_1 \leqslant k$, such that $E_{t_1} = E_{i m_1}$. The element $y_{i, m_1} \notin E_{t_1}$, hence there is a set of the covering $(E_{t_r})_{1 \leqslant r \leqslant p}$, say E_{t_2}, such that $y_{i, m_1} \in E_{t_2}$. It follows from the construction of the rooted tree that there exists an index m_2 with $1 \leqslant m_2 \leqslant f$, such that $E_{t_2} = E_{u_{m_2}}$, where $E_{u_1}, E_{u_2}, \ldots, E_{u_f}$ are those sets from E which contain the element y_{i, m_1} etc. In this way we infer that the covering $(E_{t_r})_{1 \leqslant r \leqslant p}$, is, in fact, a covering of the set X associated with the path $(x_i, x_{i, m_1}, \ldots, x_{i, m_1}, \ldots, m_p)$ which joins the root x_i to the terminal node $x_{i, m_1, \ldots, m_p}$ $(m_p = 1)$ of the rooted tree A_E.

The minimal coverings will thus correspond to the paths which join the root of the rooted tree to the terminal nodes and have a minimum number of arcs.

In many cases it is not necessary either to enumerate all irreducible solutions of the covering problem, or to determine all minimal coverings, it sufficing to obtain one minimal covering. We present below a modification of the previous algorithm, which reduces the number of branches at each node of the rooted tree and enables the detection of at least one minimal covering.

A partition $\{X_j\}_{1 \leqslant j \leqslant q}$ of the set X is said to be compatible with the family E of subsets, or simply E-compatible, if for every $j = 1, \ldots, q$ there is a set $E_{i_j} \in E$ such that $X_j \subset E_{i_j}$; in this case the covering $(E_{i_j})_{1 \leqslant j \leqslant q}$ is said to be generated by the partition $\{X_j\}_{1 \leqslant j \leqslant q}$. Conversely, given a covering of the set X with sets from the family E, it generates at least one partition compatible with the family E, by eliminating the elements common to several sets of the covering, and placing each of these elements in a class by itself. Clearly an E-compatible partition of X with a minimum number of classes generates a minimum covering of the set X, while a minimum covering of X generates at least one E-compatible partition with a minimum number of classes.

For every $X_1 \subset X$, denote by E_{X_1} the family of sets $(E_i \setminus X_1)_{1 \leqslant i \leqslant m}$. This family, obtained by removal of the elements of the set X_1 from the sets E_i, is, in general, no longer made up of sets incomparable with respect to inclusion. We now give the following algorithm. Let $x_i \in X$ be an element contained in the sets E_{i_1}, $E_{i_2}, \ldots, E_{i_k}$ from the family E. Define a rooted directed tree $A_E^* = (\tilde{X}^*, \Lambda^*)$ having the root x_i with Λ^* defined by

$$\Lambda^*(x_i) = \{x_{i,r} \mid r = 1, 2, \ldots, k\}.$$

Associate every arc of the form $(x_i, x_{i,r})$ with the set $E_{i_r} \in E$. Choose an arbitrary element $y_{i,1} \in X \setminus E_{i_1}$ which belongs to the following sets of $E_{E_{i_1}}$, maximal with respect to set inclusion: $F_1, F_2, \ldots, F_j$. Define $\Lambda^*(x_{i,1}) = \{x_{i,1,r} \mid r = 1, 2, \ldots, j\}$ associating each arc $(x_{i,1}, x_{i,1,r})$ with the set $F_r \in E_{E_{i_1}}$. Continue this procedure for the other vertices $x_{i,2}, \ldots, x_{i,k}$ of the graph, etc.

Consider the path $(x_i, x_{i,n_1}, \ldots, x_{i,n_1,\ldots,n_s})$ from the root x_i to the vertex $x_{i,n_1,\ldots,n_s}$ and denote by $E_{i,n_1,\ldots,n_s}$ the union of the sets associated with the arcs (x_i, x_{i,n_1}), $(x_{i,n_1}, x_{i,n_1,n_2})$, $\ldots$ of this path; notice that these sets are pairwise disjoint. If $X = E_{i,n_1,\ldots,n_s}$, set

$$\Lambda^*(x_{i,n_1,\ldots,n_s}) = \varnothing.$$

In this case $n_s = 1$ and the vertex $x_{i,n_1,\ldots,n_s}$ is a terminal vertex of the rooted tree. In the opposite case consider an arbitrary element $y_{i,n_1,\ldots,n_s} \in X \setminus E_{i,n_1,\ldots,n_s}$ which belongs to the following sets of $E_{E_{i,n_1,\ldots,n_s}}$, maximal with respect to set inclusion: $H_1, H_2, \ldots, H_q$. Define

$$\Lambda^*(x_{i,n_1,\ldots,n_s}) = \{x_{i,n_1,\ldots,n_s,r} \mid r = 1, 2, \ldots, q\},$$

associating each arc of the form $(x_{i,n_1,\ldots,n_s}, x_{i,n_1,\ldots,n_s,r})$ with the set $H_r \in E_{E_{i,n_1,\ldots,n_s}}$.

This construction of the rooted tree A_E^* ends when all terminal vertices have been obtained. Each path $(x_i, \ldots, x_{i,n_1,\ldots,n_s})$ of length s that joins the root x_i to the terminal vertex $x_{i,n_1,\ldots,n_s}$ $(n_s = 1)$ corresponds to an E-compatible partition of the set X consisting of the s sets associated with the arcs of this path.

PROPOSITION 2. *The set of partitions associated with the paths that join the root* x_i *to the terminal vertices of the rooted tree* A_E^* *contains at least one E-compatible partition with a minimum number of classes.*

Let $\{X_j\}_{1 \leqslant j \leqslant p}$ be an E-compatible partition of X with the minimum number of classes, say p.

The element x_i belongs to a class X_q but we can assume that $x_i \in X_1$, hence there is an index m_1 with $1 \leqslant m_1 \leqslant k$, for which $X_1 \subset E_{i_{m_1}}$, by the compatibility condition of the partition with the family E.

If $X_1 \neq E_{i_{m_1}}$ we complete the class X_1 up to $E_{i_{m_1}}$ by removing the elements of $E_{i_{m_1}} \setminus X_1$ from the remaining classes of the partition. We thus obtain a new E-compatible partition $E_{i_{m_1}}, X_2', \ldots, X_p'$ of the set X, where $X_j' \subset X_j$ and $X_j' \neq \varnothing$ for each $j = 2, \ldots, p$ because of the minimum condition for the number of classes of the partition $\{X_j\}_{1 \leqslant j \leqslant p}$. We may assume that $y_{i,m_1} \in X_2'$ and since $X_2' \subset X \setminus E_{i_{m_1}}$ there is an index m_2 with $1 \leqslant m_2 \leqslant j$ such that $X_2' \subset F_{m_2}$, if $F_1, F_2, ..., F_j$ are the maximal sets with respect to inclusion, from the family $(E_i \setminus E_{i_{m_1}})_{1 \leqslant i \leqslant m}$, where the index i_{m_1} is fixed. If $X_2' \neq F_{m_2}$, we complete the class X_2' up to F_{m_2} by removing the elements of $F_{m_2} \setminus X_2'$ from the classes $X_3', ..., X_p'$ of the partition, etc.

In this way we obtain an E-compatible partition of the set X into p classes, which is associated with a p-arc path $(x_i, x_{i, m_1}, ..., x_{i, m_1, ..., m_p})$ that joins the root x_i to the terminal node $x_{i, m_1, ..., m_p}$ of the rooted tree A_E^*.

Let us give an example of the use of Petrick's algorithm and of the above two algorithms in the case in which $X = \{1, 2, 3, 4, 5\}$ and the family E consists of the sets $A = \{1, 2, 3\}$, $B = \{1, 2, 4\}$, $C = \{1, 3, 4\}$; $D = \{1, 3, 5\}$, $E = \{1, 4, 5\}$. Writing down the algebraic covering condition for the set X, we obtain

$$(a \cup b \cup c \cup d \cup e)(a \cup b)(a \cup c \cup d)(b \cup c \cup e)(d \cup e) = (a \cup b)(c \cup ab \cup bd \cup ae \cup de)(d \cup e)$$

$$= acd \cup ace \cup abd \cup abe \cup abd \cup abde \cup aed \cup ae \cup ade \cup ade \cup bcd \cup bce \cup abd$$

$$\cup abe \cup bd \cup bde \cup abde \cup abe \cup bde \cup bde = ae \cup bd \cup acd \cup bce,$$

an expression which gives all irreducible coverings, namely

$$(A, E), (B, D), (A, C, D) \text{ and } (B, C, E),$$

among which the first two are minimal.

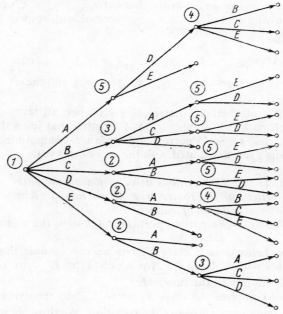

Fig. 13.2

By applying the algorithm for the construction of the rooted tree A_E, we obtain the rooted tree in Fig. 13.2, where for each branch point we have circled the element which belongs to those sets from the family E, associated with the arcs that start from the branch point in question.

For every rooted tree so constructed, each irreducible covering may appear several times, but some coverings associated with the paths that join the origin to the terminal nodes, may be reducible.

By using the following absorption rule: the covering $(E_{i_k})_{1 \leqslant k \leqslant p}$ absorbs the covering $(E_{j_l})_{1 \leqslant l \leqslant p}$ if for every k with $1 \leqslant k \leqslant p$ there exists an index l with $1 \leqslant l \leqslant q$ such that $E_{i_k} = E_{j_l}$ (and hence $p \leqslant q$), we can obtain all irreducible coverings of the set X with sets from the family E.

We can obtain a rooted tree A_E with fewer arcs and hence with fewer reducible coverings by using the following rule: if the last node of the path $(x_i, x_{i, n_1}, ..., x_{i, n_1, ..., n_s})$ is not a terminal node, then we choose the element $y_{i, n_1, ..., n_s} \in$ $\in X \backslash E_{i, n_1, ..., n_s}$ in such a way that the number of sets $E_{h_1}, E_{h_2}, ..., E_{h_q}$ from the family E that contain $y_{i, n_1, ..., n_s}$ is minimal.

By applying this rule to the given example, we obtain the rooted tree of Fig. 13.3. This rooted tree contains only 8 terminal nodes, hence 8 coverings among which 4 are irreducible instead of 21 as there were in the previous rooted tree. To determine an E-compatible partition with a minimum number of classes and hence a minimal covering, we construct the rooted tree A_E^* in Fig. 13.4 with the aid of the previous algorithm. We obtain the minimal E-compatible partitions $\{1, 2, 3\} \{4, 5\}$; $\{1, 2, 4\} \{3, 5\}$; $\{1, 4, 5\} \{2, 3\}$; $\{1, 3, 5\} \{2, 4\}$ and hence the minimal coverings (A, E) and (B, D).

In order to reduce the amount of computation in this case as well, we add the condition that the element with respect to which we perform the branching should belong to a minimum number of maximal sets:

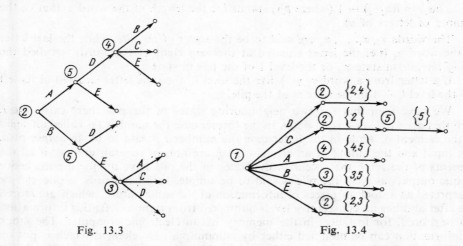

Fig. 13.3   Fig. 13.4

Considering the path $(x_i, x_{i, n_1}, ..., x_{i, n_1, ..., n_s})$ from the root x_i to the node $x_{i, n_1, ..., n_s}$ which is not a terminal node of the rooted tree, we shall choose an element $y_{i, n_1, ..., n_s} \in$ $\in X \backslash E_{i, n_1, ..., n_s}$ which belongs to the minimum number of sets from $E_{E_{i, n_1, ..., n_s}}$, maximal with respect to set inclusion. This criterion for the choice of the element with respect to which the branching is performed, will be applied starting with the root x_i.

By applying this principle we obtain the rooted tree in Fig. 13.5 which contains two minimum partitions, $\{1, 2, 3\}$ $\{4, 5\}$ and $\{1, 2, 4\}$ $\{3, 5\}$, along the paths that join the root to the terminal nodes, partitions that correspond to the two minimal coverings of the set X.

The more general covering problem which is obtained when we associate with each set E_i of the family E a positive real weight c_i and ask for a covering for which the sum of the weights corresponding to the sets of the covering is minimal, can also be solved with the aid of the rooted tree A_E by choosing from the set of irreducible coverings one which is minimal with respect to the sum of weights.

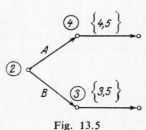

Fig. 13.5

In order to study the information furnished by a rooted tree like those occurring in the previous algorithms, as well as for a convenient organization of computations in a digital computer, we shall use the concept of a *pile*. By a *pile* over the set E we mean any finite set $A = (\alpha_0, \alpha_1, ..., \alpha_q)$, where $\alpha_0, \alpha_1, ..., \alpha_q$ are words with letters from the alphabet E, such that:

1) $\alpha_0 = \alpha_q = \Lambda$, the empty word which has no letters and is of length zero;

2) for every $i = 1, 2, ..., q$, one and only one of the following holds:

— α_{i-1} is a left factor for α_i, i.e. there is a letter $a \in E$ such that $\alpha_i = \alpha_{i-1} a$, hence $l(\alpha_i) = l(\alpha_{i-1}) + 1$;

— α_i is a left factor for α_{i-1}, i.e. there is a letter $a \in E$ such that $\alpha_{i-1} = \alpha_i a$, hence $l(\alpha_i) = l(\alpha_{i-1}) - 1$ (where $l(\alpha)$ stands for the length of the word α that is, the number of letters of α).

The words $\alpha_0, \alpha_1, ..., \alpha_q$ are said to be the *states of the pile*, while the last letter of the word α_i (i.e. the letter situated at the very right of the word) is called the *top of the pile* in state α_i, or the level 1 of the pile in state α_i.

If a letter from a word $\alpha_i \neq \Lambda$ has the level k, then the letter situated at its left has the level $k + 1$ in the state α_i of the pile.

We have seen that for two neighbouring states of the pile there exists $a \in E$ such that $\alpha_i = \alpha_{i-1}a$ or $\alpha_{i-1} = \alpha_i a$. In the former case the number i is called an *input* of the element a, while in the latter case the number i is said to be an *output* of a. The input and the output of an element $a \in E$ are not, in general, unique; if all the elements of the set E that occur in the states of the pile have unique inputs, hence unique outputs as well, the pile is said to be simple. In other words, a pile can be interpreted as a stock of pieces of information of the same nature, which are stored one after another on consecutive levels numbered from right to left; this information can be stored, for instance, in the memory of an electronic computer. The stock of information can be modified either by adjoining a new element which is put on the first level, the other elements being translated to the left, so that their levels increase by one unit each, or by removing the element on the first level, the other elements being translated to the right, so that their levels decrease by one unit each.

A model of a pile can be realized in the memory of a computer by using a list of arrays and a vector, the components of which indicate the address of the upper bound of each array; at each moment the computer memorizes a single state of the pile, the passage from one state to another being performed according to the indicated rule. Processes represented by rooted directed trees are encountered especially in problems related to information storage and retrieval, in the mathematical theory of classification (where rooted trees are used which connect the various coverings of the set of objects situated on various levels according to the number of properties common to each pair of objects) and in *SEP* algorithms ("séparation et évaluation progressives" or "branch and bound"), which are combinatorial algorithms that need the enumeration of a set of solutions which contains an optimal solution [4], [25], [39]. We mention that *SEP* algorithms can be used for the solution of integer programming problems. Memory piles are used for the syntactic analysis and compilation of programs written in various programming languages, for example in ALGOL [19], [32]. We can associate a rooted tree with a pile so that the set of states of the pile is the set of all paths that start from the root of the rooted tree, the sequence of these paths being constructed according to *Tarry's rule:* Start from the root and follow the arc at the extreme left; at each node, run along the first arc from the left among those which have not yet been followed. From a terminal node go back along the path which joins the root to that terminal node, until one reaches the first node for which there still exist incident arcs that have not yet been explored. Resume the exploration of arcs starting from this node, etc. The search ends at the moment when the root is reached in the backtracking process and all the arcs starting from the root have been explored.

We thus obtain a set of paths which, interpreted as a sequence of vertices, constitute the non-empty states of a pile associated with the vertices of the rooted tree, while interpreted as a sequence of arcs, constitute the states of a pile associated with the arcs of the rooted tree, with the convention that the set of arcs of a one-vertex path is associated with the empty word Λ. Thus, for example, if the arcs of the rooted tree in Fig. 13.3 are denoted by A, B, C, D, E, we obtain the following pile associated with the arcs of the rooted tree: $(\Lambda, A, AD, ADB, AD, ADC, AD, ADE, AD, A, AE, A, \Lambda, B, BD, B, BE, BEA, BE, BEC, BE, BED, BE, B, \Lambda)$. Denoting the terminal nodes by T and the other nodes by the corresponding circled numbers in Fig. 13.3, we obtain the pile associated with the vertices of the same rooted tree (notice that the only empty states of this pile are the initial and the final one): $(\Lambda, 2, 25, 254, 254T, 254, 254T, 254, 254T, 254, 25, 25T, 25, 2, 25, 25T, 25, 253, 253T, 253, 253T, 253, 253T, 253, 25, 2, \Lambda)$.

C. Pair [33] and J. Derniame [14] gave a procedure for associating a pile with the vertices and arcs of any graph and suggested several efficient methods for studying path problems and connectivity in a graph.

The piles used in covering problems will be associated with the vertices of the rooted trees, while the stock of information associated with each vertex will consist of families of sets or of subsets from the family E, that appear in the description of the algorithms for the construction of the rooted trees A_E and A_E^*, respectively. The sets associated with the arcs at each vertex will be marked with a star, while two stars will mark the sets associated with the arcs that belong to the path starting from the origin which is being explored at that moment. Whenever we arrive at a terminal vertex, we print all sets of the pile which are marked with two stars and we begin the exploration of another path in accordance with the rule given above. Now refer to the algorithms described in propositions 1 and 2 as algorithms 1 and 2, respectively. Using the notion of pile and taking into account that at each branching we choose a vertex which belongs to a minimum number of sets of E and to a minimum number of sets of E, maximal with respect to inclusion, respectively, algorithms 1 and 2 take the following form:

ALGORITHM 1 consists of the following instructions:

α) Store all sets E_i with $i = 1, 2, ..., m$, at the top of the pile (on the first level).

β) Determine the element of the set X that belongs to a minimum number of sets from the first level and mark with a star all the sets from the first level which contain that element.

If several elements with the above property exist, choose the element having the smallest index.

The sets marked with a star are written at the beginning of the list of sets of the first level.

γ) Mark with two stars the set (which had been marked with one star) at the beginning of the list of sets of the first level and go to $\varepsilon 1$.

δ) Check whether on the first level the set marked with two stars is followed by a set marked with one star. If not, go to θ. If so, mark with two stars the set which had been marked with one star and which is situated on the right of the set which had been marked with two stars and conversely, mark with one star the set which had been marked with two stars.

$\varepsilon 1$) Remove from the set X the elements of those sets of the pile which are marked with two stars. If the set obtained in this way is empty, go to $\eta 1$. Otherwise choose an element (namely, the element of smallest index if there are several) of the resulting set which belongs to the smallest number of sets from the last non-empty level of the pile; copy these sets onto the first level and mark them with a star. Go to γ.

$\eta 1$) Print the family of those sets from E marked with two stars: it constitutes a covering of the set X with sets from the family E. Go to δ.

θ) Eliminate the first level of the pile. If the pile is empty, stop. Otherwise go to δ.

The order of performing the above instructions is from α to θ, unless the instruction requires a jump to an instruction different from the next one.

The smallest number of sets in a covering of X is equal to the smallest number of non-empty levels of the pile from the moment of the first printing at $\eta 1$ of a covering of the set X.

ALGORITHM 2 consists of the following instructions:

$\alpha 0$) Determine an upper bound M for the number of sets in a minimal covering of X with sets from the family E. Take, for example $M = |X|$.

α) Store all sets E_i with $i = 1, 2, ..., m$ at the top of the pile (on the first level).

β) Determine the element of the set X that belongs to the smallest number of sets from the first level and mark with a star all the sets of the first level which contain that element. If several elements with the above property exist, choose the element having the smallest index. Put the sets marked with a star at the beginning of the list of sets from the first level.

γ) Mark with two stars the set (which had been marked with one star) at the beginning of the list of sets from the first level and go to $\varepsilon 2$.

δ) Check whether on the first level the set marked with two stars is followed by a set marked with one star. If not, go to θ. If so, mark with two stars the set which had been marked with one star and which is situated on the right of the set which had been marked with two stars and, conversely, mark with one star the set which had been marked with two stars.

$\varepsilon 2$) Remove the elements marked with two stars from all the other sets (marked or not) from the first level and from the sets thus obtained keep those which are maximal with respect to inclusion. If the maximal set is unique, go to $\eta 2$, otherwise check whether $l + 1 < M$, where l is the number of non-empty levels of the pile. If so, store the above obtained maximal sets on the first level and go to β. If not, go to δ.

$\eta 2$) Print the family consisting of the unique maximal set obtained at $\varepsilon 2$ and all the sets of the pile that are marked by two stars. This family of sets is an E-compatible partition of the set X. Check whether $l + 1 < M$, where l is the number of non-empty levels of the pile. If so, replace M by $l + 1$ and go to δ. If not, go to δ.

θ) Eliminate the first level of the pile. If the pile is empty, stop. Otherwise go to δ.

We have applied here a method which is frequently used in SEP algorithms and which consists of a progressive improvement of the upper bound M on the basis of the results of intermediate calculations, thus obtaining E-compatible partitions of the set X with the property that the number of classes of a partition is less than or equal to the number of classes of the previously obtained partitions, since we do not explore those paths of A_E^* starting from the root which are known to contain more than M arcs.

In the case of algorithm 2 the number of sets in a minimal covering of the set X is equal to the smallest number of non-empty levels of the pile, considered from the first moment when a unique maximal set is obtained at $\varepsilon 2$, increased by one time unit. If we are interested only in the minimum of the number of sets in a covering of X, then instead of printing the successive partitions obtained during the computation, we shall just compare the numbers of non-empty levels of the pile at the moments when unique maximal sets are obtained at $\varepsilon 2$; the smallest of these numbers, increased by one unit, is the desired result and will be printed.

The problem of finding a minimum covering arises at a certain step in the minimization of disjunctive normal forms of Boolean functions. More precisely, after all the prime implicants of the given function have been determined, we must find the smallest number of prime implicants, the disjunction of which generates the Boolean function; this corresponds to a two-level circuit which realizes the given function and consists of the smallest number of "and" operators [5], [35]. In this case, the set X is the set of those vertices of the n-dimensional hypercube in which the given Boolean function f takes on the value 1, while E is the family of prime implicants of the function f, every k-letter prime implicant corresponding to a 2^{n-k}-vertex subset of X.

The determination of matchings of a bipartite graph $G = (X, Y, \Gamma)$ (for which the necessary and sufficient condition for the existence of a matching, given in the theorem of König-Hall (Chap. 10), holds), reduces to the problem of finding all irreducible coverings of the set X by pairwise disjoint (i.e. with no common extremities) arcs of the form (x, y) with $x \in X$ and $y \in Y$. The property that the arcs of a matching are pairwise disjoint implies the property that every covering of X which corresponds to a matching of G is irreducible. Therefore all the matchings C of a bipartite graph $G = (X, Y, \Gamma)$ can be determined with the aid of algorithm 1, where E is the set of arcs of the graph, with the modification that at step $\varepsilon 1$, after the elimination of an arc (x, y), we take into consideration and write on the first level of the pile only those arcs which contain neither x nor y.

For the bipartite graph in Fig. 10.1 we obtain the rooted tree in Fig. 13.6, which contains four paths that join the root to terminal nodes; the set of arcs associated with each path corresponds to a matching of the bipartite graph. The construction of the upper branch has been interrupted because the arcs of the matching cannot have common extremities.

So by applying algorithm 1 with the above mentioned modification at $\varepsilon 1$ we either obtain all the matchings of the bipartite graph (X, Y, Γ) or come to the conclusion that there is no such matching.

In the case of the determination of the chromatic number, a problem which is related to the minimization of the number of internal states of incompletely specified sequential automata [48], the set X is the set of vertices of a graph, while E

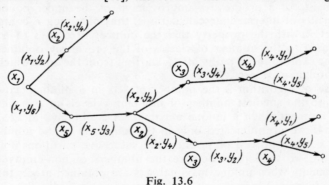

Fig. 13.6

is the family of all maximal independent sets of the graph. For this particular problem certain improvements can be made to the covering algorithms, with the effect of reducing the number of operations, and this will be discussed in the next section.

13.3. DETERMINATION OF THE CHROMATIC NUMBER OF A FINITE GRAPH

Given a finite, undirected and loop-free graph G, the chromatic number $\gamma(G)$ is the smallest number of colours which can be used for colouring the vertices of the graph in such a way that two vertices joined by an edge have different colours.

We recall that an independent set is a set of vertices such that no two vertices of the set are joined by an edge. Then the chromatic number of the graph is the smallest number of classes in a partition of X the classes of which are independent sets.

Every unoriented graph G can be associated with a symmetric binary relation R_G defined as follows: $(x, y) \in R_G$ if and only if $[x, y] \in U$, where U is the set of edges of the graph.

If the graph G has no loops, i.e. no edges of the form $[x, x]$ with $x \in X$, then the relation $R_{\bar{G}} = X \times X \setminus R_G$ is symmetric and reflexive and the problem of determining the chromatic number of the graph G reduces to the determination of an equivalence relation $R_{\bar{G}}^* \subset R_{\bar{G}} \subset X^2$ with the smallest number of classes. Every equivalence class of the relation $R_{\bar{G}}^*$ is an independent set of the graph G.

In the sequel we shall establish upper and lower bounds for the chromatic number of a graph $G = (X, U)$. Denote by $d(x_k)$ the degree of the vertex $x_k \in X$, i.e. the number of edges incident with x_k and assume that the n vertices of the graph are numbered so that the relations

$$d(x_1) \geqslant d(x_2) \geqslant \ldots \geqslant d(x_n)$$

hold. Let $\Delta = \{X_k\}_{1 \leqslant k \leqslant q}$ be a partition of X made up of independent sets and set

$$m_k = \max_{x_p \in X_k} d(x_p) \text{ for } k = 1, \ldots, q.$$

Also put

$$\alpha(G) = \max_{k=1, \ldots, n} (\min (k, 1 + d(x_k)))$$

and

$$\beta_\Delta(G) = \max_{k=1, \ldots, q} (\min (k, 1 + m_k)).$$

R. Brooks [6] established the inequality $\gamma(G) \leqslant 1 + \max_{p=1, \ldots, n} d(x_p)$, while D. Welsh and M. Powell proved [57] that

$$\gamma(G) \leqslant \alpha(G) \leqslant 1 + \max_{p=1, \ldots, n} d(x_p). \tag{13.2}$$

PROPOSITION 3. *For every partition Δ of the set of vertices into independent sets, the relation $\gamma(G) \leqslant \beta_\Delta(G)$ holds and the condition $m_1 \geqslant m_2 \geqslant \ldots \geqslant m_q$ implies the inequalities*

$$\gamma(G) \leqslant \beta_\Delta(G) \leqslant \alpha(G) \leqslant 1 + \max_{p=1, \ldots, n} d(x_p). \tag{13.3}$$

The equality $\beta_\Delta(G) = \alpha(G)$ holds for $X_1 = \{x_1\}$, $X_2 = \{x_2\}$, ..., $X_n = \{x_n\}$, while $\beta_\Delta(G) = \gamma(G)$ if, for instance, the partition Δ has the smallest number of classes, i.e. $\gamma(G) = \min_\Delta \beta_\Delta(G)$.

If X_1 is not a maximal independent set of the graph G, we can adjoin some vertices to X_1 until we get a maximal independent set X_1'. If $X_2 \setminus X_1'$ is neither empty nor a maximal independent set of the subgraph with the vertex set $X \setminus X_1'$, we adjoin in the same way new vertices to $X_2 \setminus X_1'$ until we obtain a maximal independent set X_2' of the subgraph $X \setminus X_1'$, etc. After a finite number of steps we obtain a new partition $\Delta' = \{X_k'\}_{1 \leqslant k \leqslant q'}$ of X, consisting of independent sets, such that $q' \leqslant q$ and

$$X_k \subset \bigcup_{i=1}^{\min(k, q')} X_i' \quad \text{for every} \quad k = 1, ..., q.$$

Denote by $i(x)$ the index i such that $x \in X_i'$. Let x be a vertex with $i(x) = k$. It follows from the construction of the partition Δ' that $d(x) \geqslant k - 1$, hence $i(x) \leqslant d(x) + 1$. If the vertex $x_0 \in X_k$, then $i(x_0) \leqslant k$ and

$$i(x_0) \leqslant \max_{x \in X_k} i(x) \leqslant \max_{x \in X_k} (d(x) + 1) = m_k + 1.$$

Therefore we obtain

$$\gamma(G) \leqslant \max_{x_0} i(x_0) \leqslant \max_{k=1, ..., q} \min(k, m_k + 1).$$

Furthermore, consider the relations $d(x_1) \geqslant d(x_2) \geqslant ... \geqslant d(x_n)$ and take into account

$$m_1 \geqslant m_2 \geqslant ... \geqslant m_q,$$

thus obtaining

$$d(x_1) = m_1, \ d(x_2) \geqslant m_2, ..., d(x_q) \geqslant m_q.$$

Under these conditions

$$\min(k, 1 + m_k) \leqslant \min(k, 1 + d(x_k))$$

for every $k = 1, ..., q$, hence

$$\beta_\Delta(G) \leqslant \max_{k=1, ..., q} (\min(k, 1 + d(x_k))) \leqslant \alpha(G).$$

If the partition Δ has the smallest number of classes, we get $\beta_\Delta(G) \leqslant \leqslant \max_{k=1, ..., q} k = q = \gamma(G)$.

However, there are cases in which $\gamma(G) = \beta_\Delta(G)$ although the partition Δ does not have the smallest number of classes.

To obtain a good upper bound for the chromatic number of a graph, we construct the following partition Δ:

We choose a vertex x_{i_1} of maximum degree; we then choose a vertex x_{i_2} of maximum degree within the set of vertices that are not joined by an edge to x_{i_1}; then we choose a vertex x_{i_3} of maximum degree within the vertices that are not joined by edges either to x_{i_1} or to x_{i_2}, etc.

When this process cannot be continued, that is, when each of the remaining vertices is joined by an edge to at least one of the vertices already selected, we take the set of vertices selected as the class X_1 of the partition Δ. We now take a vertex $x_{j_1} \in X \setminus X_1$ of maximum degree, then a vertex $x_{j_2} \in X \setminus X_1$ of maximum degree within the vertices of $X \setminus X_1$ that are not joined by an edge to x_{j_1} etc. If the vertices $x_{k_1}, x_{k_2}, \ldots, x_{k_r}$ belong to the class X_l of the partition, then we choose a vertex $x_{k_{r+1}}$ of maximum degree from the vertices of the set $X \setminus (\bigcup_{i=1}^{l-1} X_i)$ that are not joined by edges to any of the vertices $x_{k_1}, \ldots, x_{k_r}$. If this set is empty, we form the class X_{l+1} etc., and in this case the upper bound $\beta_\Delta(G)$ will be equal to the number of classes in the partition Δ, for this partition has been built from maximal independent sets in the corresponding subgraphs.

The above algorithm for obtaining an upper bound for the chromatic number and for a partition of the set of vertices of a graph, made up of independent sets, is due to D. Welsh and M. Powell [57]; a computer program for it, written in *FORTRAN* IV, has been published in [48]. The program has 87 instructions and prints an upper bound for the chromatic number equal to the number of classes of the partition Δ constructed above, as well as the indices of the vertices of each class of the partition Δ, each class being printed on a separate line. An example with $n = 20$ has been solved with an IBM 360/30 computer in two minutes, under the hypothesis that $d(x_1) \geqslant d(x_2) \geqslant \ldots \geqslant d(x_n)$. If this condition is not met, the program furnishes another partition of the set of vertices made up of maximal independent subsets of the corresponding subgraphs, but without observing the rule on the degrees of the selected vertices in the previous algorithm.

The isolated vertices are immaterial in the problem of determining the chromatic number of a graph, because each one can be incorporated into any class of a partition of the non-isolated vertices into independent sets.

PROPOSITION 4. *If G is a graph with no isolated vertices, having n vertices, m edges and chromatic number $\gamma(G) = k$, then*

$$m \geqslant \binom{k}{2} + \left\{ \frac{n-k}{2} \right\} \tag{13.4}$$

where $\{x\}$ stands for the least integer greater than or equal to x. If $n - k$ is even, the graph G with the minimum number of edges is unique, while if $n - k$ is odd there are two types of graphs which have a minimum number of edges but they coincide for $k = 2$.

Let X be the set of vertices of G and assume first that for every $x \in X$, the subgraph G_x obtained from G by removing the vertex x and the edges incident to it, contains isolated vertices.

Let y be an isolated vertex of G_x, hence the only edge incident to y in the graph G joins y to x. But the graph G_y also contains isolated vertices, hence $[x, y]$ is the only edge of G incident to x or to y.

By iteration of the above reasoning we deduce that if for every $x \in X$ the subgraph G_x contains isolated vertices whereas the graph G has no isolated vertices, then n is even, $\gamma(G) = 2$ and the number of edges of this graph is equal to

$$m = \frac{n}{2} = \binom{2}{2} + \frac{n-2}{2}.$$

But this number is the smallest number of edges of the graph G, because the degree $d(x)$ of any vertex x is at least equal to 1 (as G has no isolated vertices), therefore $2m = \sum_{x \in X} d(x) \geqslant n$, hence $m \geqslant \frac{n}{2}$. So in this case the proposition is established.

Now we consider the case when there is a vertex x such that G_x has no isolated vertices and prove the proposition by induction on n.

For $n=k$ the property is obvious, as the graph is complete and has $\binom{k}{2}$ edges.

Assume that the proposition is true for the graphs G' with $n' = n - 1$ vertices and with chromatic number $k \leqslant n-1$. Let G be a graph with n vertices. If $\gamma(G)=n$, then G is a complete graph with n vertices and the property is obvious. If $\gamma(G) = k \leqslant n - 1$, let $x \in X$ be a vertex such that G_x has no isolated vertices. Two cases may occur:

a) $\gamma(G_x)=k$, hence the smallest number of edges for the subgraph G_x is $\binom{k}{2} + \left\{ \frac{n-k-1}{2} \right\}$, by the inductive hypothesis. But x is not an isolated vertex, hence $d(x) \geqslant 1$, therefore the number of edges of G is greater than or equal to

$$\binom{k}{2} + \left\{ \frac{n-k-1}{2} \right\} + 1 \geqslant \binom{k}{2} + \left\{ \frac{n-k}{2} \right\}.$$

The equality $m = \binom{k}{2} + \left\{ \frac{n-k}{2} \right\}$ holds only if $n - k$ is odd, $d(x) = 1$ and G_x has the smallest number of edges.

b) $\gamma(G_x) = k - 1$. In this case there is a partition $\{x\}, C_1, ..., C_{k-1}$ of the set X, where all the classes are independent sets and the vertex x is joined by an edge to at least one vertex from each class $C_1, ..., C_{k-1}$, for in the opposite case $\gamma(G) \leqslant k - 1$, thus contradicting the hypothesis $\gamma(G) = k$. This implies $d(x) \geqslant k - 1$, hence a lower bound for the number of edges of the graph G is the number of edges of G_x plus $k-1$, which implies, by the inductive hypothesis,

$$m \geqslant \binom{k-1}{2} + \left\{ \frac{n-1-(k-1)}{2} \right\} + k - 1 = \binom{k}{2} + \left\{ \frac{n-k}{2} \right\};$$

the equality holds only if $d(x) = k - 1$ and the subgraph G_x has the smallest number of edges.

Following through this proof in the cases in which the inequalities become equalities, we obtain by induction the following characterization of the graphs G with no isolated vertices, with n vertices and $\gamma(G) = k$, for which the number of edges is a minimum:

If $n - k$ is even, the graph G with the smallest number of edges is unique (up to an isomorphism) and consists of a complete k-subgraph and of $n - k$ vertices which are pairwise joined by $\dfrac{n-k}{2}$ edges. If $n - k$ is odd, there are two non-isomorphic graphs with a minimum number of edges. One graph consists of a complete k-subgraph, of $n - k - 1$ vertices that are pairwise joined by $\dfrac{n-k-1}{2}$ edges and of another vertex which is joined by an edge to an arbitrary vertex of the complete k-subgraph. The other graph consists of a complete k-subgraph, of $n - k - 1$ vertices that are pairwise joined by $\dfrac{n-k-1}{2}$ edges and of another vertex which is joined by an edge to a vertex which does not belong to the complete k-subgraph. These two types of graphs coincide for $k = 2$.

If no supplementary restriction is imposed upon the graph G with n vertices and chromatic number k, then the smallest number of edges is equal to $\dbinom{k}{2}$, because between any two classes of a partition of X consisting of k independent sets there is at least one edge, for otherwise $\gamma(G) < k$. We can prove by induction on n that the only graph which has this minimum number of edges consists of a complete k-subgraph and of $n - k$ isolated vertices. So for the case with no isolated vertices we obtain $m \geqslant \dfrac{k^2 - k}{2} + \dfrac{n - k}{2}$ or $k^2 - 2k + n - 2m \leqslant 0$, hence $k \leqslant 1 + \sqrt{2m - n + 1}$.

COROLLARY. *If the graph G with n vertices and m edges has no isolated vertices, its chromatic number satisfies the inequality*

$$\gamma(G) \leqslant 1 + \sqrt{2m - n + 1}. \tag{13.5}$$

This inequality becomes an equality for the complete graph with n vertices.

A. Ershov and G. Kuzhukhin obtained [18] the following upper bound for the chromatic number of a connected graph with n vertices and m edges:

$$\gamma(G) \leqslant \left[\frac{3 + \sqrt{9 + 8(m - n)}}{2} \right] \tag{13.6}$$

which implies that if the graph G is connected and $\gamma(G) = k$, the number of edges $m \geqslant \dbinom{k}{2} + n - k$. This lower bound for the number of edges of a connected graph with n vertices and chromatic number equal to k can be obtained by induction, with a proof similar to that of proposition 4. This also enables us to reveal the structure of the graphs with the smallest number of edges in the above class: for $k = 2$ they are the trees with n vertices, for $k = 3$ they consist of an odd cycle with p vertices $(3 \leqslant p \leqslant n)$ and of $n - p$ vertices such that the graph obtained by identifying the vertices of the cycle is a tree, while for $k \geqslant 4$ they consist of a complete k-subgraph and of $n - k$ vertices, such that the graph obtained by identifying the vertices of the complete k-subgraph is a tree [50].

By counting the minimal connected graphs with n labelled vertices and chromatic number k, for $k = 2$ one obtains n^{n-2} graphs with the aid of the Cayley formula, for $k = 3$ one obtains

$$\frac{1}{2} \sum_{\substack{3 \leqslant p \leqslant n \\ p \text{ odd}}} [n-1]_{p-1} n^{n-p}$$

while for $4 \leqslant k \leqslant n$ one obtains $\binom{n-1}{k-1} n^{n-k}$ graphs with labelled vertices by virtue of proposition 14 in Chap. 6.

To obtain a lower bound for the chromatic number, we shall use Turán's theorem (proposition 6, Chap. 3) which states that for the graphs G with n vertices which do not contain complete subgraphs with k vertices $(2 \leqslant k \leqslant n)$, the maximum number of edges is given by the formula

$$M(n, k) = \frac{k-2}{k-1} \cdot \frac{n^2 - r^2}{2} + \frac{r(r-1)}{2},$$

where r is the remainder in the division of n by $k - 1$. The graph G for which this upper bound on the number of edges is attained is unique up to isomorphism and consists of $k - 1$ classes of vertices, where r classes contain $t + 1$ vertices each, t being defined by $n = (k-1)t + r$ where $0 \leqslant r \leqslant k - 2$, while the other classes contain t vertices each and each of the vertices of G is joined by edges to all the vertices which do not belong to the class of that vertex. So G is a complete $(k-1)$-partite graph.

PROPOSITION 5. *The maximum number of edges of the graphs G with n vertices and chromatic number $\gamma(G) = k - 1$ is equal to $M(n, k)$ and the structure of the maximal graph with $M(n, k)$ edges is given by Turán's theorem.*

Notice that

$$\{G | \gamma(G) = k - 1\} = \bigcup_{i=2}^{k-1} \{G | c(G) = i, \ \gamma(G) = k - 1\},$$

where $c(G)$ stands for the maximum number of elements of a complete subgraph of G, so that $c(G) \leqslant \gamma(G)$. But for every i

$$\{G | c(G) = i, \ \gamma(G) = k - 1\} \subset \{G | c(G) = i\},$$

hence we have the following upper bound for the number of edges of the graphs G with $\gamma(G) = k - 1$:

$$\max_{i=2,\ldots,k-1} M(n, i + 1) = M(n, k).$$

But the maximal graph with $M(n, k)$ edges has chromatic number $k - 1$.

This proposition yields an implicit lower bound for the chromatic number of the graph G: determine the greatest number k such that the number of edges $m \geqslant M(n, k) + 1$, hence $\gamma(G) \geqslant c(G) \geqslant k$; if $r = 0$ we obtain an explicit but weaker lower bound.

PROPOSITION 6 (D. Geller). *The chromatic number of a graph G with n vertices and m edges satisfies the inequality*

$$\gamma(G) \geqslant \left\{ \frac{n^2}{n^2 - 2m} \right\}. \tag{13.7}$$

We obtain $\dfrac{k-2}{k-1} \cdot \dfrac{n^2}{2} \geqslant M(n, k)$, because the difference of these numbers is equal to $\dfrac{r(k-1-r)}{2(k-1)} \geqslant 0$, since $0 \leqslant r \leqslant k - 2$. Therefore, if

$$m > \frac{k-2}{k-1} \cdot \frac{n^2}{2} \geqslant M(n, k),$$

proposition 5 implies $\gamma(G) \geqslant k$. But

$$m > \frac{k-2}{k-1} \cdot \frac{n^2}{2} \text{ implies } k < \frac{2n^2 - 2m}{n^2 - 2m},$$

hence we can take

$$k = \left\{ \frac{2n^2 - 2m}{n^2 - 2m} - 1 \right\} = \left\{ \frac{n^2}{n^2 - 2m} \right\},$$

thus obtaining $\gamma(G) \geqslant \left\{ \dfrac{n^2}{n^2 - 2m} \right\}.$

Notice that this value is a lower bound for the number $c(G)$ as well.

This lower bound is attained for example for the complete graph K_n or for a tree; for the latter $m = n - 1$ and

$$\gamma(G) \geqslant \left\{ \frac{n^2}{n^2 - 2n + 2} \right\} = 2 \text{ for every } n \geqslant 2.$$

From the equality $\sum_{i=1}^{n} d(x_i) = 2m$, we obtain

$$\gamma(G) = \frac{n^2}{n^2 - \sum_{i=1}^{n} d(x_i)} \geqslant \frac{n^2}{n^2 - n \cdot \min_{i=1, \ldots, n} d(x_i)} = \frac{n}{n - \min_{i=1, \ldots, n} d(x_i)},$$

ie. the lower bound obtained in [16].

For every integer $n \geqslant 2$, there are graphs G for which (13.7) yields $\gamma(G) \geqslant 2$, but for which the chromatic number $\gamma(G)$ is, in reality, equal to n, because W. Tutte [15] and A. Zykov [60] have shown that for every integer $n \geqslant 2$ there is a graph G with no complete 3-subgraph and having chromatic number equal to n.

To determine exactly the chromatic number, we shall construct first the family E of maximal independent sets and then apply the covering algorithm 2.

Now $\gamma(G) \leqslant \beta_\Delta(G)$, where $\beta_\Delta(G)$ is an upper bound for the chromatic number of the graph which can be determined with the aid of the above methods, and moreover, $\gamma(G) s(G) \geqslant n$, where $s(G)$ is the independence number of the graph G and n is the number of vertices of G. Hence $\gamma(G) \geqslant \left\{ \dfrac{n}{s(G)} \right\}$. If we are at a non-terminal vertex $x_{i, n_1, \ldots, n_s}$ of the rooted tree constructed in algorithm 2 ($n_s > 1$), then the chromatic number of the graph G, which is equal to the smallest length of the paths that join the root to the terminal vertices of the rooted tree, is less than or equal to the smallest length of the paths that join the root to the terminal nodes and pass through the vertex $x_{i, n_1, \ldots, n_s}$. But the minimum length of the latter paths is equal to the number s of arcs contained in the path which joins the root to $x_{i, n_1, \ldots, n_s}$ plus the minimum number of arcs contained in the paths that join $x_{i, n_1, \ldots, n_s}$ to the terminal vertices of the rooted tree, i.e. it is equal to s plus the chromatic number of the subgraph generated by the vertex set $X \setminus E_{i, n_1, \ldots, n_s}$. We have denoted by $E_{i, n_1, \ldots, n_s}$ the union of the independent sets associated with the arcs of the path from the root to the node $x_{i, n_1, \ldots, n_s}$.

Denoting by G' the subgraph of the graph G which is generated by the vertex set $X \setminus E_{i, n_1, \ldots, n_s}$, we obtain the lower bound $\left\{ \dfrac{n'}{s(G')} \right\}$ for $\gamma(G')$, where $n' = |X \setminus E_{i, n_1, \ldots, n_s}|$ i.e. the number of elements which belong to the union of the independent sets from the first level of the pile, while $s(G')$ is the independence number of G', i.e. the greatest number of elements in a set from the first level of the pile. The number of arcs contained in the path from the root of A_E^* to the vertex $x_{i, n_1, \ldots, n_s}$ is equal to s, which represents the number of non-empty levels of the pile at step β of algorithm 2. It follows that if $s + \left\{ \dfrac{n'}{s(G')} \right\} > \beta_\Delta(G)$, we shall not explore the rooted subtree having the root $x_{i, n_1, \ldots, n_s}$, because it cannot contain a path of minimum length joining the root to the terminal nodes.

With the above modification, algorithm 2 leads to the chromatic number of a graph and explores fewer paths of the rooted tree A_E^*. When we are interested in obtaining a great number of partitions of the set X we disregard the above remarks and apply algorithm 2 without modification.

Now since $s + \left\{ \dfrac{n'}{s(G')} \right\} > M$, where $\{x\}$ stands for the least integer greater than or equal to x and M is an upper bound for the chromatic number of the graph, then algorithm 2, described in pile language, runs as follows:

$\alpha 0$) Determine an upper bound M for the chromatic number of the graph G, for example $M = |X|$.

α) Determine all maximal independent sets of the graph, for example with the aid of the Bednarek — Taulbee algorithm, and store these sets at the top of the pile.

β) Check whether $s + \left\{ \dfrac{n'}{s(G')} \right\} > M$, where s is the number of non-empty levels of the pile, n' is the number of elements contained in the union of the independent sets from the first level of the pile and $s(G')$ is the greatest number of elements contained in a set from the first level. If so, go to θ; if not, determine the vertex of the graph which belongs to the smallest number of sets from the first level and mark with a star all independent sets of the first level which contain that vertex. If there are several vertices with this property, choose the vertex having the least index. Store the sets marked with a star at the beginning of the list of sets from the first level.

γ) Mark with two stars the set situated at the beginning of the list of sets from the first level, previously marked with one star, and go to ε.

δ) Check whether on the first level the set marked with two stars is followed by a set marked with one star. If not, go to θ; if so, mark with two stars the set previously marked with one star, located at the right of the set marked with two stars, and mark with one star the set previously marked with two stars.

ε) Eliminate the elements of the set marked with two stars from the other sets of the first level, whether or not they are marked, and from the sets thus obtained keep those which are maximal with respect to set inclusion. If the maximal set is unique, go to η; otherwise check whether $s + 1 < M$, where s is the number of non-empty sets of the pile. If so, write the maximal sets on the first level and go to β. If not, go to δ.

η) Print the family consisting of the unique maximal set found at ε and all the sets of the pile marked with two stars. This family of sets is a partition of the set X, made up of independent sets. Check whether $s + 1 < M$, where s is the number of non-empty levels of the pile. If so, take M equal to $s + 1$ and go to δ. If not, go to δ.

θ) Remove the first level of the pile. If the resulting pile is empty, stop. Otherwise go to δ.

The set E over which the pile has been constructed, is the union of the family of independent sets of the graph and the set $\{*, **\}$. The family of partitions printed during the computation contains at least one partition with the smallest number of independent sets; this number is precisely the chromatic number of the graph. The chromatic number is therefore equal to one more than the smallest number of non-empty levels of the pile, obtained at the moment when, at step ε of the algorithm one obtains a unique maximal set.

If we are only interested in the chromatic number of the graph G and not in a partition which realizes it, then it is not necessary to print the partitions obtained during the computations, but we must compare the numbers of non-empty levels of the pile at the moments when a unique maximal independent set is obtained at step ε; the computer should finally print the smallest of these numbers plus one unit.

For a greater number of vertices of the graph G, the difficulty with the algorithm consists of the fact that the number of maximal independent sets increases rapidly, in the most unfavourable case this number being of the order of $3^{n/3}$ for a graph with n vertices [30]. For this reason a variant of the above algorithm can be applied in which for every node of the rooted tree we only determine the family of those maximal independent sets which contain a certain vertex x of the graph for which the branching is being performed. This family of independent sets can be obtained in the following way: the vertex x is incorporated in each maximal independent set of the subgraph obtained from G by deleting the vertex x and all the vertices joined to x by an edge.

Methods for the approximate determination of a chromatic decomposition of the set of vertices of a graph have been suggested by D. C. Wood [59] and by R. Wilkov and W. Kim [58]; these methods can also be applied to graphs with a large number of vertices.

Problems similar to that of the determination of a partition with a minimum number of independent sets occur in the *problem of classifying on one level a set of objects with respect to a set of properties* [17].

Let us give an example of an application of the algorithm for the determination of the chromatic number of a graph in the case of a graph G with seven nodes the edges of which are the following: [1, 2], [1, 3], [1, 5], [1, 7], [2, 3], [2, 4], [3, 4], [3, 5], [4, 5], [4, 6], [4, 7], [5, 7] and [6, 7]. It follows that the degrees are the following: $d(1) = d(3) = d(5) = d(7) = 4$, $d(2) = 3$, $d(4) = 5$ and $d(6) = 2$. By applying the algorithm for obtaining an upper bound $\beta_\Delta(G)$ of the chromatic number, we obtain

$$X_1 = \{4, 1\}, \ X_2 = \{3, 7, 2\} \text{ and } X_3 = \{5, 6\},$$

hence $\beta_\Delta(G) = 3$ and $\gamma(G) \leqslant 3$. The same result is obtained if we choose the following four-class partition made up of independent sets:

$$X_1 = \{1, 4\}, \ X_2 = \{5, 2\}, \ X_3 = \{7, 3\} \text{ and } X_4 = \{6\}.$$

We obtain $m_1 = 5$, $m_2 = 4$, $m_3 = 4$, $m_4 = 2$ and, by applying proposition 3,

$$\gamma(G) \leqslant \max(\min(1, 6), \min(2, 5), \min(3, 5), \min(4, 3)) = \max(1, 2, 3, 3) = 3.$$

Let us determine the maximal independent sets with the aid of the Bednarek—Taulbee algorithm:

$L_1 = \{1\}$, $I_1 = I_1' = \varnothing$, $L_2 = \{1\}$, $\{2\}$, $I_2 = I_2' = \varnothing$, $L_3 = \{1\}$, $\{2\}$, $\{3\}$, $I_3 = I_3' = \{1\}$, $L_4 = \{1, 4\}$, $\{2\}$, $\{3\}$, $I_4 = I_4' = \{2\}$; $L_5 = \{1, 4\}$, $\{2, 5\}$, $\{3\}$, $I_5 = I_5' = \{1\}$, $\{2, 5\}$, $\{3\}$; $L_6 = \{1, 4\}$, $\{1, 6\}$, $\{2, 5, 6\}$, $\{3, 6\}$; $I_6 = I_6' = \{2\}$, $\{3\}$; $L_7 = \{1, 4\}$, $\{1, 6\}$, $\{2, 5, 6\}$, $\{2, 7\}$, $\{3, 6\}$, $\{3, 7\}$.

Starting from the family L_7 of maximal independent sets of the graph, the pile constructed as indicated in the algorithm for the determination of the chromatic number, has the following states, where the levels are numbered from top to bottom:

1)	
2a)	{1, 4}, {1, 6}, {2, 5, 6}, {2, 7}, {3, 6}, {3, 7}
2b)	* {1, 4}, {1, 6}, {2, 5, 6}, {2, 7}, {3, 6}, {3, 7}
2c)	** {1, 4}, {1, 6}, {2, 5, 6}, {2, 7}, {3, 6}, {3, 7}
3a)	{2, 5, 6}, {2, 7}, {3, 6}, {3, 7}
	** {1, 4}, {1, 6}, {2, 5 ,6}, {2, 7}, {3, 6}, {3, 7}
3b)	* {2, 5, 6}, {2, 7}, {3, 6}, {3, 7}
	** {1, 4}, {1, 6}, {2, 5, 6}, {2, 7}, {3, 6}, {3, 7}
3c)	** {2, 5, 6}, {2, 7}, {3, 6}, {3, 7}
	** {1, 4}, {1, 6}, {2, 5, 6}, {2, 7}, {3, 6}, {3, 7}
3d)	The pile is the same as at 3c); at this moment the partition {1, 4}, {2, 5, 6}, {3, 7} is printed
4)	** {1, 4}, {1, 6}, {2, 5, 6}, {2, 7}, {3, 6}, {3, 7}
5)	

The pile is empty at states 1) and 5). Since the maximal class, obtained at 3d) by elimination of the vertices 2, 5, 6 from the other sets belonging to the first level, is unique, namely {3, 7}, this set was printed together with all the sets of the pile marked with two stars at that moment. This led to following partition of the set of vertices into independent sets:

$$\{1, 4\}, \ \{2, 5, 6\}, \ \{3, 7\}.$$

Since there is a unique partition and it contains three classes, it follows that the chromatic number of the graph is equal to 3 and the printed partition realizes a colouring of the vertices of the graph G with the smallest number of colours. In the case in which several partitions are printed, we must choose the partition with the smallest number of classes.

This algorithm has been programmed in FORTRAN IV for an IBM 360/30 computer; the program contains 367 instructions among which the first 150 determine the maximal independent sets of the graph with the aid of the Bednarek-Taulbee algorithm [48]. The program prints the various partitions which consist of independent sets, as well as the chromatic number of the graph.

Two examples were solved, namely a graph with 20 vertices and chromatic number 6, for which a single minimum partition was printed (duration: 3 minutes) and a graph with 40 vertices, 652 edges and chromatic number 13, for which 7 partitions were printed, among which were 4 minimum partitions, the other 3 partitions having 14 classes each (duration: 5 minutes). The 13-class partitions are the following:

38	39			28	39	40		38	39	40		38	39	40	
35	36	37		22	23	24		35	36	37		35	36	37	
32	33	34		25	26	27		22	23	24		32	33	34	
29	30	31		28	29	30		25	26	27		22	23	24	
26	27	28		31	32	33		28	29	30		25	26	27	
21	23	24	25	21	34	35	36	21	31	32	33	21	28	29	30
20	22	40		18	19	20		18	19	20		18	19	20	
17	18	19		17	37			15	16	17		15	16	17	
14	15	16		14	15	16		14	34			12	13	14	
11	12	13		11	12	13		11	12	13		11	31		
8	9	10		8	9	10		8	9	10		8	9	10	
5	6	7		5	6	7		5	6	7		5	6	7	
1	2	3	4	1	2	3	4	1	2	3	4	1	2	3	4

The 40-vertex graph is the loop-free graph complementary to that in Fig. 13.7. For this graph the upper bound obtained by D. Welsh and M. Powell for the chromatic number is $\alpha(G) = 34$, whereas the partition Δ constructed as above has 14 classes, therefore $\beta_\Delta (G) = 14$.

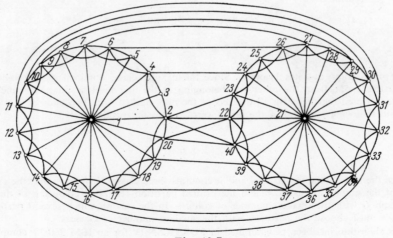

Fig. 13.7

However, there exist 9 edges which can be adjoined to the graph of Fig. 13.7 in such a way that the computer will print more than 200 partitions for the complementary graph with the upper bound 14.

Whereas the problem was solved by the computer in five minutes, the writing of the matrix with 1600 binary components lasted more than one hour.

In the case of problems of greater dimensions, the use of some listing procedures is necessary, for instance, a certain subroutine which searches in the list for the indices and the values of the non-zero elements of a square matrix.

Similarly one can write a program for the solution of any covering problem with the aid of algorithm 2 (without checking the condition at step β).

In order to avoid the printing of all the partitions associated with the paths of the rooted tree, one can proceed as follows: determine first the smallest number of sets in a covering, then explore again the rooted tree and print a partition (or all the partitions) with this smallest number of classes.

As concerns the number of paths of the rooted tree which must be explored in the suggested algorithm, this number may equal $\left[\dfrac{n}{2}\right]!$ for a graph with n vertices. The same result is valid for the number of solutions that must be compared in any of the known algorithms which determine exactly the chromatic number of a graph by enumerating a set of coverings which includes all minimal coverings.

Indeed, consider a graph G with n vertices consisting of two complete subgraphs G_1 and G_2 that contain $\left[\dfrac{n}{2}\right]$ and $\left\{\dfrac{n}{2}\right\}$ vertices, respectively, a vertex being linked by edges only to the vertices which belong to the same complete subgraph.

This graph has the chromatic number $\left\{\dfrac{n}{2}\right\}$, while the rooted tree A_E^* constructed in the above algorithm has the property that the vertices situated at the same distance from the root have the same outdegrees $\Big($ namely, outdegrees equal to $\left[\dfrac{n}{2}\right], \left[\dfrac{n}{2}\right] - 1, ..., 2, 1, 0$ if n is even and equal to $\left[\dfrac{n}{2}\right], \left[\dfrac{n}{2}\right] - 1, ..., 2, 1, 1, 0$ if n is odd $\Big)$ when we run along any path from the root to the terminal node. All the paths of this rooted tree have length $\left\{\dfrac{n}{2}\right\}$ and their number is

$$1 \cdot 2 \cdot ... \cdot \left(\left[\frac{n}{2}\right] - 1\right)\left[\frac{n}{2}\right] = \left[\frac{n}{2}\right]!.$$

Since to different paths correspond different partitions, the number of all possible partitions of the vertices of the graph into $\left\{\dfrac{n}{2}\right\}$ classes such that two vertices joined by an edge belong to different classes, is greater than or equal to $\left[\dfrac{n}{2}\right]!$, two partitions being considered identical if they are obtained from each other by a permutation of the classes or by a permutation of the elements in a class.

The number of irreducible coverings by maximal independent sets of this graph, which is complementary to the complete bipartite graph $K_{\left\{\frac{n}{2}\right\}, \left[\frac{n}{2}\right]}$ is greater than or equal to the number of surjections of a set with $\left\{\dfrac{n}{2}\right\}$ elements onto a set with $\left[\dfrac{n}{2}\right]$ elements; the latter number is $s_{\left\{\frac{n}{2}\right\}, \left[\frac{n}{2}\right]}$. For any surjection of the vertex set of the graph G_2 with $\left\{\dfrac{n}{2}\right\}$ elements onto the vertex set of the graph G_1 with $\left[\dfrac{n}{2}\right]$ elements determines an irreducible covering of the graph G by maximal independent sets, every set from the covering containing a vertex and its image under the surjection. So we get

$$s_{\left\{\frac{n}{2}\right\}, \left[\frac{n}{2}\right]} \geqslant s_{\left[\frac{n}{2}\right], \left[\frac{n}{2}\right]} = \left[\frac{n}{2}\right]!$$

and $\left[\dfrac{n}{2}\right]!$ is a strict lower bound for every $n \geqslant 6$. In this case the number of irreducible coverings of the graph G by maximal independent sets is equal to the number of mappings of a set with $\left\{\dfrac{n}{2}\right\}$ elements into the family of subsets of a set with $\left[\dfrac{n}{2}\right]$ elements, such that the associated graph contains neither isolated vertices nor chains having 4 vertices (the existence of a chain with at least 4 vertices would contradict the irreducibleness hypothesis). In the case when our means of computation are insufficient for the enumeration of all the solutions, we must obtain certain lower and upper bounds for the chromatic number of the graph and use methods giving approximate solutions, as above.

In the sequel we indicate a few applications of the chromatic number of a graph in certain optimization problems.

Consider a production process consisting of n activities $A_1, A_2, ..., A_n$, which are associated with an undirected graph with vertices $x_1, ..., x_n$, vertices x_i and x_j being linked by an edge if and only if the activities A_i and A_j cannot be carried out simultaneously. The problem of determining a minimum chromatic decomposition of this graph is precisely the problem of organizing the production process in a minimum number of stages such that all the activities of the same stage are performed simultaneously [39]. The problem of optimizing time-tables in schools can be given a similar formulation [57].

Another problem which reduces to the determination of the chromatic number of a graph is the following [37]: determine the minimum number of channels that can be used by n TV stations, if two neighbouring stations cannot use the same channel. We introduce a graph, the vertices of which represents the n stations.

Two vertices x_i and x_j are joined by an edge if the stations i and j cannot use the same channel. In this case an independent set of the graph represents a set of stations which can use the same channel.

As another example, consider the storage problem [10]. Certain goods (chemicals, say) are to be stored in several compartments, but, for safety reasons certain chemicals cannot be places in the same compartment; determine the smallest number of compartments in which the chemicals can be stored under the above conditions. In this case we must find the chromatic number of the graph, the vertices of which represent the chemicals, with two vertices linked by an edge if the corresponding chemicals cannot be stored in the same compartment.

Problems

1.   A cover of a set S is a collection of different non-empty subsets of S, the union of which is S.

Let $C(n)$ denote the number of covers of a set with finite cardinality n.

Prove that the numbers $C(n)$ are generated by the recurrence:

$$C(n) = 2^{2^n - 1} - 1 - \sum_{j=1}^{n-1} \binom{n}{j} C(j)$$

and using the inverse binomial formulae prove that

$$C(n) = \sum_{j=0}^{n} (-1)^j \binom{n}{j} 2^{2^{n-j} - 1}.$$

(Comtet, 1966)

H i n t: Use the fact that any collection of non-empty subsets of S covers some subset of S.

2.   A cover C of a set S consisting of k non-empty subsets of S will be called a k-member irreducible cover of S if the union of any $k - 1$ subsets from C is a proper subset of S.

Let $M(n, k)$ enumerate the k-member irreducible covers of an n-set.

Prove that

$$M(n, k) = \sum_{i=k}^{n} \binom{n}{i} (2^k - k - 1)^{n-i} S(i, k)$$

where $S(i, k)$ is the Stirling number of the second kind.

From elementary properties of the Stirling numbers deduce that:

$$M(n, n - 1) = \frac{1}{2} n(2^n - n - 1),$$

$$M(n, 2) = S(n + 1, 3).$$

(Hearne, Wagner, 1973)

3.   Prove that the chromatic number of a graph satisfies

$$\gamma(G) \leqslant 1 + \max_{G' \subset G} \min_{v \in G'} d(v)$$

where $d(v)$ is the degree of the vertex x and $G' \subset G$ means that G' is an induced subgraph of G.

(Szekeres, Wilf, 1968)

4.   The degree of balance of a nonoriented graph G, denoted by $\delta(G)$, is the smallest number of edges to be deleted or inserted between the vertices of G so that G becomes a union of two disjoint cliques C_1 and C_2 (C_1 may be empty) with the property that there is no edge joining a vertex of C_1 to a vertex of C_2.
      Prove Abelson-Rosenberg's conjecture:

$$\max \delta(G) = \left\{ \frac{n(n-2)}{4} \right\}$$

where the maximum is taken over all graphs with n vertices and the graphs attaining this maximum are bipartite graphs $K_{p,\,q}$ with $p, q \geqslant 0, p + q = n$.

(Tomescu, 1973)

H i n t: Use induction on n.

5.   A collection $F \subset \mathscr{P}_r(X)$ where $|X| = n$ is called a uniform cover of weight λ for the q-subsets of X, $q \geqslant r$, if each q-set includes the same number λ of r-sets from F. F is a minimal uniform cover if no proper subset of F is a uniform cover and is called non-trivial if F is itself a proper subset of the family of r-sets.
      Prove that there are no non-trivial minimal uniform covers for $q > r$, if the numbers $\binom{n}{r}$ and $\binom{q}{r}$ are relatively prime.

(Simmons, 1972)

H i n t: The number of ways an r-set can be extended to be a q-set is equal to $\binom{n-r}{q-r}$. Derive from this that $\lambda \binom{n}{q} = |F| \binom{n-r}{q-r}$, i.e. $\lambda \binom{n}{r} = |F| \binom{q}{r}$, where $|F| \leqslant \binom{n}{r}$.

6.   For a graph G let $A = \{a_{ij}\}_{i,\,j=1,\,\ldots,\,n}$ be its adjacency matrix. Every set of vertices S can be described by its characteristic vector $(x_1, \ldots, x_n)$ defined as follows: for each $i = 1, 2, \ldots, n$, $x_i = 1$ if the vertex indexed by i belongs to S and $x_i = 0$ otherwise.
      Prove that:

i) A set S of vertices is internally stable (or independent) if and only if its characteristic vector satisfies the Boolean equation

$$\bigcup_{i=1}^{n} \bigcup_{j=i}^{n} (a_{ij} \cup a_{ji}) x_i x_j = 0.$$

(Maghout, 1963)

ii) A set S of vertices is a maximal internally stable set of vertices if and only if its characteristic vector satisfies the system of Boolean equations

$$x_i = \bar{a}_{ii} \prod_{\substack{j=1 \\ j \neq i}}^{n} (\bar{a}_{ij}\,\bar{a}_{ji} \cup \bar{x}_j) \quad (i = 1, 2, ..., n).$$

(Rudeanu, 1966)

7.   If $\bar{G}$ is the complementary graph of a graph G with n vertices, prove that:

$$2\sqrt{n} \leqslant \gamma(G) + \gamma(\bar{G}) \leqslant n + 1,$$

$$n \leqslant \gamma(G)\,\gamma(\bar{G}) \leqslant \left[\left(\frac{n+1}{2} \right)^2 \right].$$

(Nordhaus, Gaddum, 1956)

H i n t: Prove by induction on n that $\gamma(G) + \gamma(\bar{G}) \leqslant n + 1$. In order to show that $\gamma(G)\gamma(\bar{G}) \geqslant n$ use the inequalities $\gamma(G)\,s(G) \geqslant n$ and $\gamma(\bar{G}) \geqslant s(G)$, the number of internal stability of G. Apply the inequality $4ab \leqslant (a+b)^2$ also.

8.   For a graph G having n vertices and m edges prove that its number of internal stability $s(G)$ satisfies

$$s(G) \geqslant \frac{n^2}{2m+n}.$$

H i n t: Applying Turán's theorem show that Proposition 6 gives a lower bound for the maximum number of vertices $c(G)$ of a complete subgraph of G. Pass to the complementary graph $\bar{G}$.

9.   Show that the minimum number of elementary chains which cover all the vertices of a tree T is equal to $\left\{ \dfrac{p}{2} \right\}$, where p is the number of endvertices of T.

(Tomescu, 1970)

10.   Prove that for n odd the edges of the complete graph K_n may be covered by $\dfrac{n-1}{2}$ edge-disjoint Hamiltonian cycles.

H i n t: If the vertices of K_n are $0, 1, 2, ..., 2k$, let us consider the following Hamiltonian cycle

$$[0, 1, 2, 2k, 3, 2k - 1, 4, 2k - 2, ..., k + 3, k, k + 2, k + 1, 0].$$

Add 1 modulo $2k$ to all nonzero numbers from this sequence $k - 1$ times.

11.   Let P be a partially ordered set. A subset S of P will be called a chain if any two elements in S are comparable; it will be called an antichain if no two distinct elements in S are comparable. If m is a natural number, prove that if P possesses no chain of cardinal $m + 1$, then it can be expressed as the union of m antichains. Deduce from this Erdös-Szekeres' result (Chap. I).

(Mirsky, 1971)

H i n t: Use induction with respect to m.

BIBLIOGRAPHY

1. Abelson, R., Rosenberg, M., *Symbolic psycho-logic: A model of attitudinal cognition*, Behavioral Sci., **3**, 1958, 1—13.
2. Bazerque, G., *Implantation des arborescences: Applications aux tableaux*, Rev. Française Informat. Recherche Opérationnelle, B—3, 1970, 27—107.
3. Bednarek, A., Taulbee, O., *On maximal chains*, Rev. Roumaine Math. Pures Appl., **11**, 1966, 23—25.
4. Bertier, P., Roy, B., *Procédure de résolution pour une classe de problèmes pouvant avoir un caractère combinatoire*, Cahiers Centre Études Recherche Opér., Bruxelles, **6**, 1964, 202—208.
5. Breuer, M., *Simplification of the covering problem with application to Boolean expressions*, J. Assoc. Comput. Mach., **17**, 1970, 166—181.
6. Brooks, R., *On colouring the nodes of a network*, Proc. Cambridge Philos. Soc., **37**, 1941, 194—197.
7. Carvallo, M., *Principes et applications de l'analyse booléenne*, Gauthier-Villars, Paris, 1965.
8. Chinal, J., *Techniques booléennes et calculateurs arithmétiques*, Dunod, Paris, 1967.
9. Choudhury, A., Basu, A., Desarkar, S., *On the determination of the maximum compatibility classes*, IEEE Trans. Computers, CT—18, 1969, 665.
10. Christophides, N., *An algorithm for the chromatic number of a graph*, Comput. J., **14**, 1971, 38—39.
11. Comtet, L., *Recouvrements, bases de filtre et topologies d'un ensemble fini*, C. R. Acad.Sci. Paris, Sér. A, **262**, 1966, 1091—1094.
12. Constantinescu, P., *The classification of a set of elements with respect to a set of properties*, Comput. J., **8**, 1966, 352—357.
13. Das, S., Sheng, C., *On finding maximum compatibles*, Proc. IEEE, **57**, 1969, 694.
14. Derniame, J., *Étude d'algorithmes pour les problèmes de cheminement dans les graphes finis*, Thèse, Fac. Sciences, Nancy, 1966.
15. Descartes, B., (W. Tutte's pseudonym), *Solution of advanced problem 4526*, Amer. Math. Monthly, **61**, 1954, 352.
16. Dodu, J. C., Ludot, J. P., Pouget, J., *Sur le regroupement optimal des sommets dans un réseau électrique*, Rev. Française Informat. Recherche Opérationnelle, V—1, 1969, 17—37.
17. Drăghici, M., *Au sujet d'un problème de classification sur un niveau*, Bull. Math. Soc. Sci. Math., R. S. Roumanie, **12** (60), 1968, 65—74.

18. Ershov, A., Kuzhukhin, G., *Estimates of the chromatic number of connected graphs*, Dokl. Akad. Nauk SSSR, **142**, 1962, 270−273.
19. Genuys, F., *Commentaires sur le langage Algol*, Rev. Chiffres, **5**, 1962, 29.
20. Hammer, P., Rudeanu, S., *Boolean methods in operations research and related areas*, Springer, Berlin, 1968.
21. Harary, F., Ross, I., *A procedure for clique detection using the group matrix*, Sociometry, **20**, 1957, 205−215.
22. Hearne, T., Wagner, C., *Minimal covers of finite sets*, Discrete Math., **5**, 1973, 247−251.
23. Ilzinia, I., *On finding the cliques of a graph* (in Russian), Avtom. i Vyčisl. Tehn. (Riga), **2**, 1967, 7−11.
24. Kaufmann, A., *Introduction à la combinatorique en vue des applications*, Dunod, Paris, 1968.
25. Lawler, E., Wood, D., *Branch and bound methods: a survey*, Operations Res. **14**, 1966, 699−719.
26. Maghout, K., *Sur la détermination des nombres de stabilité et du nombre chromatique d'un graphe*, C. R. Acad. Sci. Paris, **248**, 1959, 3522−3523.
27. Maghout, K., *Application de l'algèbre de Boole à la théorie des graphes et aux programmes linéaires et quadratiques*, Cahiers Centre Études Recherche Opér., **5**, 1963, 21−99.
28. Malgrange, Y., *Recherches des sous-matrices premières d'une matrice à coefficients binaires: Applications à certains problèmes de graphe*, Deuxième Congrès de l'AFCALTI, Gauthier-Villars, Paris, 1962, 231−242.
29. Mirsky, L., *A dual of Dilworth's decomposition theorem*, Amer. Math. Monthly, **78**, 1971, 876−877.
30. Moon, J. W., Moser, L., *On cliques in graphs*, Israel J. Math., **3**, 1965, 23−28.
31. Nordhaus, E. A., Gaddum, J. W., *On complementary graphs*, Amer. Math. Monthly, **63**, 1956, 176−177.
32. Pair, C., *Justification théorique de l'utilisation des piles en compilation*, C. R. Acad. Sci., Paris, **257**, 1963, 3278−3281.
33. Pair, C., *Étude de la notion de pile. Application à l'analyse syntaxique*, Thèse, Fac. Sciences, Nancy, 1966.
34. Paull, M. C., Unger, S. H., *Minimizing the number of states in incompletely specified sequential switching functions*, IRE Trans. Electronic Computers, EC−8, 1959, 356−367.
35. Petrick, S. R., *A direct determination of the irredundant forms of a Boolean function from the set of prime implicants*, AFCRC-TR-56-110, Air Force Cambridge Research Center, 1956.
36. Pichat, E., *Contribution à l'algorithmique non numérique dans les ensembles ordonnés*, Thèse, Grenoble, 1970.
37. Read, R. C., *An introduction to chromatic polynomials*, J. Combinatorial Theory, **4**, 1968, 52−71.
38. Roth, R., *Computer solutions to minimum-cover problems*, Operations Res., **17**, 1969, 455.
39. Roy, B., *Algèbre moderne et théorie des graphes (orientés vers les sciences économiques et sociales)*, Vol. II, Dunod, Paris, 1970.
40. Rudeanu, S., *Note sur l'existence et l'unicité du noyau d'un graphe* II, Rev. Française Recherche Opérationnelle, **10**, 1966, 301−310.
41. Rudeanu, S., *On solving Boolean equations in the theory of graphs*, Rev. Roumaine Math. Pures Appl., **11**, 1966, 653−664.
42. Rudeanu, S., *Boolean equations for the chromatic decomposition of graphs*, An. Univ. Bucureşti Mat.-Mec., **18**, 1969, 119−126.
43. Schneider, M., *Méthodes pour récenser toutes les cliques maximales et θ-maximales d'un graphe*, Rev. Française Automat. Informat. Recherche Opérationnelle, V-3, 1973, 21−33.
44. Simmons, G. J., *An extremal problem in the uniform covering of finite sets*, J. Combinatorial Theory (A), **13**, 1972, 183−197.
45. Szekeres, G., Wilf, H., *An inequality for the chromatic number of a graph*, J. Combinatorial Theory, **4**, 1968, 1−3.
46. Tomescu, I., *Sur un problème de partition avec un nombre minimal de classes*, C. R. Acad. Sci. Paris, Sér. A, **265**, 1967, 645−648.
47. Tomescu, I., *Sur le problème du coloriage des graphes généralisés*, C. R. Acad. Sci. Paris, Sér. A, **267**, 1968, 250−252.
48. Tomescu, I., *Méthodes combinatoires dans la théorie des automates finis*, Logique, Automatique, Informatique, Ed. Acad. R.S.R., Bucureşti, 1971, 269−423.

49. Tomescu, I., *An algorithm for determining the chromatic number of a finite graph*, Econom. Comp. Econom. Cybernet. Stud. Res. (Bucharest), **1**, 1969, 69—81.
50. Tomescu, I., *Le nombre des graphes connexes k-chromatiques minimaux aux sommets étiquetés*, C. R. Acad. Sci. Paris, **273**, 1971, 1124—1126.
51. Tomescu, I., *Une caractérisation des graphes k-chromatiques minimaux sans sommet isolé*, Rev. Française Automat. Informat. Recherche Opérationnelle, R—1, 1972, 88—91.
52. Tomescu, I., *Note sur une caractérisation des graphes dont le degré de déséquilibre est maximal*, Math. Sci. Humaines, **42**, 1973, 37—40.
53. Tomescu, I., *A combinatorial algorithm for solving covering problems*, IEEE Trans. Computers, C—22, 1973, 218—220.
54. Tomescu, I., *Une démonstration du théorème de Dilworth et applications à un problème de couverture des graphes*, Calcolo (Pisa), **7**, 1970, 289—294.
55. Vitaver, L., *The determination of the minimal colourings of the vertices of a graph with the aid of the Boolean powers of the adjacency matrix* (in Russian), Dokl. Akad. Nauk SSSR, **147**, 1962, 758—759.
56. Weissman, J., *Boolean algebra, map coloring and interconnections*, Amer. Math. Monthly, **69**, 1962, 608—613.
57. Welsh, D. J., Powell, M., *An upper bound for the chromatic number of a graph and its application to timetabling problems*, Comput. J., **3**, 1967, 85—86.
58. Wilkov, R., Kim, W., *A practical approach to the chromatic partition problem*, J. Franklin Inst., **289**, 1970, 333—349.
59. Wood, D. C., *A technique for colouring a graph applicable to large scale timetabling problems*, Comput. J., **12**, 1969, 317—319.
60. Zykov, A. A., *On certain properties of linear complexes* (in Russian), Mat. Sb., **24**, 1949, 163—188.

CHAPTER 14

Maximum and Minimum Numbers of Colourings of a Graph

In this chapter we obtain the maximum and minimum numbers of colourings of a graph; we recall that a colouring is a partition of the vertex set such that each class of the partition contains nonadjacent vertices only.

In the previous chapter we have defined the chromatic number $\gamma(G)$ of the graph G as the smallest number of classes of a colouring.

This problem has a combinatorial character, as will be seen in the sequel.

PROPOSITION 1. *The greatest number of minimum colourings of an n-vertex graph having the chromatic number k, is equal to k^{n-k}.*

The only graph which reaches this maximum number of colourings consists of a complete k-subgraph and $n - k$ isolated vertices.

Here the vertices of the graph are labelled but the order of the classes in a colouring is immaterial.

The proof is by induction on the number n of vertices. For $n = 1, 2$ the result is immediate. Now assume the property to be true for all graphs with at most $n-1$ vertices.

Let G be a graph with n vertices and chromatic number $k(1 \leqslant k \leqslant n)$. Let x be a vertex of the graph and denote by G_x the subgraph obtained by removing vertex x; also let $C_m(G)$ stand for the number of minimum colourings of the graph G. If $\gamma(G_x) = k$, we obtain

$$C_m(G) \leqslant k C_m(G_x) \leqslant k k^{n-k-1} = k^{n-k},$$

the inequalities becoming equalities only in the case when x is an isolated vertex and the subgraph G_x has the maximum number of minimum colourings, because the vertex x can be adjoined to a minimum partition of the graph G_x in at most k different ways.

If $\gamma(G_x) = k - 1$, a minimum colouring of G is the following: $\{x\}, C_1, ..., C_{k-1}$, where the sets $C_1, ..., C_{k-1}$ are independent sets of vertices and there exist $k - 1$ vertices $x_1 \in C_1, ..., x_{k-1} \in C_{k-1}$ each of which is linked by an edge to x, because in the contrary case $\gamma(G) = k - 1$.

If X is the set of vertices of the graph G, every minimum covering of G has a class which contains the vertex x and includes a subset of $X \setminus \{x_1, ..., x_{k-1}\}$. But $|X \setminus \{x, x_1, ..., x_{k-1}\}| = n - k$, hence the number of minimum partitions of X which contain in the same class the vertex x and r vertices from the set $X \setminus \{x, x_1, ..., x_{k-1}\}$, where $0 \leqslant r \leqslant n - k$, has the upper bound $\binom{n-k}{r}(k-1)^{n-k-r}$, by virtue of the inductive hypothesis. For r vertices can be chosen from of the $n - k$ vertices in $\binom{n-k}{r}$ different ways and the greatest number of minimum partitions of a graph G with $n - k - r$ vertices and $\gamma(G) = k - 1$ is equal to $(k-1)^{n-k-r}$. Therefore

$$C_m(G) \leqslant \sum_{r=0}^{n-k} \binom{n-k}{r} (k-1)^{n-k-r} = k^{n-k}$$

and equality holds only in the case when $\{x, x_1, ..., x_{k-1}\}$ is a complete k-subgraph and the remainder of the vertices are isolated. We have thus proved by induction that $C_m(G) \leqslant k^{n-k}$ for every graph G with n vertices and chromatic number k, the maximum being reached only in the case in which G consists of a complete k-subgraph and $n - k$ isolated vertices. Notice that this is the only graph with $\gamma(G) = k$ and the smallest number $\binom{k}{2}$ of edges.

COROLLARY. *The greatest number of minimum colourings of a graph with n vertices equals* $\max_{r=[x],\{x\}} (r^{n-r})$, *where x is the real number which satisfies the equation* $x(1 + \ln x) = n$.

The greatest number of minimum colourings of an n-vertex graph equals $\max_{k=1,...,n} (k^{n-k})$ and the equation $x(1 + \ln x) = n$ is obtained by equating to zero the derivative of the function x^{n-x}. This number is also equal to the maximum number of equivalence relations $R^* \subset R \subset X \times X$ with the smallest number of classes, where $|X| = n$ and R is a symmetric and reflexive binary relation. The maximum number of minimum colourings of a graph with n vertices increases very rapidly with n, as seen from Table 14.1.

TABLE 14.1

n	max $C_m(G)$	n	max $C_m(G)$
1	1	9	1 024
2	1	10	4 096
3	2	11	16 384
4	4	12	78 125
5	9	13	390 625
6	27	14	1 953 125
7	81	15	10 077 696
8	256	16	60 466 176

A $(k + r)$-colouring of a graph G with n vertices and chromatic number k is a partition of its set of vertices into $k + r$ classes, where $1 \leqslant r \leqslant n - k$, such that two vertices which belong to the same class are not joined by an edge.

From example 2, Chap. 4, it follows that the number $C(n, k, r)$ of $(k + r)$-colourings of a graph consisting of a complete k-subgraph and $n - k$ isolated vertices is given by $C(n, k, r) = \sum_{p=r}^{n-k} \binom{n - k}{p} S(p, r) k^{n-k-p}$ for every r with $1 \leqslant r \leqslant n - k$, where the $S(p, r)$ are Stirling numbers of the second kind. The numbers $C(n, k, r)$ satisfy the recurrence relations

$$C(n, k, r) = C(n - 1, k, r - 1) + (k + r) C(n - 1, k, r)$$

and   (14.1)

$$C(n, k, r) = \sum_{q=0}^{n-k-r} \binom{n - k}{q} C(n - 1 - q, k - 1, r)$$

for $n > k$ and $1 \leqslant r \leqslant n - k$. The former recurrence relation is obtained from the number of colourings of a graph consisting of a complete k-subgraph and $n-k-1$ isolated vertices by adjoining an isolated vertex, while the latter relation is obtained from the number of colourings of a graph consisting of a complete $(k - 1)$-subgraph and $n - k$ isolated vertices by adjoining a new vertex which is linked to all the vertices of the complete $(k - 1)$-subgraph, because in this way we obtain without repetitions all $(k + r)$-colourings, the number of which has been denoted by $C(n, k, r)$. It was shown in [9], with a proof similar to that of proposition 1 and taking into account (14.1), that the greatest number of $(k + r)$-colourings of a graph G with n vertices and chromatic number k, is equal to $C(n, k, r)$ for every $1 \leqslant r \leqslant n - k$ and the only graph with this maximum number of colourings is the graph which consists of a complete k-subgraph and $n - k$ isolated vertices. Obviously $C(n, k, r) = 0$ for every $r > n - k$, since there is no partition of the set of n vertices into more than n classes, because any class is by definition non-empty.

The *chromatic polynomial* $P(G; \lambda)$ of a graph $G = (X, U)$ is a polynomial in λ, the value of which for $\lambda = \lambda_0$, i.e. $P(G; \lambda_0)$, represents the number of colourings of the vertices of the graph with λ_0 colours, that is, the number of mappings $f : X \to \{1, 2, ..., \lambda_0\}$ such that $[x, y] \in U \Rightarrow f(x) \neq f(y)$ [6].

The distinction between these colourings and those used so far consists of the fact that in the present definition two partitions differ also in respect of the order of classes, each $(k + r)$-class partition in the previous sense generating $(k+r)!$ partitions in the new sense. For example, the chromatic polynomial of a complete k-graph is equal to $\lambda(\lambda - 1) ... (\lambda - k + 1)$, since the colour of the first vertex can be chosen from λ colours, the colour of the second vertex can then be chosen from $\lambda - 1$ colours, ..., the colour of the last vertex can be chosen from the $\lambda - - k + 1$ remaining colours. Obviously $P(G; \lambda) = 0$ for every $\lambda < \gamma(G)$. If G_n stands for a graph with n vertices, then

$$\max_{\gamma(G_n)=k} P(G ; \lambda) = \lambda(\lambda - 1) ... (\lambda - k + 1)\lambda^{n-k}$$

for every natural number λ and for every $\lambda \geqslant k$ the maximum is reached only for that graph with n vertices and chromatic number k which consists of a complete k-subgraph and $n - k$ isolated vertices. Indeed, for every $\lambda < \gamma(G_n)$ we obtain $P(G_n; \lambda) = 0$ and for $\lambda \geqslant \gamma(G_n)$ we get

$$P(G_n; \lambda) = \sum_{j=\gamma(G_n)}^{\min(n, \lambda)} j! \binom{\lambda}{j} C_j(G_n) \leqslant \sum_{j=\gamma(G_n)}^{\min(n, \lambda)} j! \binom{\lambda}{j} C(n, \gamma(G_n), j - \gamma(G_n))$$

$$= \lambda(\lambda - 1)...(\lambda - k + 1)\lambda^{n-k}$$

where $\gamma(G_n) = k$ and $C_j(G_n)$ stands for the number of colourings with j classes, interpreted as partitions of G_n for which the order of classes is immaterial. For j colours can be chosen from the λ colours in $\binom{\lambda}{j}$ different ways and a partition into j classes generates $j!$ colourings in the new sense (which takes into account the order of classes), while the chromatic polynomial of a graph consisting of a complete k-subgraph and $n - k$ isolated vertices equals $\lambda(\lambda - 1) ... (\lambda - k + 1)\lambda^{n-k}$.

We now return to the previous definition of a colouring, as being a partition of the set of vertices into independent sets, for which the order of the classes is immaterial. We shall determine the greatest number of 3-colourings of a connected graph G for which $\gamma(G) = 2$ and $\gamma(G) = 3$, respectively.

A graph G is said to be connected if for every pair of vertices there is at least one chain which connects them. Proposition 1 is not restricted to connected graphs, because the graph with the maximum number of colourings is not connected (as it contains isolated vertices) unless $n = k$.

Let $\text{Col}_3(G)$ stand for the number of 3-colourings of a graph G.

PROPOSITION 2. *If the graph G with $n \geqslant 3$ vertices is connected and has chromatic number $\gamma(G) = 2$, then $\text{Col}_3(G) \leqslant 2^{n-2} - 1$. The only connected graphs having the maximum number $2^{n-2}-1$ of 3-colourings are the trees with n vertices.*

The proof is by induction on n. For $n = 3$ the proposition is obvious. Assume it to be true for n.

A connected graph G with $n + 1$ vertices ($n + 1 \geqslant 3$) has at least two vertices that are not separating vertices [1], i.e. vertices with the property that the deletion of any of them results in a subgraph which is not connected.

Let x be such a vertex and G_x the connected graph obtained from G by removal of x. A connected graph G having chromatic number 2 has a unique colouring with 2 classes, say C_1 and C_2, as we now show: since the graph G is connected, it follows that if C_1 stands for the class of an arbitrary but fixed vertex y, the vertices joined by edges to y will belong to the class C_2, the vertices linked by edges to the vertices of C_2 will belong to the class C_1, etc. Hence

$$\text{Col}_3(G) \leqslant 1 + 2\,\text{Col}_3(G_x) \leqslant 1 + 2(2^{n-2} - 1) = 2^{n-1} - 1,$$

because there is a unique 3-colouring of G which has a class consisting of the single vertex x, since G_x has a unique 2-colouring and vertex x can be adjoined to a 3-colouring of G_x in at most two different ways. This is because the graph is connected, so that x is linked by an edge to at least one vertex of G_x which belongs to a class of a 3-colouring of G_x.

Notice also that the equality holds only in the case in which x is linked to exactly one vertex of the graph G_x (which is a tree), so that G is a tree as well.

For example, given an elementary chain $[x_1, x_2, ..., x_n]$, if x_1 and x_2 are in distinct classes, there are 2 possibilities for incorporating x_3 into one of the two classes which do not contain x_2, there are 2 possibilities for including x_4 in one of the two classes which do not contain x_3, etc., thus there are $2^{n-2} - 1$ different colourings with 3 classes.

PROPOSITION 3. *The number of 3-colourings of an odd cycle C_{2k+1} with $2k+1$ vertices equals $\dfrac{1}{3}(4^k - 1)$.*

Notice first that an odd cycle has chromatic number 3 and $\mathrm{Col}_3(C_{2k+1}) = 2N_1(k) + N_2(k) + 1$, if $N_1(k)$ is the number of those 3-colourings of the elementary chain $[x_1, ..., x_{2k}]$ which include the vertices x_1 and x_{2k} in the same class, while $N_2(k)$ is the number of those 3-colourings of the elementary chain $[x_1, ..., x_{2k}]$ for which the vertices x_1 and x_{2k} belong to different classes. The vertex x_{2k+1} of the cycle C_{2k+1} is connected only to vertices x_1 and x_{2k}, hence the number of 3-colourings of C_{2k+1} which have a class consisting of x_{2k+1} alone is 1, while the number of 3-colourings of C_{2k+1} which do not have the above property is $2N_1(k) + N_2(k)$. We have taken into account that the vertex x_{2k+1} can be added to any of the $N_1(k)$ colourings of the chain in two ways and the same vertex can be added to any of the $N_2(k)$ colourings of the chain in exactly one way, namely in the class which contains neither x_1 nor x_{2k}. But $N_1(k) + N_2(k)$ is the number of 3-colourings of the elementary chain $[x_1, ..., x_{2k}]$, so that the previous proposition implies

$$N_1(k) + N_2(k) = 2^{2k-2} - 1,$$

therefore

$$\mathrm{Col}_3(C_{2k+1}) = 2^{2k-2} + N_1(k).$$

We remark that $N_1(k) = \mathrm{Col}_3(C_{2k-1})$, i.e. the number of 3-colourings of the elementary chain $[x_1, ..., x_{2k}]$ which contain vertices x_1 and x_{2k} in the same class equals the number of 3-colourings of the cycle C_{2k-1} and this is obtained by identifying the vertices x_1 and x_{2k}. We thus obtain

$$\mathrm{Col}_3(C_{2k+1}) = 2^{2k-2} + \mathrm{Col}_3(C_{2k-1}) = 2^{2k-2} + 2^{2k-4} + \mathrm{Col}_3(C_{2k-3}) =$$

$$= ... = 2^{2k-2} + ... + 2^2 + \mathrm{Col}_3(C_3) = \frac{1}{3}(4^k - 1)$$

since $\mathrm{Col}_3(C_3) = 1$.

This result can also be deduced from the fact that the chromatic polynomial of an elementary cycle with n vertices is equal to $(\lambda - 1)^n + (-1)^n (\lambda - 1)$ [6].

PROPOSITION 4. *The greatest number of 3-colourings of a connected graph G with n vertices and chromatic number $\gamma(G) = 3$ is* $\dfrac{1}{3}(2^{n-1} - 1)$ *for odd n and* $\dfrac{2}{3}(2^{n-2} - 1)$ *for even n. If n is odd, the only connected graph for which this maximum number of 3-colourings is reached is the odd polygon C_n, while if n is even, the only graph which realizes the above maximum is the graph consisting of the polygon C_{n-1} and another vertex which is joined by an edge to a vertex of C_{n-1}.*

Set $g(n) = \dfrac{1}{3}(2^{n-1} - 1)$ for odd n and $g(n) = \dfrac{2}{3}(2^{n-2} - 1)$ for even n.

Now the proof is by induction on n. For $n=3$ the single graph G with n vertices and $\gamma(G) = 3$ is the complete 3-graph, for which there is a unique 3-colouring, so that the proposition is verified. Assume the proposition to be true for n. Let G be a graph having $n + 1$ vertices. There exists at leats one vertex x such that the subgraph G_x with n vertices is connected. If $\gamma(G_x) = 3$, then

$$\mathrm{Col}_3(G) \leqslant 2\,\mathrm{Col}_3(G_x) \leqslant 2g(n) \leqslant g(n + 1)$$

because $2g(n) = g(n + 1)$ for odd n and $2g(n) + 1 = g(n + 1)$ for even n and since the graph G is connected, the vertex x is linked by at least one edge to a vertex which belongs to a class of a 3-colouring of G_x, hence there are at most two possibilities for incorporating vertex x into a class of a 3-colouring of G_x. According to the induction hypothesis, the equality between $\mathrm{Col}_3(G)$ and $g(n+1)$ holds only in the case in which n is odd and the vertex x is joined by an edge to a single vertex of the odd cycle C_n.

If $\gamma(G_x) = 2$, there is a 3-colouring of the graph G having the following form: $\{x\}$, H_1, H_2, where H_1 and H_2 are independent sets of vertices.

If the graph G is the complement of a perfect graph, it contains by definition a triangle, i.e. a cycle with 3 vertices: x, u_1, v_1, where $u_1 \in H_1$ and $v_1 \in H_2$. In the contrary case, according to a theorem of Berge [1], it contains an odd elementary cycle $[x, u_1, v_1, u_2, v_2, \ldots, u_k, v_k]$, where $u_i \in H_1$ and $v_i \in H_2$ for $1 \leqslant i \leqslant k$ and $k \geqslant 2$.

Therefore in every case the graph G contains an odd elementary cycle $[x, u_1, v_1, \ldots, u_k, v_k]$ with $u_i \in H_1$, $v_i \in H_2$ and $k \geqslant 1$.

We shall remove those edges which are diagonals of this odd polygon so that the number of 3-colourings will increase. We also notice that the subgraph generated by an odd polygon has chromatic number equal to 3, hence the chromatic number of the graph does not decrease and the graph remains connected.

It follows from the connectedness of the graph G that there is at least one vertex y of the graph which does not belong to this cycle with $2k + 1$ vertices and which is joined by an edge to at least one vertex of the cycle. If several edges exist which join vertex y to vertices of the cycle, we can retain but one edge and the graph thus obtained is also connected and has chromatic number 3, while the number of its 3-colourings is greater. Then we find a vertex z which is linked by an edge to at least one vertex from the set $\{x, y, u_1, v_1, \ldots, u_k, v_k\}$ and we repeat the above construction, etc. Hence

$$\mathrm{Col}_3\,(G) \leqslant 2^{n-2k}\,\mathrm{Col}_3\,(C_{2k+1}) \leqslant g(n+1)$$

since

$$2g(3) = g(4); \ 2g(4) < g(5); \ 2g(5) = g(6);$$

$$2g(6) < g(7); \ \ldots$$

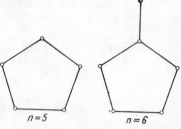

Fig. 14.1

and in view of the connectedness of the graph, for each new vertex there are at most two possibilities for incorporating it into a class of a 3-colouring already obtained.

By using the inductive hypothesis and the definition of the numbers $g(n)$, we deduce that the graph which realizes this maximum number of 3-colourings of a connected graph G with chromatic number $\gamma(G) = 3$ is unique and has the form indicated in the statement of the proposition.

For $n = 5$ and $n = 6$ these graphs are represented in Fig. 14.1.

The smallest number of edges of a connected graph G with n vertices and chromatic number equal to k is $\binom{k}{2} + n - k$, which can be inferred for example from the upper bound (13.6) for the chromatic number of a connected graph in terms of the numbers of vertices and of edges, or it can be proved directly by analogy with the proof of proposition 4, Chap. 13.

For $k = 3$, the minimum number of edges of a connected graph G with n vertices and chromatic number 3 is equal to n, but there is more than one graph realizing this smallest number of edges. One such graph is that in the statement of proposition 4, while another one consists of a triangle and $n - 3$ other vertices each of which is joined by an edge to a vertex of the triangle; this graph has several 3-colourings equal in number to $2^{n-3} < g(n)$ for every $n \geqslant 5$.

These results can be expressed in terms of the chromatic polynomial of a graph G with n vertices, because the number of t-colourings is equal to $\dfrac{P(G;t)}{t!}$ if $\gamma(G) = t$.

It is quite difficult to obtain similar results for connected graphs with $\gamma(G) > 3$, because no characterization is available for the vertex-critical t-chromatic connected graphs, if $t > 3$.

In the sequel we indicate how to obtain the minimum number of colourings of a graph [11]. To this end we shall make use of certain properties of the Stirling numbers of the second kind, given in the following two propositions.

PROPOSITION 5. *If $p \geqslant q$ and $u \geqslant v$, then*

$$S(p, u)S(q, v) \geqslant S(p, v)S(q, u).$$

The proof is by induction on the indices p, q, u, v. The result is valid for $p = u$ or $p = q$ or $q = v$ or $u = v$, while

$$S(p, u) S(q, v) - S(p, v) S(q, u) = S(p - 1, u - 1) S(q - 1, v - 1)$$

$$- S(p - 1, v - 1) S(q - 1, u - 1)$$

$$+ uv(S(p - 1, u) S(q - 1, v) - S(p - 1, v) S(q - 1, u))$$

$$+ u(S(p - 1, u) S(q - 1, v - 1) - S(p - 1, v - 1) S(q - 1, u))$$

$$+ v(S(p - 1, u - 1) S(q - 1, v) - S(p - 1, v) S(q - 1, u - 1)).$$

To obtain this expression, we have used the recurrence relation

$$S(p, q) = S(p - 1, q - 1) + qS(p - 1, q).$$

PROPOSITION 6. *If $p > q$ and $s \geqslant 3$, then the inequality*

$$\sum_{i+j=s} S(p + 1, i) S(q, j) > \sum_{i+j=s} S(p, i) S(q + 1, j)$$

holds.

The proof follows from proposition 5, taking into account the previous recurrence relation for the Stirling numbers of the second kind, since

$$\sum_{i+j=s} S(p + 1, i)S(q, j) - \sum_{i+j=s} S(p, i) S(q + 1, j)$$

$$= \sum_{j=1}^{\left[\frac{s}{2}\right]} (s - 2j) (S(p, s - j) S(q, j) - S(p, j) S(q, s - j)).$$

The inequality is strict because of the term corresponding to $j = 1$.

A complete k-partite graph is a graph consisting of k independent sets $I_1, ..., I_k$ such that two vertices are joined by an edge if and only if they belong to different sets I_j, I_h.

The Turán graph $T(n, k)$ is a complete k-partite graph with n vertices for which m parts contain $t + 1$ vertices each, while the other $k - m$ parts contain t vertices each, where m is the remainder in the division of n by k ($n = kt + m$ with $0 \leqslant m \leqslant k - 1$).

According to proposition 5 in §13.3, the graph $T(n, k)$ is the unique graph (up to an isomorphism) having the maximum number of edges among the graphs with n vertices and chromatic number equal to k.

PROPOSITION 7. *The number of $(k + r)$-colourings of the Turán graph $T(n, k)$ is given by*

$$D(n, k, r) = \sum_{\substack{n_1, \ldots, n_k \geqslant 1 \\ n_1 + \ldots + n_k = k + r}} S(t + 1, n_1) \ldots S(t + 1, n_m) S(t, n_{m+1}) \ldots S(t, n_k).$$

If n_1 is the number of classes of the partition of the $t + 1$ vertices of the first independent set of the graph $T(n, k)$, ..., n_k is the number of classes of the partition of the t vertices of the last independent set of the graph $T(n, k)$, it follows that $n_1 + \ldots + n_k = k + r$ and $n_i \geqslant 1$ for $i = 1, \ldots, k$. Notice that all colourings of the graph $T(n, k)$ with $k + r$ classes are obtained without repetitions from the partitions of $k + r$ into k parts, two partitions differing in the order of terms as well. We also obtain $D(n, k, r) = 1$ for $r = 0$ and $r = n - k$ and $D(n, k, r) = 0$ for $r > n - k$.

PROPOSITION 8. *The smallest number of $(k + r)$-colourings of a graph G with n vertices and chromatic number k is equal to $D(n, k, r)$ and for every r with $0 < r < n - k$, the only graph which has this minimum of colourings is the graph $T(n, k)$.*

To prove this, consider a partition of the set of vertices of G into k independent sets. If there exist two vertices x and y which belong to different classes of this partition and which are not joined by an edge, we can link vertices x and y by a new edge and denote the graph obtained in this way by $G_{[x, y]}$.

The chromatic number of the graph $G_{[x, y]}$ is equal to k.

The number of $(k + r)$-colourings of the graph $G_{[x, y]}$ is strictly less than that for G for $0 < r < n - k$, because in $G_{[x, y]}$ there cannot be a class consisting only of vertices x and y, whereas the $(k + r)$-colourings of the graph $G_{[x, y]}$ are $(k + r)$-colourings of the graph G as well.

If we repeat the procedure, we obtain a complete k-partite graph which has a number of $(k + r)$-colourings less than that of G.

If there exist two independent sets I_1 and I_2 of this complete k-partite graph such that $|I_1| = p + 1$ and $|I_2| = q$ with $p > q$, we can obtain a new complete k-partite graph with n vertices by choosing a vertex $x \in I_1$ and defining the new subsets:

$$I_1' = I_1 \setminus \{x\}, \quad I_2' = I_2 \cup \{x\} \text{ and } I_j' = I_j \text{ for } j = 3, \ldots, k.$$

For $0 < r < n - k$, the number of $(k + r)$-colourings of this graph decreases strictly, since the two sides of the inequality which occurs in proposition 6 represent the number of s-colourings of the graph consisting of the subsets I_1 and I_2 and the number of s-colourings of the graph consisting of the parts I_1' and I_2', respectively.

If we repeat this procedure, we obtain the graph $T(n, k)$ which has the minimum number $D(n, k, r)$ of colourings with $k + r$ classes and which is unique with respect to this property for $0 < r < n - k$.

If $r = 0$, we can delete certain suitably chosen edges of the graph $T(n, k)$ for $n > k$, thus obtaining graphs each of which also has chromatic number k and has a unique k-colouring. If $r = n - k$, every graph has a unique n-colouring with a unique vertex in each class.

The number of $(k + r)$-colourings of a complete k-partite graph, the parts of which are $I_1, ..., I_k$ with $|I_1| = n - k + 1$ and $|I_j| = 1$ for $j = 2, ..., k$, is equal to $S(n - k + 1, r + 1)$. It follows from the proof of the previous proposition that $D(n, k, r) \leqslant S(n - k + 1, r + 1)$, the inequality being strict only for $0 < r < n - k$.

If $P_{T(n, k)}(\lambda)$ stands for the chromatic polynomial of the graph $T(n, k)$ and $P(G_n; \lambda)$ denotes the chromatic polynomial of a graph G_n with n vertices, we obtain

$$\min_{\gamma(G_n) = k} P(G_n; \lambda) = P_{T(n, k)}(\lambda)$$

for every natural number λ, the minimum being attained for each $\lambda > k$ only for the graph $T(n, k)$.

The proof follows from the fact that $P(G_n; \lambda) = 0$ for every $\lambda < \gamma(G_n)$, while for $\lambda \geqslant \gamma(G_n)$ we have

$$P(G_n; \lambda) = \sum_{j=\gamma(G_n)}^{\min(n, \lambda)} j! \binom{\lambda}{j} C_j(G_n) \geqslant \sum_{j=\gamma(G_n)}^{\min(n, \lambda)} j! \binom{\lambda}{j} D(n, \gamma(G_n), j - \gamma(G_n)) = P_{T(n, k)}(\lambda)$$

or $\gamma(G_n) = k$, if $C_j(G_n)$ denotes the number of j-colourings of the graph G_n.

The chromatic polynomial of the graph $T(n, k)$ can be computed with the aid of the method of R. C. Read [6], since the chromatic polynomial of the graph consisting of m isolated vertices is $x^m = \sum_{k=1}^{m} S(m, k) [x]_k$.

For example, in the case $n = 8$ and $k = 3$ one obtains

$$P_{T(8,3)}(\lambda) = ([\lambda]_1 + 3[\lambda]_2 + [\lambda]_3)^2([\lambda]_1 + [\lambda]_2) =$$

$$= \lambda(\lambda - 1)(\lambda - 2)(\lambda^5 - 18\lambda^4 + 136\lambda^3 - 529\lambda^2 + 1047\lambda - 836)$$

where $[\lambda]_p[\lambda]_q = [\lambda]_{p+q}$ by definition.

Problems

1. For a graph G with n vertices and m edges prove that $P(G; \lambda) = \lambda^n - m\lambda^{n-1} + ...$ and $P(G; 0) = 0$.

2. Show that a graph G with n vertices is a tree if and only if $P(G; \lambda) = \lambda(\lambda - 1)^{n-1}$.

H i n t: To prove necessity use the property that a tree T_n with n vertices is obtained from a tree T_{n-1} by adding a new vertex joined to exactly one old vertex. For sufficiency use the property that $P(G; \lambda) = P(C_1; \lambda) \dots P(C_p; \lambda)$ if G contains p connected components $C_1, \dots, C_p$.

3. Prove that the chromatic polynomial of an n-cycle C_n is $P(C_n; \lambda) = (\lambda - 1)^n + (-1)^n (\lambda - 1)$.

4. If u and v are nonadjacent vertices of a graph G, let G' be the graph obtained from G by identifying u and v as a single vertex (and by replacing any double edges which result in this way by single edges) and let G'' be the graph obtained from G by joining u and v by an edge.

Prove that $P(G; \lambda) = P(G'; \lambda) + P(G''; \lambda)$.

5. Using the previous property, prove that the coefficients of $P(G; \lambda)$ are alternating in sign.

6. Let α_r denote the coefficient of λ^r in the expansion of the chromatic polynomial of a connected graph. Prove that

$$|\alpha_r| \geqslant \binom{n-1}{r-1}.$$

(Read, 1968)

H i n t: Use the property that a connected graph with n vertices is obtained from a tree T_n by adding some new edges and also use problems 2 and 4.

7. Prove that the smallest nonzero coefficient of λ^r in $P(G; \lambda)$ equals the number of connected components of G.

H i n t: Use the property that for a connected graph G we have $|\alpha_1| \geqslant \binom{n-1}{0} = 1$.

8. Prove that the number of 3-colourings of the cycle C_n with n vertices is equal to $\frac{1}{3} (2^{n-1} - 1)$ for n odd and to $\frac{1}{3} (2^{n-1} - 2)$ for n even.

9. Prove that the number of 3-colourings of the graph consisting of two even cycles C_{2s} and C_{2t}, having a common vertex x, is equal to

$$\frac{1}{9} (2^{2s+2t-1} + 2^{2s} + 2^{2t} - 7).$$

10. The maximum number of 3-colourings of a Hamiltonian graph G with n vertices and chromatic number $\gamma(G) = 3$ is equal to $\frac{1}{3} (2^{n-1} - 1)$ for n odd (for an odd cycle C_n), and is equal to $\frac{1}{3} (2^{n-1} - 2) - \frac{1}{9} (2^{n-1} + 2^{2s} - 2^{n-2s} - 7)$

where $s = \left[\dfrac{n}{4}\right]$ for n even $\left(\text{for a cycle } C_n \text{ having a chord joining two vertices at}\right.$ a distance equal to $2\left[\dfrac{n}{4}\right]\Big)$.

(Tomescu, 1973)

11. Let α_r denote the coefficient of λ^r in the expansion of the chromatic polynomial $P(G; \lambda)$ of a graph G with n vertices and k edges. Show that

$$|\alpha_r| \leqslant \binom{k}{n-r}.$$

(Meredith, 1972)

H i n t: Use induction upon n and k and the property stated in problem 4 in the form $P(G'; \lambda) = P(G; \lambda) - P(G''; \lambda)$.

12. For a graph $G = (X, U)$ define its extended chromatic polynomial $V_k(G; \lambda)$ as the number of mappings of X into the set $\{1, 2, ..., \lambda\}$ which are not constant on the vertices of any complete k-subgraph of G.

Prove that if G is any finite graph, then $V_k(G; \lambda)$ can be expressed as a polynomial in λ with the following properties:

i) The coefficient a_i of λ^i is an integer for all i.

ii) The coefficient $a_i \neq 0$ only if $m \leqslant i \leqslant n$, where m is the number of components of G, and $n = |X|$.

iii) The coefficient $a_n = 1$.

(Sobczyk, Gettys, 1972)

H i n t: We have $V_k(G; \lambda) = \sum\limits_{i=1}^{n} \binom{\lambda}{i} a_{i,k}$ where $a_{i,k}$ is the number of mappings of X onto $\{1, 2, ..., i\}$ which are not constant on any complete k-subgraph of G. For (ii) use the property $V_k(G; \lambda) = \prod\limits_{i=1}^{m} V_k(H_i; \lambda)$ where $H_1, ..., H_m$ are the connected components of G.

13. For the complete graph K_3 prove that

$$V_3(K_3; \lambda) = \lambda^3 - \lambda.$$

14. Prove that the chromatic polynomial of a complete bipartite graph $K_{p,q}$ is equal to

$$P(K_{p,q}; \lambda) = \sum_{r=1}^{p} \sum_{s=1}^{q} S(p, r) S(q, s) [\lambda]_{r+s}.$$